Fundamentals of Petroleum and Petrochemical Engineering

CHEMICAL INDUSTRIES

A Series of Reference Books and Textbooks

Founding Editor

HEINZ HEINEMANN
Berkeley, California

Series Editor

JAMES G. SPEIGHT
CD & W, Inc.
Laramie, Wyoming

1. *Fluid Catalytic Cracking with Zeolite Catalysts,* Paul B. Venuto and E. Thomas Habib, Jr.
2. *Ethylene: Keystone to the Petrochemical Industry,* Ludwig Kniel, Olaf Winter, and Karl Stork
3. *The Chemistry and Technology of Petroleum,* James G. Speight
4. *The Desulfurization of Heavy Oils and Residua,* James G. Speight
5. *Catalysis of Organic Reactions,* edited by William R. Moser
6. *Acetylene-Based Chemicals from Coal and Other Natural Resources,* Robert J. Tedeschi
7. *Chemically Resistant Masonry,* Walter Lee Sheppard, Jr.
8. *Compressors and Expanders: Selection and Application for the Process Industry,* Heinz P. Bloch, Joseph A. Cameron, Frank M. Danowski, Jr., Ralph James, Jr., Judson S. Swearingen, and Marilyn E. Weightman
9. *Metering Pumps: Selection and Application,* James P. Poynton
10. *Hydrocarbons from Methanol,* Clarence D. Chang
11. *Form Flotation: Theory and Applications,* Ann N. Clarke and David J. Wilson
12. *The Chemistry and Technology of Coal,* James G. Speight
13. *Pneumatic and Hydraulic Conveying of Solids,* O. A. Williams
14. *Catalyst Manufacture: Laboratory and Commercial Preparations,* Alvin B. Stiles
15. *Characterization of Heterogeneous Catalysts,* edited by Francis Delannay
16. *BASIC Programs for Chemical Engineering Design,* James H. Weber
17. *Catalyst Poisoning,* L. Louis Hegedus and Robert W. McCabe
18. *Catalysis of Organic Reactions,* edited by John R. Kosak
19. *Adsorption Technology: A Step-by-Step Approach to Process Evaluation and Application,* edited by Frank L. Slejko

20. *Deactivation and Poisoning of Catalysts,* edited by Jacques Oudar and Henry Wise
21. *Catalysis and Surface Science: Developments in Chemicals from Methanol, Hydrotreating of Hydrocarbons, Catalyst Preparation, Monomers and Polymers, Photocatalysis and Photovoltaics,* edited by Heinz Heinemann and Gabor A. Somorjai
22. *Catalysis of Organic Reactions,* edited by Robert L. Augustine
23. *Modern Control Techniques for the Processing Industries,* T. H. Tsai, J. W. Lane, and C. S. Lin
24. *Temperature-Programmed Reduction for Solid Materials Characterization,* Alan Jones and Brian McNichol
25. *Catalytic Cracking: Catalysts, Chemistry, and Kinetics,* Bohdan W. Wojciechowski and Avelino Corma
26. *Chemical Reaction and Reactor Engineering,* edited by J. J. Carberry and A. Varma
27. *Filtration: Principles and Practices: Second Edition,* edited by Michael J. Matteson and Clyde Orr
28. *Corrosion Mechanisms,* edited by Florian Mansfeld
29. *Catalysis and Surface Properties of Liquid Metals and Alloys,* Yoshisada Ogino
30. *Catalyst Deactivation,* edited by Eugene E. Petersen and Alexis T. Bell
31. *Hydrogen Effects in Catalysis: Fundamentals and Practical Applications,* edited by Zoltán Paál and P. G. Menon
32. *Flow Management for Engineers and Scientists,* Nicholas P. Cheremisinoff and Paul N. Cheremisinoff
33. *Catalysis of Organic Reactions,* edited by Paul N. Rylander, Harold Greenfield, and Robert L. Augustine
34. *Powder and Bulk Solids Handling Processes: Instrumentation and Control,* Koichi Iinoya, Hiroaki Masuda, and Kinnosuke Watanabe
35. *Reverse Osmosis Technology: Applications for High-Purity-Water Production,* edited by Bipin S. Parekh
36. *Shape Selective Catalysis in Industrial Applications,* N. Y. Chen, William E. Garwood, and Frank G. Dwyer
37. *Alpha Olefins Applications Handbook,* edited by George R. Lappin and Joseph L. Sauer
38. *Process Modeling and Control in Chemical Industries,* edited by Kaddour Najim
39. *Clathrate Hydrates of Natural Gases,* E. Dendy Sloan, Jr.
40. *Catalysis of Organic Reactions,* edited by Dale W. Blackburn
41. *Fuel Science and Technology Handbook,* edited by James G. Speight
42. *Octane-Enhancing Zeolitic FCC Catalysts,* Julius Scherzer
43. Oxygen in Catalysis, Adam Bielanski and Jerzy Haber
44. *The Chemistry and Technology of Petroleum: Second Edition, Revised and Expanded,* James G. Speight
45. *Industrial Drying Equipment: Selection and Application,* C. M. van't Land
46. *Novel Production Methods for Ethylene, Light Hydrocarbons, and Aromatics*, edited by Lyle F. Albright, Billy L. Crynes, and Siegfried Nowak
47. *Catalysis of Organic Reactions*, edited by William E. Pascoe

48. *Synthetic Lubricants and High-Performance Functional Fluids*, edited by Ronald L. Shubkin
49. *Acetic Acid and Its Derivatives*, edited by Victor H. Agreda and Joseph R. Zoeller
50. *Properties and Applications of Perovskite-Type Oxides*, edited by L. G. Tejuca and J. L. G. Fierro
51. *Computer-Aided Design of Catalysts*, edited by E. Robert Becker and Carmo J. Pereira
52. *Models for Thermodynamic and Phase Equilibria Calculations*, edited by Stanley I. Sandler
53. *Catalysis of Organic Reactions*, edited by John R. Kosak and Thomas A. Johnson
54. *Composition and Analysis of Heavy Petroleum Fractions*, Klaus H. Altgelt and Mieczyslaw M. Boduszynski
55. *NMR Techniques in Catalysis*, edited by Alexis T. Bell and Alexander Pines
56. *Upgrading Petroleum Residues and Heavy Oils*, Murray R. Gray
57. *Methanol Production and Use*, edited by Wu-Hsun Cheng and Harold H. Kung
58. *Catalytic Hydroprocessing of Petroleum and Distillates*, edited by Michael C. Oballah and Stuart S. Shih
59. *The Chemistry and Technology of Coal: Second Edition, Revised and Expanded*, James G. Speight
60. *Lubricant Base Oil and Wax Processing*, Avilino Sequeira, Jr.
61. *Catalytic Naphtha Reforming: Science and Technology*, edited by George J. Antos, Abdullah M. Aitani, and José M. Parera
62. *Catalysis of Organic Reactions*, edited by Mike G. Scaros and Michael L. Prunier
63. *Catalyst Manufacture*, Alvin B. Stiles and Theodore A. Koch
64. *Handbook of Grignard Reagents*, edited by Gary S. Silverman and Philip E. Rakita
65. *Shape Selective Catalysis in Industrial Applications: Second Edition, Revised and Expanded*, N. Y. Chen, William E. Garwood, and Francis G. Dwyer
66. *Hydrocracking Science and Technology*, Julius Scherzer and A. J. Gruia
67. *Hydrotreating Technology for Pollution Control: Catalysts, Catalysis, and Processes*, edited by Mario L. Occelli and Russell Chianelli
68. *Catalysis of Organic Reactions*, edited by Russell E. Malz, Jr.
69. *Synthesis of Porous Materials: Zeolites, Clays, and Nanostructures*, edited by Mario L. Occelli and Henri Kessler
70. *Methane and Its Derivatives*, Sunggyu Lee
71. *Structured Catalysts and Reactors*, edited by Andrzej Cybulski and Jacob A. Moulijn
72. *Industrial Gases in Petrochemical Processing*, Harold Gunardson
73. *Clathrate Hydrates of Natural Gases: Second Edition, Revised and Expanded*, E. Dendy Sloan, Jr.
74. *Fluid Cracking Catalysts*, edited by Mario L. Occelli and Paul O'Connor
75. *Catalysis of Organic Reactions*, edited by Frank E. Herkes
76. *The Chemistry and Technology of Petroleum: Third Edition, Revised and Expanded*, James G. Speight

77. *Synthetic Lubricants and High-Performance Functional Fluids: Second Edition, Revised and Expanded,* Leslie R. Rudnick and Ronald L. Shubkin
78. *The Desulfurization of Heavy Oils and Residua, Second Edition, Revised and Expanded,* James G. Speight
79. *Reaction Kinetics and Reactor Design: Second Edition, Revised and Expanded,* John B. Butt
80. *Regulatory Chemicals Handbook*, Jennifer M. Spero, Bella Devito, and Louis Theodore
81. *Applied Parameter Estimation for Chemical Engineers*, Peter Englezos and Nicolas Kalogerakis
82. *Catalysis of Organic Reactions,* edited by Michael E. Ford
83. *The Chemical Process Industries Infrastructure: Function and Economics,* James R. Couper, O. Thomas Beasley, and W. Roy Penney
84. *Transport Phenomena Fundamentals*, Joel L. Plawsky
85. *Petroleum Refining Processes*, James G. Speight and Baki Özüm
86. *Health, Safety, and Accident Management in the Chemical Process Industries*, Ann Marie Flynn and Louis Theodore
87. *Plantwide Dynamic Simulators in Chemical Processing and Control,* William L. Luyben
88. *Chemical Reactor Design*, Peter Harriott
89. *Catalysis of Organic Reactions*, edited by Dennis G. Morrell
90. *Lubricant Additives: Chemistry and Applications*, edited by Leslie R. Rudnick
91. *Handbook of Fluidization and Fluid-Particle Systems*, edited by Wen-Ching Yang
92. *Conservation Equations and Modeling of Chemical and Biochemical Processes*, Said S. E. H. Elnashaie and Parag Garhyan
93. *Batch Fermentation: Modeling, Monitoring, and Control,* Ali Çinar, Gülnur Birol, Satish J. Parulekar, and Cenk Ündey
94. *Industrial Solvents Handbook, Second Edition*, Nicholas P. Cheremisinoff
95. *Petroleum and Gas Field Processing*, H. K. Abdel-Aal, Mohamed Aggour, and M. Fahim
96. *Chemical Process Engineering: Design and Economics*, Harry Silla
97. *Process Engineering Economics*, James R. Couper
98. *Re-Engineering the Chemical Processing Plant: Process Intensification*, edited by Andrzej Stankiewicz and Jacob A. Moulijn
99. *Thermodynamic Cycles: Computer-Aided Design and Optimization*, Chih Wu
100. *Catalytic Naphtha Reforming: Second Edition, Revised and Expanded*, edited by George T. Antos and Abdullah M. Aitani
101. *Handbook of MTBE and Other Gasoline Oxygenates*, edited by S. Halim Hamid and Mohammad Ashraf Ali
102. *Industrial Chemical Cresols and Downstream Derivatives*, Asim Kumar Mukhopadhyay
103. *Polymer Processing Instabilities: Control and Understanding*, edited by Savvas Hatzikiriakos and Kalman B. Migler
104. *Catalysis of Organic Reactions*, John Sowa
105. *Gasification Technologies: A Primer for Engineers and Scientists*, edited by John Rezaiyan and Nicholas P. Cheremisinoff
106. *Batch Processes*, edited by Ekaterini Korovessi and Andreas A. Linninger

107. *Introduction to Process Control*, Jose A. Romagnoli and Ahmet Palazoglu
108. *Metal Oxides: Chemistry and Applications*, edited by J. L. G. Fierro
109. *Molecular Modeling in Heavy Hydrocarbon Conversions*, Michael T. Klein, Ralph J. Bertolacini, Linda J. Broadbelt, Ankush Kumar and Gang Hou
110. *Structured Catalysts and Reactors, Second Edition*, edited by Andrzej Cybulski and Jacob A. Moulijn
111. *Synthetics, Mineral Oils, and Bio-Based Lubricants: Chemistry and Technology*, edited by Leslie R. Rudnick
112. *Alcoholic Fuels*, edited by Shelley Minteer
113. *Bubbles, Drops, and Particles in Non-Newtonian Fluids, Second Edition*, R. P. Chhabra
114. *The Chemistry and Technology of Petroleum, Fourth Edition*, James G. Speight
115. *Catalysis of Organic Reactions*, edited by Stephen R. Schmidt
116. *Process Chemistry of Lubricant Base Stocks*, Thomas R. Lynch
117. *Hydroprocessing of Heavy Oils and Residua*, edited by James G. Speight and Jorge Ancheyta
118. *Chemical Process Performance Evaluation*, Ali Cinar, Ahmet Palazoglu, and Ferhan Kayihan
119. *Clathrate Hydrates of Natural Gases, Third Edition*, E. Dendy Sloan and Carolyn Koh
120. *Interfacial Properties of Petroleum Products*, Lilianna Z. Pillon
121. *Process Chemistry of Petroleum Macromolecules*, Irwin A. Wiehe
122. *The Scientist or Engineer as an Expert Witness*, James G. Speight
123. *Catalysis of Organic Reactions*, edited by Michael L. Prunier
124. *Lubricant Additives: Chemistry and Applications, Second Edition*, edited by Leslie R. Rudnick
125. *Chemical Reaction Engineering and Reactor Technology*, Tapio O. Salmi, Jyri-Pekka Mikkola, and Johan P. Warna
126. *Asphaltenes: Chemical Transformation during Hydroprocessing of Heavy Oils*, Jorge Ancheyta, Fernando Trejo, and Mohan Singh Rana
127. *Transport Phenomena Fundamentals, Second Edition*, Joel Plawsky
128. *Advances in Fischer-Tropsch Synthesis, Catalysts, and Catalysis*, edited by Burton H. Davis and Mario L. Occelli
129. *Advances in Fluid Catalytic Cracking: Testing, Characterization, and Environmental Regulations*, edited by Mario L. Occelli
130. *Fundamentals of Petroleum and Petrochemical Engineering*, Uttam Ray Chaudhuri

Fundamentals of Petroleum and Petrochemical Engineering

Uttam Ray Chaudhuri
University of Calcutta
Calcutta, India

CRC Press
Taylor & Francis Group
Boca Raton London New York

CRC Press is an imprint of the
Taylor & Francis Group, an **informa** business

CRC Press
Taylor & Francis Group
6000 Broken Sound Parkway NW, Suite 300
Boca Raton, FL 33487-2742

Printed in the United States of America on acid-free paper
10 9 8 7 6 5 4 3 2 1

International Standard Book Number: 978-1-4398-5160-9 (Hardback)

Library of Congress Cataloging-in-Publication Data

Chaudhuri, Uttam Ray.
Fundamentals of petroleum and petrochemical engineering / Uttam Ray Chaudhuri.
p. cm. -- (Chemical industries ; 130)
Includes bibliographical references and index.
ISBN 978-1-4398-5160-9 (hardback)
1. Petroleum engineering. I. Title.

TN870.C5117 2010
665.5--dc22 2010032407

Visit the Taylor & Francis Web site at
http://www.taylorandfrancis.com

and the CRC Press Web site at
http://www.crcpress.com

Dedication

This book is dedicated to the memory of my parents

Inspired by wife Sampa, daughter Aratrika, and brother Utpal

Contents

Preface xxi
Author xxiii
Introduction xxv

Chapter 1 Crude Petroleum Oil 1

1.1 Composition of Crude Oil 1
1.1.1 Hydrocarbon Groups 1
1.1.1.1 Complex Hydrocarbons 3
1.1.1.2 Non-Hydrocarbons or Hetero-Atomic Compounds 4
1.2 Physical Properties of Crude Oil 6
1.3 Origin of Hydrocarbons 7
1.4 Exploration Techniques 9
1.4.1 Gravimetric Method 9
1.4.2 Magnetometric Method 10
1.4.3 Seismic Survey 10
1.4.4 Remote Sensing Method 10
1.4.5 Geochemical Methods 11
1.4.6 Stratigraphy 11
1.5 Resource Estimation 11
1.5.1 Effect of Pressure 12
1.5.2 Connate Water 13
1.5.3 Effect of Temperature 13
1.5.4 Effect of Viscosity 13
1.6 Oil Field Development 14
1.7 Well Logging 16
1.8 Oil Production Processes 16
1.9 Crude Conditioning and Storage 19
1.10 Transportation and Metering of Crude Oil 20
1.11 Gas Hydrates 21
1.11.1 Production Method 22
1.12 Coal Bed Methane 22
Questions 23

Chapter 2 Petroleum Products and Test Methods 25

2.1 Crude Oil Analysis 25
2.1.1 API Gravity 27
2.1.2 Characterisation Factor 27
2.1.3 Bottom Sediment and Water 28

2.2 Domestic Fuels ... 28
2.2.1 Liquified Petroleum Gas ... 28
2.2.2 Kerosene ... 30
2.2.2.1 Smoke Point ... 30
2.2.2.2 Flash Point ... 31
2.2.2.3 Char Point and Bloom ... 31
2.2.2.4 Distillation Test ... 32
2.2.2.5 Sulfur Content and Corrosion ... 32
2.3 Automotive Fuels ... 32
2.3.1 Motor Spirit ... 32
2.3.1.1 American Standard for Testing Material Distillation ... 33
2.3.1.2 Octane Number ... 34
2.3.1.3 Corrosion ... 35
2.3.1.4 Reid Vapour Pressure ... 36
2.3.1.5 Oxidation Stability ... 36
2.3.1.6 Additives ... 36
2.3.2 High Speed Diesel ... 36
2.3.2.1 Cetane Number ... 37
2.3.2.2 Diesel Index ... 37
2.3.2.3 Sulfur ... 38
2.3.2.4 Corrosion ... 38
2.3.2.5 Flash Point ... 38
2.3.2.6 Flame Length ... 38
2.3.2.7 Pour Point ... 38
2.3.2.8 Viscosity ... 38
2.4 Aviation Fuels ... 39
2.5 Furnace Fuels ... 39
2.5.1 Gaseous Fuels ... 39
2.5.2 Liquid Fuels ... 40
2.6 Lubricating Oils ... 40
2.6.1 Viscosity ... 41
2.6.2 Saybolt Method ... 41
2.6.3 Redwood Method ... 41
2.6.4 Brookfield Method ... 42
2.6.5 Viscosity Index ... 42
2.6.6 Cloud Point ... 43
2.6.7 Pour Point ... 43
2.7 Miscellaneous Products ... 44
2.7.1 Jute Batching Oil ... 44
2.7.2 Mineral Turpentine Oil ... 44
2.7.3 Carbon Black Feed Stock ... 46
2.7.4 Bitumen ... 46
2.7.5 Petroleum Coke ... 47
2.7.6 Wax ... 48
Questions ... 48

Chapter 3 Processing Operations in a Petroleum Refinery 49

3.1 Crude Oil Receiving 49
3.2 Desalting of Crude Oil 49
3.3 Distillation and Stripping 51
3.3.1 Atmospheric Distillation 52
3.4 Stabilisation 54
3.5 Amine Absorption 55
3.6 De-Ethaniser 55
3.7 Meroxing and Caustic Wash 56
3.8 Liquified Petroleum Gas Splitter 58
3.9 Naphtha Redistillation 58
3.10 Naphtha Pretreatment 59
3.11 Naphtha Platinum Reforming (Platforming) 60
3.12 Kerosene Hydrodesulfurisation 62
3.13 Diesel Hydrodesulfurisation 63
3.14 Vacuum Distillation 64
3.15 Solvent Extraction 66
3.16 Propane Deasphalting 67
3.17 Solvent Dewaxing 67
3.18 Hydrofinishing 69
3.19 Catalytic Processes for Lube Oil Base Stock Manufacture 70
3.20 Hydrocracking 71
3.21 Mild Hydrocracking 73
3.22 Hydrogen Generation 74
3.22.1 Feed Desulfurisation 74
3.22.2 Primary Reforming 74
3.22.3 Secondary Reforming 75
3.22.4 Shift Reactors 75
3.22.5 Hydrogen Purification 76
3.23 Fluid Catalytic Cracking 76
3.24 Bitumen Blowing 79
3.25 Vis-Breaking 79
3.26 Coking 80
Questions 81

Chapter 4 Lubricating Oil and Grease 83

4.1 Components of Finished Lubricating Oils 83
4.2 Automotive Oils 84
4.3 Industrial Lubricants 85
4.3.1 Bearing Lubricants 85
4.3.2 Hydraulic Lubricants 86
4.3.3 Compressor Lubricants 86
4.3.4 Pump Lubricants 86
4.4 Aviation Lubricants 87

4.5 Marine Lubricants 88
4.6 Greases 88
4.7 Lube Blending and Grease Manufacture 89
4.8 Environmental Impact of Lubricants 90
4.9 Reclamation of Used Lubricants 92
4.10 Power Consumption in a Blending Tank 94
Questions 98

Chapter 5 Petrochemicals 101

5.1 Defintions of Petrochemicals 101
5.1.1 Feedstocks 101
5.1.2 Intermediates 101
5.1.3 Finished Products 102
5.2 Naphtha Cracking 103
5.2.1 Primary Fractionator or Stabiliser 105
5.2.2 Hydrogen Separator 105
5.3 Conversion Processes for Selected Petrochemicals 107
5.3.1 Polyethylene 107
5.3.1.1 Low Density Polyethylene 108
5.3.1.2 High Density Polyethylene 109
5.3.1.3 Linear Low Density Polyethylene 110
5.3.2 Polypropylene 111
5.3.3 Polyethylene Terephthalate 111
5.3.3.1 Terephthalic Acid 112
5.3.3.2 Ethylene Glycol 113
5.3.4 Polyvinyl Chloride 115
5.3.5 Polystyrene 115
5.3.6 Polybutadiene 116
5.3.7 Acrylonitrile Butadiene Styrene 116
5.3.8 Styrene–Butadiene Rubber 116
5.3.9 Poly Methyl Metha Acrylate 117
5.3.10 Polytetrafluoroethylene 117
5.3.11 Nylons 117
5.3.12 Phenol Formaldehyde 120
5.3.13 Urea Formaldehyde 120
5.3.14 Melamine Formaldehyde 120
5.3.15 Polyurethane 121
5.3.15.1 Toluene Diisocyanate 121
5.3.16 Silicone 122
5.4 Petrochemical Complex 122
5.4.1 Downstream Units 122
5.4.2 Petrochemicals' Hub 122
5.5 Processing of Plastic, Rubber, and Fibre 123
5.5.1 Moulding of Plastics 124
5.5.2 Extrusion Moulding 124

5.5.3 Blow Moulding ... 126
5.5.4 Compression Moulding ... 126
5.5.5 Thermal Moulding ... 126
5.5.6 Injection Moulding ... 126
5.5.7 Rubber Compounding ... 127
Questions ... 128

Chapter 6 Offsite Facilities, Power and Utilities ... 131

6.1 Layout of Petroleum and Petrochemical Plants ... 131
6.2 Processing Units ... 131
6.3 Offsite Facilities ... 134
6.3.1 Floating Roof Tank ... 134
6.3.2 Fixed Roof Tank ... 135
6.3.3 Pressure Vessels ... 136
6.3.4 Horton Sphere ... 136
6.3.5 Accessories ... 136
6.3.6 Blending Operations ... 139
6.3.7 Filling, Loading, and Despatch Operations ... 139
6.3.8 Pipeline Transport ... 139
6.3.9 Effluent Water Treatment ... 140
6.3.10 Off Gas Treatment ... 141
6.3.11 Internal Fuel Oil Circulation ... 144
6.4 Power and Steam Generating Plant ... 144
6.5 Cooling Tower ... 147
6.6 Water Conditioning Plant ... 147
Questions ... 150

Chapter 7 Material and Energy Balances ... 151

7.1 Measurement of Quantity of Crude Oil and Products ... 151
7.1.1 Tank Dipping ... 151
7.1.2 Volume Correction ... 153
7.1.3 Density Correction ... 153
7.2 Measurement of Gases in Closed Vessels ... 158
7.3 Material Balance in a Plant ... 158
7.3.1 Flow Meter Readings ... 160
7.3.2 Fuel Consumption ... 161
7.3.3 Steam Consumption ... 161
7.3.4 Overall Material Balance ... 161
7.4 Energy Balance in a Plant ... 164
7.4.1 Heat Balance ... 164
7.4.2 Energy Balance in a Heat Exchanger ... 164
7.4.3 Energy Balance in a Furnace ... 165
7.4.4 Energy Balance in a Distillation Column ... 166
7.4.5 Overall Energy Balance ... 168
Questions ... 170

Chapter 8 Heat Exchangers and Pipe-Still Furnaces 171

8.1 Heat Exchangers 171
8.2 Theory of Heat Exchange 171
8.2.1 Heat Balance 173
8.2.2 Rate of Heat Transfer 174
8.3 Fouling 177
8.4 Plate Type Heat Exchanger 185
8.5 Extended Surface Exchanger 185
8.6 Scraped Surface Exchanger 185
8.7 Heat Exchanger Train 186
8.8 Pipe-Still Furnace 188
8.9 Pipe-Still Furnace Elements 190
8.9.1 Heater Pipes or Tubes 190
8.9.2 Refractories 191
8.9.3 Burners 191
8.9.4 Convection Zone 192
8.9.5 Radiant Section 192
8.9.6 Stack or Chimney 193
8.10 Operation of a Furnace 193
8.11 Draught in a Furnace 194
8.12 Furnace Design by the Wilson, Lobo and Hottel Method 194
8.12.1 Furnace Design by the Lobo and Evans Method 205
Questions 206

Chapter 9 Distillation and Stripping 207

9.1 Processes of Distillation and Stripping 207
9.2 Batch Distillation 207
9.3 Boiling Point and Equilibrium Diagrams 208
9.4 Theory of Distillation 208
9.5 Continuous Distillation 210
9.5.1 Top Reflux Drum 211
9.5.2 Rectification Section 212
9.5.2.1 Streams Leaving the Envelope under Study 212
9.5.2.2 Streams Entering the Envelope under Study 212
9.5.2.3 Reboiler 213
9.5.2.4 Stripping Section 213
9.5.2.5 Feed Plate or Flash Zone 214
9.5.2.6 Evaluation of Fraction Vaporised (f) from the Quality of the Feed 215
9.6 McCabe–Thiele Method 217
9.6.1 Operating Line for the Feed Section or Feed Line 218
9.6.2 Operating Line and Plates for the Stripping Section 218

9.7 Enthalpy Balance Method 219
9.7.1 Reflux Drum 219
9.7.2 Top Plate 220
9.7.3 Reboiler 221
9.7.4 Numerical Solution 223
9.7.5 Types of Refluxes 229
9.7.6 Internal Reflux 230
9.7.7 Minimum Reflux 230
9.8 Gap and Overlap 230
9.9 Packie's Correlation 231
Question 234

Chapter 10 Extraction 235

10.1 Extraction Principle 235
10.2 Extraction Process 236
10.3 Definition of Terms Related to Extraction 236
10.3.1 Partition Coefficient 236
10.3.2 Partial Solubility 237
10.3.3 Solvent to Feed Ratio 237
10.3.4 Solvent Recovery 237
10.3.5 Separation of Phases 237
10.3.6 Selectivity 237
10.3.7 Solvent Power 238
10.3.8 Critical Solution Temperature 238
10.4 Phase Equilibrium in the Extraction Process 238
10.5 Batch Extraction 239
10.6 Continuous Extraction 240
10.6.1 Computation of Number of Plates 240
Questions 248

Chapter 11 Reactor Calculations 249

11.1 Reactors in Refineries and Petrochemical Plants 249
11.2 Reaction Stoichiometry, Mechanism, and Pathways 249
11.3 Rate of Reaction and Kinetic Equations 250
11.4 Batch, Continuous Stirred Tank Reactor, and Plug Flow Reactor Concepts 252
11.5 Naphtha Reformer Calculations 269
11.6 Calculations for a Fluidised Catalytic Cracking Reactor 271

Chapter 12 Elements of Pipeline Transfer Facilities 275

12.1 Pipes and Tubes 275
12.2 Fittings and Supports 275
12.2.1 Corrosion Protection 276

12.3 Crude Oil Transfer Lines ... 277
12.3.1 Design Steps for Crude Pipes ... 277
12.3.2 Economic Pipe Diameter ... 279
12.4 Product Transfer Lines ... 279
12.5 Gas Transfer Lines ... 280
12.6 Pumps and Compressors ... 281
12.6.1 Centrifugal Pumps ... 281
12.6.1.1 Priming ... 283
12.6.1.2 Specific Speed ... 285
12.6.2 Positive Displacement Pumps ... 287
12.6.3 Rotary Pumps ... 288
12.6.4 Compressors ... 289
12.7 Power Calculations for Pumping and Compression ... 290

Chapter 13 Instrumentation and Control in a Refinery ... 297

13.1 Control Hardware ... 297
13.1.1 Hardware ... 297
13.1.2 Cables ... 298
13.2 Control Loops ... 299
13.3 The Process Piping and Instrumentation Diagram ... 301
13.4 Control Software ... 301
13.5 Distributed Control System ... 304
13.6 The Control Room ... 305
13.7 Crude Throughput Control ... 305
13.8 Desalter Control ... 306
13.9 Atmospheric Distillation Column Control ... 308
13.9.1 Reflux Drum Pressure Control ... 308
13.9.2 Reflux Drum Level Control ... 308
13.9.3 Top Plate Temperature ... 310
13.9.4 Draw Plate Temperature ... 310
13.9.5 Overflash Rate ... 312
13.9.6 Flash Zone Pressure and Temperature ... 312
13.9.7 Bottom Temperature ... 312
13.9.8 Furnace Control ... 312
13.10 Vacuum Distillation Control ... 312
13.11 Reformer Unit Control ... 314
13.12 Fluid Catalytic Cracking Unit Control ... 314
13.12.1 Reactor Outlet Temperature Control ... 316
13.12.2 Level Control of the Catalyst Bed in the Stripper Section of the Reactor ... 319
13.12.3 Pressure Balance between the Reactor and the Regenerator ... 319
13.13 Fail-Safe Devices ... 319
13.13.1 Normal Running Conditions ... 319
13.13.2 During Planned Shutdown ... 321

13.13.3 During Accidents or Emergency Shutdown 322
13.13.4 Power Plant Failure 323
13.14 Standard Signals in Process Control 323

Chapter 14 Miscellaneous 325

14.1 Startup 325
14.1.1 Power Plant Startup 325
14.1.2 Startup of a Crude Distillation Unit 326
14.1.3 Starting a Naphtha Pretreatment Plant 327
14.1.4 Starting a Naphtha Reforming Plant 328
14.1.5 Starting a Fluid Catalytic Cracking Plant 328
14.2 Shutdown 329
14.2.1 Shutdown of a Crude Distillation Unit 330
14.2.2 Shutdown of a Naphtha Pretreatment Unit 330
14.2.3 Regeneration of the Catalyst 330
14.2.4 Shutdown of a Naphtha Reforming Unit 331
14.2.5 Regeneration of Reforming Catalyst 331
14.3 Maintenance of Plant and Equipments 332
14.4 Fire and Explosion 333
14.4.1 Pyrophoric Iron 334
14.5 Factories Act 335
14.6 Safety Analysis 337

Chapter 15 Plant Management and Economics 343

15.1 Cost of Equipment 343
15.1.1 Capacity Ratio Method 343
15.1.2 Purchased Price 344
15.1.3 First Cost of the Equipment 344
15.1.4 Depreciation 344
15.2 Cost of a Plant 344
15.3 Operating Cost 344
15.4 Product Cost 344
15.5 Profit and Product Price 345
15.6 Taxes and Duties 345
15.7 Breakeven Point, Payout Period, and Rate of Return 345
15.7.1 Payout Period or Payback Period 346
15.7.2 Rate of Return 346
15.8 Linear Programming 349
15.9 Material Audit 352
15.9.1 Category of Materials 352
15.9.2 Papers to Be Maintained 353
15.9.2.1 Tank Dip Register 353
15.9.2.2 Pass Out Vouchers 353
15.9.2.3 Tank Dip Memos 353

15.9.2.4 Daily Stock Report 355
15.9.2.5 Daily Pumping Record 355
15.9.2.6 Daily Operation Record 355
15.9.3 Material Audit of Capital Goods 366
15.10 Energy Audit 367
15.10.1 Electricity Audit 367
15.10.2 Thermal Audit 368
15.10.3 Steam Balance 368

Appendix 375

Index 381

Preface

Modern civilisation cannot think of a day without petroleum and petrochemicals. Petroleum fuels, such a gasoline and diesel, are the major fuels for all transportation vehicles. Commodities manufactured from petrochemicals, for example, plastics, rubbers and synthetic fibres derived from petroleum, have become part and parcel of our daily life. The absence of petroleum will cause an end to our modern civilization unless alternative means are available. In fact, petroleum is a non-renewable fossilised mass, the amount of which is being exhausted with our increasing consumption. Future crude oil will be heavier and contaminated with unwanted salts and metals. Production and processing will be costlier than ever before. Therefore, it is inevitable to make use of this dwindling natural resource more judiciously and efficiently for the sustenance of our civilisation. The contents of this book have been prepared to provide a holistic working knowledge about petroleum and petrochemical technology. Chapter 1 presents the essential preliminaries for the exploration and production of crude petroleum oil and gas. This chapter is an introduction for beginners who may be entering the profession of oil and gas exploration and production. Chapter 2 is an analysis of crude oil and petroleum products. This will help scientists entering the profession as chemists in a refinery. The processing of petroleum in refineries is discussed in Chapter 3 and may be useful for apprentice engineers in a refinery. The fundamentals of lubricating oil and grease are dealt with in Chapter 4, which is useful for engineers and scientists entering the lubricants industries. Chapter 5 discusses the fundamentals of petrochemicals, their raw materials, and the end products, along with the manufacturing principles of some of the industrially important products. This chapter may be important for the engineer who is likely to follow a profession in petrochemical plants. The rest of the book, from Chapters 6 through 15, will be of common interest to engineers in refineries and petrochemical plants. Chapters 6 to 12 deal with the theories and problems of unit operations and the processes involved in refineries and petrochemical plants. The essential knowledge of automatic operations in a plant is dealt with in Chapter 13. Without this knowledge, engineers will not be successful in operating any plant. Chapter 14 deals with various miscellaneous activities, like start up, shutdown, maintenance, fire, and safety operations, which are essential to the running of any plant. Chapter 15 discusses the commercial and managerial activities that any engineer has to know for the ultimate success of refining or manufacturing businesses.

Author

Uttam Ray Chaudhuri is an associate professor in the Department of Chemical Technology, University of Calcutta. He holds a PhD in chemical engineering from the Indian Institute of Technology, Kharagpur, India. He received his graduate and postgraduate degrees in chemical engineering from Jadavpur University. He has 30 years of experience in industry, research, and teaching in the field of chemical engineering and technology. He has a number of research publications in foreign and Indian journals to his credit. He also served as a chemical engineer for more than ten years in the Indian Oil Corporation Ltd. (Refineries and Pipeline Division).

Introduction

Petroleum is a fossilised mass that has accumulated below the earth's surface from time immemorial. Raw petroleum is known as crude (petroleum) oil or mineral oil. It is a mixture of various organic substances and is the source of hydrocarbons, such as methane, ethane, propane, butane, pentane, and various other paraffinic, naphthenic, and aromatic hydrocarbons, the building blocks of today's organic industry. Various petroleum products, such as gaseous and liquid fuels, lubricating oil, solvents, asphalts, waxes, and coke, are derived from refining crude oil. Many lighter hydrocarbons and other organic chemicals are synthesised by thermal and catalytic treatments of these hydrocarbons. The hydrocarbon processing industry is basically divided into three distinct activities—petroleum production, petroleum refining, and petrochemical manufacture. Refineries produce cooking gas (liquified petroleum gas or LPG), motor spirit (also known as petrol or gasoline), naphtha, kerosene, aviation turbine fuel (ATF), high speed diesel (HSD), lubricating base oils, wax, coke, bitumen (or asphalt), etc., which are mostly a mixture of various hydrocarbons (the organic compounds made of carbon and hydrogen as the major constituent elements). In a petrochemical plant (where one or more petrochemicals are produced) or in a petrochemical complex (where many petrochemical products are produced), pure hydrocarbons or other organic chemicals with a definite number and type of constituent element or compound are produced from the products in refineries. Thus, petrochemicals are derived from petroleum products obtained from refineries. Products from a petrochemical complex are plastics, rubbers, synthetic fibres, raw materials for soap and detergents, alcohols, paints, pharmaceuticals, etc. Since petroleum is the mixture of hundreds of thousands of hydrocarbon compounds, there is a possibility of synthesising many new compounds. In fact, due to the advancement of new technology, new petrochemicals are being invented and will continue to be added to this industry in the near future. Hence, the petrochemical industry is still a growing industry. The manufacture of valuable petrochemicals from low-valued petroleum products has been the main attractive option for the refining industry investing in the petrochemical industry. Thus, modern refineries are, in fact, refinery cum petrochemical complexes.

1 Crude Petroleum Oil

1.1 COMPOSITION OF CRUDE OIL

The compounds in crude petroleum oil are essentially hydrocarbons or substituted hydrocarbons in which the major elements are carbon at 85%–90% and hydrogen at 10%–14%, and the rest with non-hydrocarbon elements—sulfur (0.2%–3%), nitrogen (< 0.1–2%), oxygen (1%–1.5%), and organo-metallic compounds of nickel, vanadium, arsenic, lead, and other metals in traces (in parts per million or parts per billion concentration). Inorganic salts of magnesium chloride, sodium chlorides, and other mineral salts are also accompanied with crude oil from the well either because of water from *formation* or water and chemicals injected during drilling and production.

1.1.1 HYDROCARBON GROUPS

Compounds solely made of carbon and hydrogen are called hydrocarbons. These hydrocarbons are grouped as paraffins, naphthenes, aromatics, and olefins. Crude oil contains these hydrocarbons in different proportions, except olefins, which are produced during processing.

Paraffins are saturated hydrocarbons. A saturated hydrocarbon is a compound where all four bonds of a carbon atom are linked to four separate atoms. Examples are methane, ethane, propane, butane, pentane, hexane, with the generic molecular formula of C_nH_{2n+2}, where n is the number of carbon atoms in that compound. The homologous series of these hydrocarbons are called alkanes (Figure 1.1).

The series starts with methane, which has the chemical formula CH_4. Alkanes are relatively unreactive as compared to aromatics and olefins. At room temperature, alkanes are not affected by concentrated fuming sulfuric acid, concentrated alkalies, or powerful oxidising agents such as chromic acid. They carry out substitution reactions slowly with chlorine in sunlight and with bromine in the presence of a catalyst. Paraffins are available both as normal and iso-paraffins. Normal paraffins are straight chain compounds and iso-paraffins are branched compounds. Normal and iso-paraffins have the same formula (i.e., same number of carbon and hydrogen atoms), but they differ widely in their physical and chemical properties because of isomerism. The number of isomers of normal paraffins increases with the number of carbon atoms in the paraffin. For example, paraffins with carbon numbers of five, six, and eight will have iso-paraffins of three, five, and eighteen, respectively. Iso-paraffins are more reactive than normal paraffins and are desirable in motor spirit. Normal paraffins can be converted to iso-paraffins by thermal or catalytic processes. This is known as the isomerisation reaction.

Olefins are unsaturated hydrocarbons, i.e., the double bond is present between the two carbon atoms in the formula. The generic formula is C_nH_{2n}, and the lowest

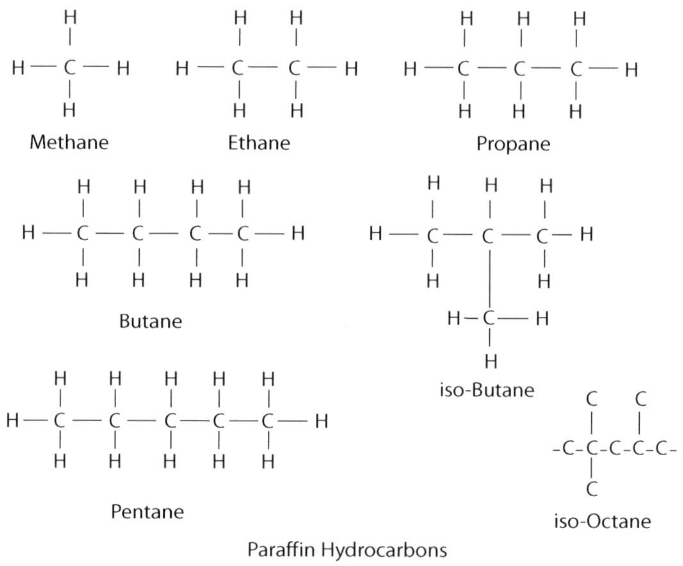

FIGURE 1.1 Common saturated hydrocarbons or paraffins.

member of this homologous series is ethylene, C_2H_4. This series is known as alkenes. These are highly reactive and can react to themselves to mono olefins (Figure 1.2).

Olefins react readily with acids, alkalies, halogens, oxidizing agents, etc. Olefins are not present in crude oil, but they are produced by thermal and catalytic decomposition or dehydrogenation of normal paraffins. Like paraffins, olefins may be straight (normal) chain or branched chain (iso-) hydrocarbons. Olefins can be determined by the bromine or iodine number in reaction with bromine or iodine. They are readily converted to

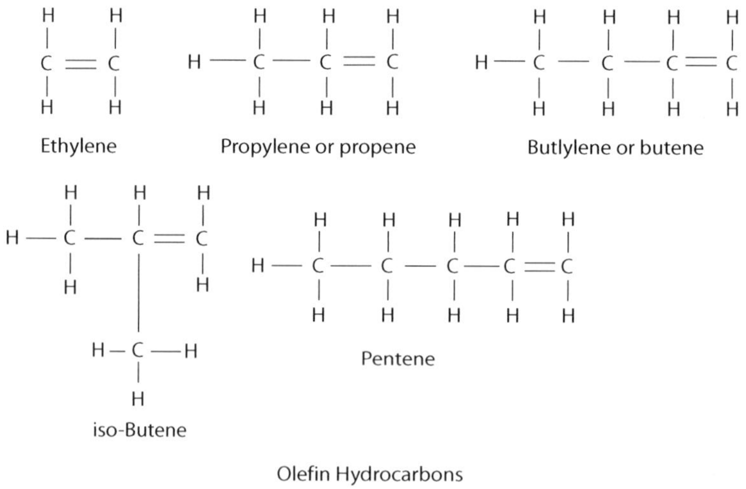

FIGURE 1.2 Common unsaturated hydrocarbons or alkenes.

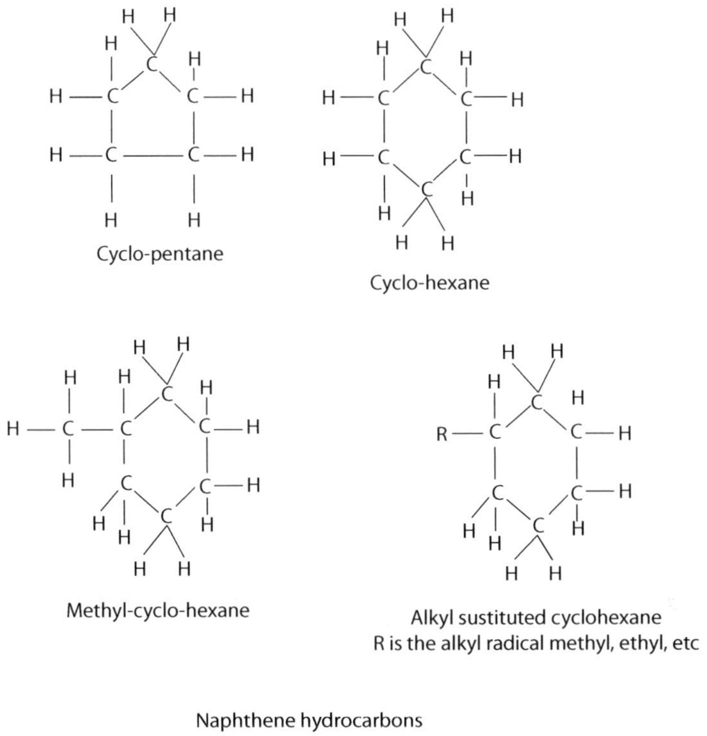

FIGURE 1.3 Common cyclic saturated hydrocarbons or cyclo alkanes.

diolefins in the presence of oxygen and form a gum-like substance. Olefins present in petroleum products can be removed by absorption in sulfuric acid.

Naphthenes are cyclic saturated hydrocarbons with the general formula, like olefins, of C_nH_{2n}, also known as cyclo-alkanes. Since they are saturated, they are relatively inactive, like paraffins. Naphthenes are desirable compounds for the production of aromatics and good quality lube oil base stocks. Some of these are shown in (Figure 1.3).

Aromatics, often called benzenes, are chemically very active as compared to other groups of hydrocarbons. Their general formula is C_nH_{2n-6}. These hydrocarbons in particular are attacked by oxygen to form organic acids. Naphthenes can be dehydrogenated to aromatics in the presence of a platinum catalyst. Lower aromatics, such as benzene, toluene, and xylenes, are good solvents and precursors for many petrochemicals. Aromatics from petroleum products can be separated by extraction with solvents such as phenol, furfurol, and diethylene glycol. Some of these are presented in (Figure 1.4).

1.1.1.1 Complex Hydrocarbons

Crude oil also contains a large number of hydrocarbons that do not fall into the category of paraffins, olefins, naphthenes, or aromatics, but may be the combined group of any two or more groups of paraffins, naphthenes, or aromatic hydrocarbons. By joining two or more naphthene rings or combining naphthene and aromatic rings, paraffin chains with aromatic rings (alkyl-aromatics), etc., a vast array of complex

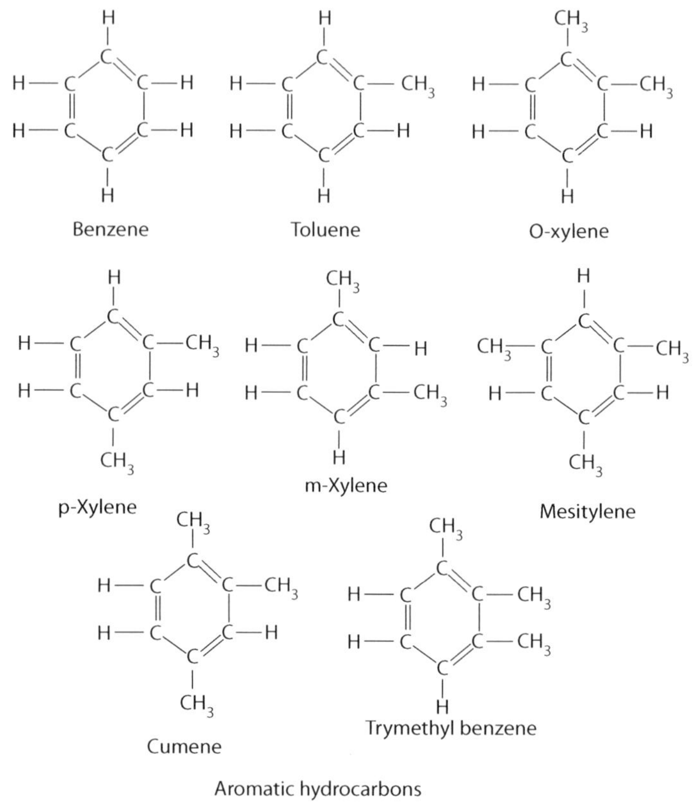

FIGURE 1.4 Common cyclic unsaturated hydrocarbons or aromatics.

hydrocarbons may be formed. Examples of these compounds are decalin, naphthalene, and diphenyl. Heavier fractions of crude oil contain these types of hydrocarbons. Multinuclear (multi ring) aromatics or polynuclear aromatics (PNA) are well known in crude oil and its residual products. PNAs are the precursors of coke, which forms due to thermal effect. These cannot be decomposed easily even by severe hydro-cracking (Figure 1.5).

1.1.1.2 Non-Hydrocarbons or Hetero-Atomic Compounds

Common hetero atoms in hydrocarbons are sulfur, oxygen, nitrogen, and metallic atoms. Sulfur compounds are present in crude oil as mercaptans, mono- and disulfides with the general formula R-SH, R-S-R1, R-S-S-R1, where R and R1 are the alkyl radicals. Mercaptans are very corrosive whereas mono- and disulfides are not. Examples of cyclic sulfur compounds are thiophenes and benzothiophene. *Hydrogen sulfide* (H_2S) gas is associated with crude oil in dissolved form and is released when heated. H_2S is corrosive at high temperatures and in the presence of moisture. Crude oil that contains large amounts of H_2S is called sour crude. Sulfur present in petroleum fuel products also forms various oxides of sulfur (SO_x) during combustion, which are strong environmental pollutants. H_2S can be removed from gases by

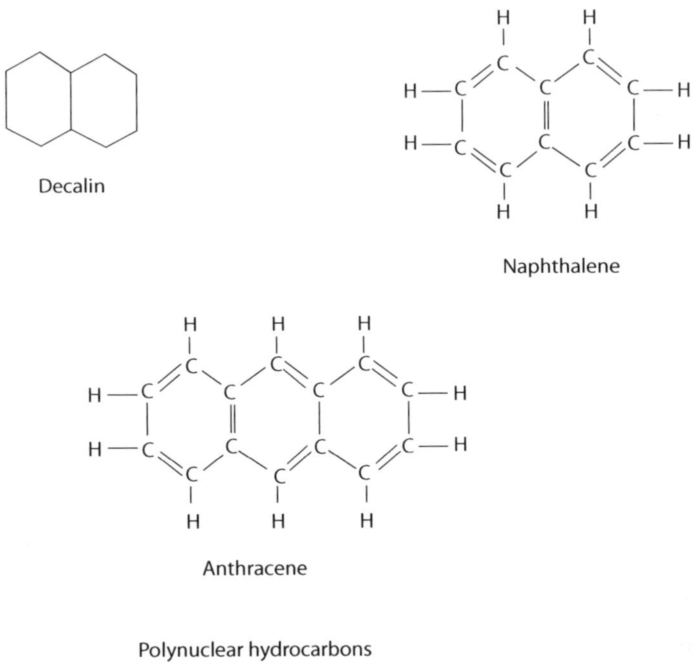

FIGURE 1.5 Structural examples of polynuclear aromatics.

absorption in an amine solution. In the light distillates, sulfur may be present as H_2S, mercaptans, and thiophenes, but in the heavier fractions of crude oil, 80%–90% of the sulfur is usually present in the complex ring structure of hydrocarbons. In this combination, the sulfur atom is very stable and non-reactive. As a result, sulfur from heavier petroleum cannot be removed without a destructive reaction, such as severe thermal or catalytic reactions. Nowadays, sulfur is recovered during refining and sold as a product. Sulfur also has a poisoning effect on various catalysts.

Nitrogen compounds in hydrocarbons are usually found in the heavier parts of the crude oil. These are responsible for colour and colour instability and poisoning of certain catalysts. Nitrogen in petroleum fuels causes the generation of oxides of nitrogen (NO_x), which are also strong pollutants of the atmosphere. Nitrogen can be eliminated from petroleum products by catalytic hydrogenation. Like sulfur, nitrogen in the heavier parts of petroleum cannot be removed without severe cracking or hydrogenation reactions.

Oxygen compounds: crude oil may contain oxygen containing compounds, such as naphthenic acids, phenols, and cresols, which are responsible for corrosive activities. Oxygen also acts as a poison for many catalysts. This can be removed by catalytic hydrogenation. Excess oxygen compounds may even lead to explosion.

Metallic compounds of vanadium, nickel, lead, arsenic, etc., are also found in crude oil. Vanadium and nickel are found in the form of organo-metallic compounds mostly in the heavier fractions of crude oil where the metal atoms are distributed within the compound in a complex form called porphyrins. Petroleum fuels containing these metallic compounds may damage the burners, lines, and walls of the combustion chambers. Some of the hetero-atomic hydrocarbons are shown in Figure 1.6.

1.2 PHYSICAL PROPERTIES OF CRUDE OIL

Crude oil is sometimes classified as paraffinic base, naphthenic base, or asphaltic base, according to the prevalence of the hydrocarbon groups. But various physical properties are required in addition to these classification in order to characterise a crude oil.

API gravity is expressed as the relation developed by the American Petroleum Institute, as

$$\text{API} = 141.5/s - 131.5, \tag{1.1}$$

where "s" is the specific gravity of oil measured with respect to water, both at 60°F (15.5°C). Since oil is lighter than water, API gravity is always greater than 10. The lighter the oil, the larger the API gravity. However, gravity is not the only measurement of crude oil, but a mere indicator of lightness. Since crude oil is, in fact, a mixture of various hydrocarbons varying from gases to semi-solid asphalts, it is convenient to separate these into various boiling fractions rather than as individual chemical species. Crude is distilled in a laboratory distillation apparatus and the boiling fractions are collected. Boiling fractions are a mixture of hydrocarbons

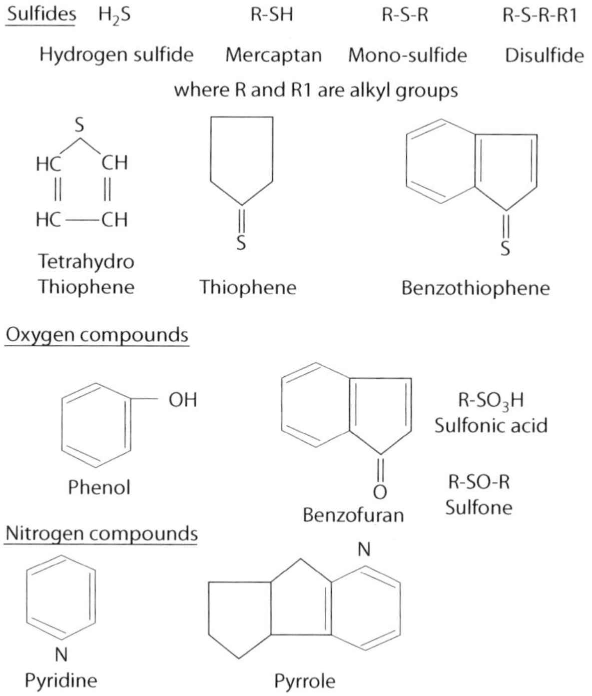

FIGURE 1.6 Some of the hetero-atomic hydrocarbons.

boiling in a certain range of temperatures. For a particular crude oil, each boiling fraction separated has a certain average boiling point. A *characterisation factor* of crude oil has been related with the average (molal average) boiling point (T_B in Rankine) of all the fractions separated and its specific gravity "s", as

$$C_F = (T_B)^{1/3} /s. \tag{1.2}$$

Characterisation factor (C_F) is universally accepted as the identity of a crude oil and its products. Various other properties, such as molecular weight, density, viscosity, and thermodynamic properties, are available for any oil product if its characterisation factor is determined. Since crude oil is always associated with water and settleable solids, it is essential to determine the relative amount of bottom sludge and water (BSW) after the necessary settling period. Water is separated by the solvent extraction method in the laboratory. *Ultimate analysis* of crude oil is a method to determine the amount of carbon, hydrogen, and other constituent elements in it. Combustion of crude oil yields ashes of metallic oxides that are analysed for the metallic components present in crude oil.

1.3 ORIGIN OF HYDROCARBONS

The word *petroleum* is derived from the Latin words for rock (petra) and oleum (oil). It is found in the form of gas and/or liquid phases in porous rock structures. Both gases and liquids are rich mixtures of organic components consisting of carbon and hydrogen and hence are known as hydrocarbons in general. Usually, these are available in the sub-surface of Earth in the porous rocks known as sedimentary basins. In the majority of the basins, gas, oil, and water coexist under pressure with methane gas at the cap and oil is sandwiched between the gas and water. Dissolved and liquified gases are usually present in liquid petroleum oil. Heavy, carbon-rich or bituminous hydrocarbons are also available in the shallow depth in the shales (oil shales) or on the surface sands (tar sands). The most abundant hydrocarbon gas in nature, methane, is also available in large quantities from the coal bed (known as coal bed methane). Large quantities of methane are also available as hydrates under the sea bed in the Arctic region and are known as gas hydrates. There are many hypotheses about the origin of

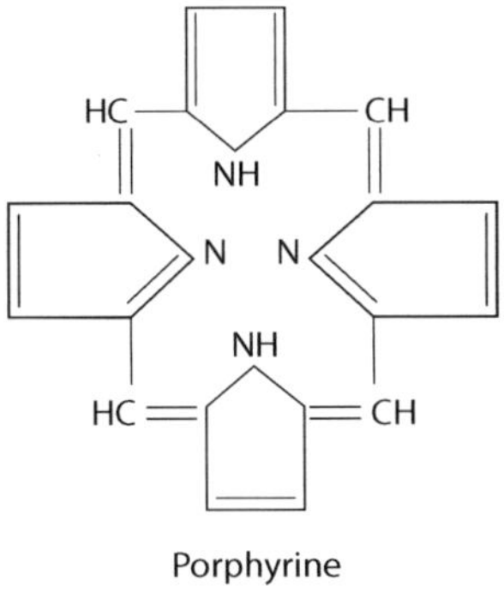

FIGURE 1.7 Complex structure of porphyrins present in asphalt.

the formation of crude oil. To date, it is generally agreed that crude petroleum oil was formed from decaying plants and vegetables and dead animals and converted to oil by the action of high pressure and high temperature under the earth's surface, and by the action of the biological activities of micro-organisms. Organic materials of plant or animal origin accumulate in the lowest places, usually in the crevices, low-lying land, sea bed, coral reefs, etc., and are gradually buried under the surface of Earth. Thus, huge amounts of organic matter are trapped layer after layer in the earth's crust and rock. Rocks that bear these organic layers are called sedimentary rocks. Several kilometres below the earth's surface, organic sediments are decayed biologically to a mass, known as *kerogen*, which has a very high mass of organic-to-inorganic ratio favourable for conversion to hydrocarbon. The temperature of Earth increases with depth (geothermal gradient) at the rate of approximately 30°C per kilometre. Thus, at a depth of 4–5 km, called *kitchen* by geologists, temperatures of 120°C–150°C exist where kerogen is converted to hydrocarbon oil under very high pressure of rocks and soil. But this conversion takes millions of years (geological time period) to complete. Methane is also formed thermogenically (i.e., thermal conversion of kerogen) along with biogenic methane already present before the formation of crude oil. Migration of oil with gas occurs within the rock layers by the pressure gradient from high to low pressure zones. The formation of crude (or crude deposit) oil has been found in the sedimentary porous rock layers trapped under the hard and impervious igneous rock layers. Crude oil and gas accumulate in the pores of the sedimentary rocky layer as shown in Figure 1.8. This formation may be found from a few kilometers (as deep as 2 km and as deep as 7 km) below the earth's surface. The first oil deposit is known as the Drake Well, discovered in the United States (near Titusville) in 1859.

Some of the common terms used in petroleum exploration and production are *source rock*, *migration*, and *reservoir*. Sedimentary rocks are the rocky layer where organics are converted to oil and gas due to high temperature and pressure over

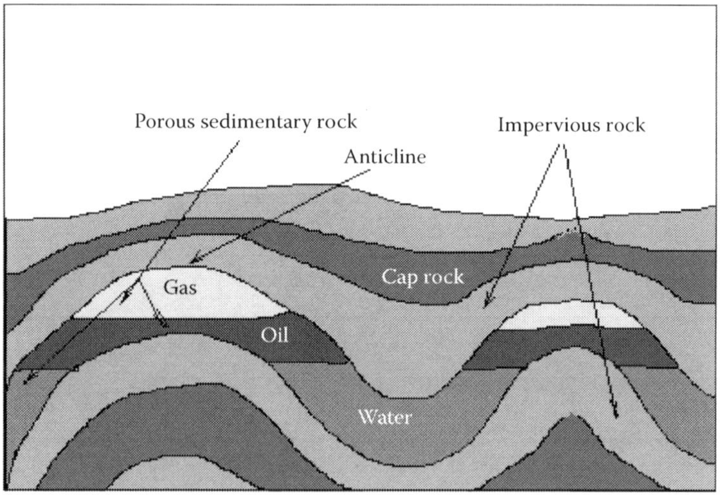

FIGURE 1.8 A typical anticline oil and gas reservoir.

millions of years. From the source rock, oil and gas then migrate to areas or traps that have a structure favourable for storing oil and gas. Traps are usually anticline or domed or faulted areas having oil and gas trapped in a porous rocky area covered by impermeable rock (seal or cap rocks) layers that do not allow further migration or escape to another area. Such an area that traps oil and gas is known as a *reservoir* or *basin*.

A *prospect* of hydrocarbon deposits is declared by the geologist when the area under study satisfies the above geological structure and conditions. The area where oil and gas are stored is known as *formation*. Drilling is started only in the *prospect* area as declared by the geologists. Oil reserves are classified into three categories, namely, *proven*, *probable*, and *possible* reserves. Proven reserves are worth for economic exploitation. Probable reserve has a certain degree of probability (about 50%) for economic exploitation. Possible reserve has very little probability (about <10%) for economic exploitation with current technology. Commercially viable formation is also known as *pay* or *pay zone*.

1.4 EXPLORATION TECHNIQUES

The selection of a drilling site is a tricky and costly affair. Though some visible evidence of a hydrocarbon source, like seepage of oil and gas from the surface, the visual appearance of surface and vegetation, the presence of oil or gas in fountains or rivers, etc., are sometimes used in locating oil and gas reserves, and many ancient oil fields were discovered by these events. But, today, such fortunate events are very rare and sometimes may not always be suitable for commercial exploitation. Modern exploration techniques use geophysical, geochemical, and geotechnical methods. Exploration of the surface of Earth can be useful for imaging or mapping sub-surface structures favourable for oil and gas accumulation. In the geophysical methods, gravimetric, magnetometric, seismic, radioactive, and stratigraphic studies of the surface are gathered. Chemical analysis of the surface soil and rocks are carried out by geochemical methods. Geotechnical methods, such as the mechanical properties of rocks and surface, are measured. Remote sensing from satellite is the most recent development for a low cost geological survey. Usual geophysical methods include gravimetric, magnetometric, and seismic methods. Geochemical methods employ chemical analysis of the cuttings (rock samples cut by drilling bit) and core (a narrow column of rock taken from the wall of a drilled hole) of the drilled site.

1.4.1 Gravimetric Method

Gravity of the earth's surface varies with distance from the surface of Earth and the type of material, such as salt, water, oil, gas, or mineral matter. The measurement of a small variation of gravity or acceleration due to gravity is recorded with accuracy and the data are converted to retrieve a geological structure of the sub-surface of Earth. A gravimeter is a very sensitive instrument, usually a spring-type balance with high resolution and accuracy capable of detecting a minute variation in gravity. Porous and oil-containing rock layers and salt have lower density compared to the surrounding non-porous and hard rock layers. Thus, a gravimetric curve is acquired and analysed for the location of deposit.

1.4.2 Magnetometric Method

Earth has its own magnetic field that varies from one location to another owing to the different structural materials of rocks and also the presence of solar-charged particles received by Earth. A variation of magnetic field strength is recorded by a sensitive instrument, called a magnetometer. Igneous non-porous rocks are found to be magnetic as compared to sedimentary rocks containing organic deposits. Thus, a magnetometric survey can also be used to locate oil deposits. Both the gravimetric and magnetometric methods are done simultaneously to predict a reproducible sub-surface structure. After the zone is confirmed by gravimetric and magnetometric surveys, a seismic survey is carried out for a clear image of the sub-surface structure.

1.4.3 Seismic Survey

This technique uses a sonic instrument over a desired site to correctly locate the prospective basin structure. In this method, a sound signal generated by the explosion method (explorers call them mini-earthquakes, which are artificially created by explosives) is transmitted through the earth's surface under study and reflected signals are detected by geophones located at specified positions. The frequency and time of the reflected signal varies with the density, porosity, and the type of reflecting surface. Various rock deposits at different depths vary with density and porosity. Seismic reflection can measure this change as it travels below the surface. Computer simulation software is used for imaging the sub-surface structure. This is applied to all the surveys for fast and accurate prediction about the oil and gas reserve location, well before a site is finally selected for drilling operations. It is to be noted that exploration has to be deterministic, but the availability of oil and gas is estimated based on probability.

1.4.4 Remote Sensing Method

Solar radiation from the Earth's surface varies in intensity and frequency depending on the sub-surface property. This observation is collected via satellite to predict the sub-surface structure. In order to image the sub-surface structure, historical geological data collected previously by gravimetric, magnetometric, and seismic surveys are used. The final image is obtained by geological imaging software (GIS). However, the remote sensing method is not applicable during nighttime or places incapable of reflecting solar radiation, like the ocean surface, which absorbs substantial amounts of solar radiation. However, extrapolation from the land surface in the vicinity of the sea can be accurately predicted, but is not applicable for the deep sea area. A radioactive or gamma-ray survey is also used in the exploration.

1.4.5 Geochemical Methods

Inorganic contents of surface or shallow cuttings or core are sampled and analysed for inorganic materials, such as salts and carbonates, which are frequently associated with hydrocarbons. *Organic contents* or the presence of organic matter is detected by heating a sample in a crucible and the loss of mass of the sample is an indication of the presence of organic matter. The ratio of organic mass to inorganic matter in a sample is used to ascertain the presence of hydrocarbons. *Total organic carbon* is defined as the carbon present in the organic matter in the sample which is different in inorganic carbon from carbonates. Core samples are examined for porosity, permeability, salt content, organic content, and many other physical and chemical properties.

1.4.6 Stratigraphy

Correlations are established between wells, fossils, rock and mud properties, before and during drilling operations for the final prediction, and this technique is known as stratigraphy. But it is important to remember that prediction from exploration may not be correct as far as the location and amount of deposit are concerned. It may happen that the drilling operation may not yield oil or the yield may not be sufficient at the explored site and that the expenditure borne by this work is irrecoverable. Hence, a more accurate determination of the location and economic deposit should be done before investing money in well construction. After confirmation from the test drilled hole, final construction is carried out.

1.5 RESOURCE ESTIMATION

The oil potential of a deposit depends on the pressure and temperature of the formation, the surface tension, the density and viscosity of the oil, the porosity and permeability of the rock, and so forth. The quantum of oil and/or gas present in the reservoir pores is called oil and/or gas *in place*. The amount of hydrocarbon oil that can be economically produced and marketed is called reserve. The oil and gas volume/quantities can be estimated by the volumetric method.

Volumetric oil *in place* in million metric tons is given by the relation:

$$AH\,\theta(1 - s_w)\rho_0 / b_0, \tag{1.3}$$

where:

A: area of oil pool in square kilometres
H: oil pay thickness in metres
θ: porosity of the reservoir rocks
ρ_0: density of oil
b_0: volume fraction of oil in the formation
s_w: fraction saturated by water in the pores

For gas *in place*, the following relation is used:

$$AH\,\theta(1 - s_w)\, p_r T_r / (Z p_s T_s), \tag{1.4}$$

where:

A: area of oil pool in square kilometres
H: oil pay thickness in metres
θ: porosity of the reservoir rocks
s_w: fraction saturated by water in the pores
p_r: reservoir pressure in the formation
p_s: pressure at the surface of earth
T_r: absolute temperature in the formation
T_s: absolute temperature at the surface

1.5.1 Effect of Pressure

At high reservoir pressure, the gas density is high and is dissolved in the liquid oil, and is thus amenable for the production of oil and gas without the aid of additional means of power for external pressurisation. In many reservoirs, methane is the major constituent of gas, which has a tendency to form hydrates with water at high pressure. Once formed, methane hydrates are difficult to disperse in the well and may damage the well piping due to abrasion. The formation of methane hydrates is also responsible for a reduction in the oil pressure of the reservoir. Methanol may be injected into such wells to disperse methane hydrates. With the production of oil and gas, the pressure of the well falls with time (years), and to maintain production, water or high pressure inert gas is injected into the surrounding wells to maintain the pressure of the producing well. Pressure in the bottom of the well at the formation can be measured with a remote access pressure gauge lowered through the well piping. Figure 1.9 presents a pressure profile, rate of production, and water injection rate with age of the producing well. There are four stages of production with ages. Stage 1 is the *baby well*, in which production is gradually rising and reaches its maximum. Stage 2 is the *young well*, which produces the oil at the maximum rate. Stage 3 is the *middle-aged* well when production starts decreasing, and finally stage 4 is the *old*

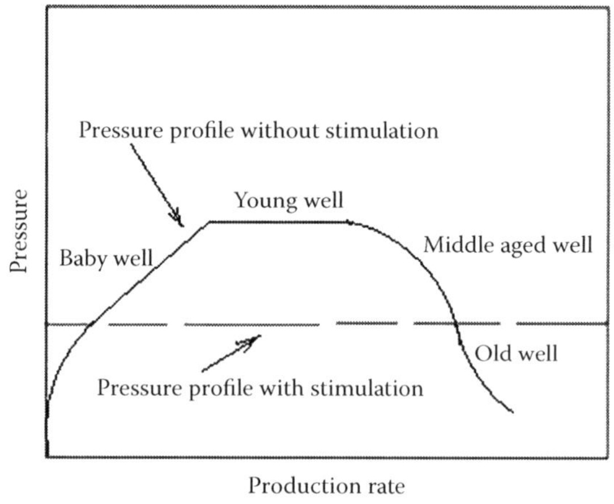

FIGURE 1.9 Pressure profiles for a producing well.

well, while production is very low and water cut is more than oil cut, it will continue until it becomes uneconomic to continue production. However, external pressurisation with water or inert gas should be planned from the beginning of stage 2 to maintain maximum pressure with controlled production. Water injection should be gradually raised to maintain the same pressure up to stage 4 to compensate for the fall in well pressure.

1.5.2 Connate Water

Oil, gas, and water are distributed in the reservoir according to their densities. Gas is also dissolved in the oil phase. Gas occupies the upper space, followed by oil in the next layer, and water in the lower part. Water occupies the major space of the reservoir. Oil, gas, and water in the reservoir are also present in the interstices of the porous rocks simultaneously. This water in the pores is called connate or interstitial water. It is important to take account of this connate water in estimating crude oil in the reserve. The greater the amount of this water, the lower the permeability for oil. Water in the reservoir usually contains mineral salts. Improper selection of sites in the surrounding wells for injection of external water or gas to pressurise the producing well may result in more water cut in the production.

1.5.3 Effect of Temperature

The temperature under the surface of Earth increases with depth. The rate of increase in temperature per 34 m of depth is called the *geothermal gradient*, which may be less than or greater than 1°C per 34 m. This may vary from well to well. The greater the thermal gradient, the more permeability of oil. Recently, attempts have been made to increase the bottom hole pressure by partial combustion of oil or injection of steam or hot gas into the surrounding wells.

1.5.4 Effect of Viscosity

Reservoir crude oil is classified as a viscoelastic fluid that exerts normal stress in addition to tangential stress developed while in the flowing condition. Thus, the flow behaviour of this type of fluid cannot be directly expressed in terms of Newtonian viscosity. However, for steady state flow, the relation for pseudoplastic fluid may be more applicable.

$$\tau = k\,(-\,du\,/\,dx)^n, \tag{1.5}$$

where τ is the shear stress, u and x are the velocity and distance, du/dx is the corresponding shear rate, k is the consistency factor (but not Newton's viscosity), and $n < 1$. However, k can be related with Newton's viscosity and can be used for the effect of temperature and other factors affecting viscosity. Production rate is inversely proportional to the viscosity of oil and directly to the pressure of the reservoir. In the well, recoverability of oil with respect to water is measured by the ratio (ξ) of viscosities of oil to water, which is

$$\xi = \mu_{oil} / \mu_{water}, \tag{1.6}$$

where μ_{oil} and μ_{water} are the viscosities of oil and water, respectively. The lower the value of ξ, the greater the oil cut and vice versa. This value increases with the age of the well and thus increases the water cut in the production. Attempts are made to inject polymer or high viscous compounds, which is readily soluble in water and increases the viscosity of water in the well. It is common to maintain a ξ value below 3 to have a greater oil cut in the production.

1.6 OIL FIELD DEVELOPMENT

Drilling is done to fracture and penetrate the rocky layers to reach the oil formation below the Earth's surface. A hollow steel pipe containing the drill bit with perforations at its mouth is used for drilling. Mud fluid is pumped through the top end of the drill pipe through a hose which moves down with the pipe as the drilling progresses. The drill pipe and the hose are suspended from the crown of a pyramidal structure called a rig. Figure 1.10 depicts a typical rig for drilling operations. A high pressure pump is employed to pump the mud solution from the mud pit through the hose such that the cuttings at the drill bit are washed out through the mouth of the drill bit and returned to the top surface through the annular space between the drill pipe and the hole developed. Cuttings with the mud solution are collected and separated from each other. Clarified mud along with fresh mud are pumped back to the drill pipe continuously. Mud is consumed due to absorption and seepage through the pores and crevices of the layers. Monitoring of the level in the mud pit is essential to assess the consumption and generation pattern of cuttings and water. An alarming decrease in the level indicates leakage through the layers due to seepage in crevices or channels. While an increase in the level indicates ingress of

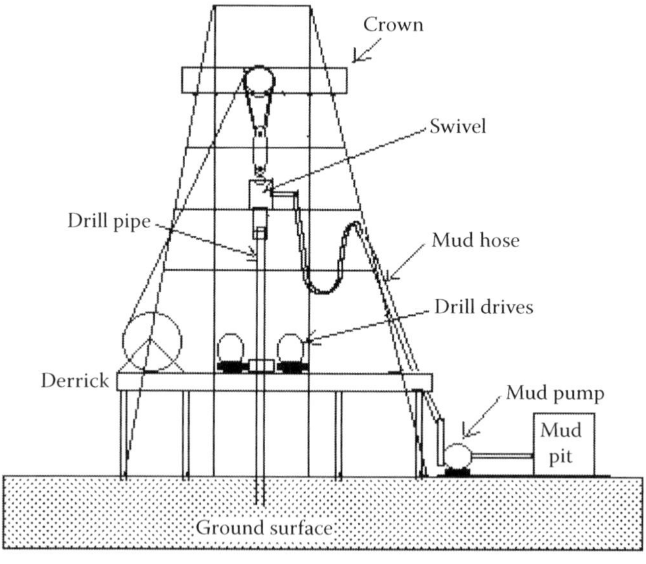

FIGURE 1.10 Schematic diagram of an oil rig for exploration.

underground water. Samples of drill cuttings are useful for surveying and assessing the direction of the drilling operation as they carry valuable information about the layers and formation. Continuous well logging is then carried out using a modern system of data acquisition and analysis. After drilling to a depth of 30–40 ft, a steel pipe is introduced into the hole to protect the wall of the hole formed. This is called the *casing string*, which is then cemented to the wall of the hole by pumping a fast-setting cement solution (usually Portland cement without sand) to the annular space between the pipe casing and the wall of the hole. This casing helps prevent caving of the wall and seepage of water from the layers. The drilling arrangement with a casing pipe is shown in Figure 1.11. An additional drill pipe is then joined of sufficient strength to withstand the various static and dynamic stresses for the increasing dead weight of piping, torsional stresses due to rotation, for upward and downward movement, abrasion from sand, fluid friction from mud fluid with cutting and corrosion, etc. The drilling operation is then continued and an additional casing pipe of a reduced diameter from the previous one is inserted and cemented at strategically located positions (for easy recovery of casing pipes after the well life is exhausted) until the target depth is reached. The final casing diameter may reach as small as 5–8 in.

At this stage, the top of the well (well head) along with the casing hanger are fitted with the necessary piping and collection headers. A pipe riser is inserted in the well to lift the oil and is connected to the well-head piping and valves. The diameter and design of the pipe riser (tubing) may differ depending on the facility of the oil lifting mechanism. The well-head connection consists of a tubing header and a Christmas tree header for collection of oil, gas, and water to the respective storage tanks. Such a complete well is presented in Figure 1.12. The surface of the casing pipe at the desired target depth is punctured by bullet or missile firing by experts. The hydrostatic pressure of the mud fluid in the well hole balances the reservoir pressure, thereby preventing spouting of the well from the formation.

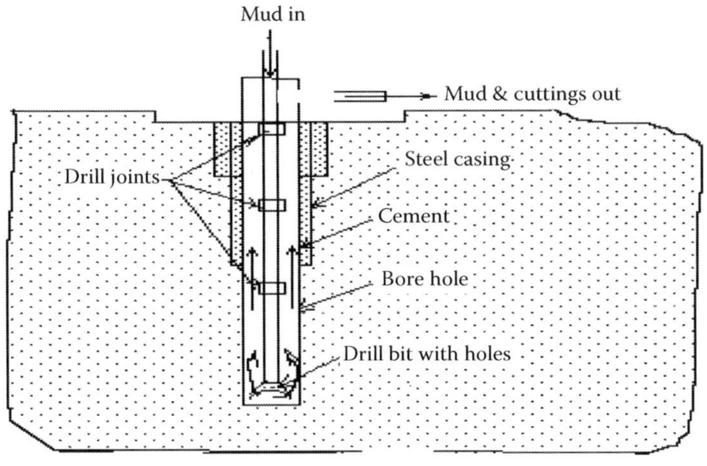

FIGURE 1.11 Drilling operation in a well.

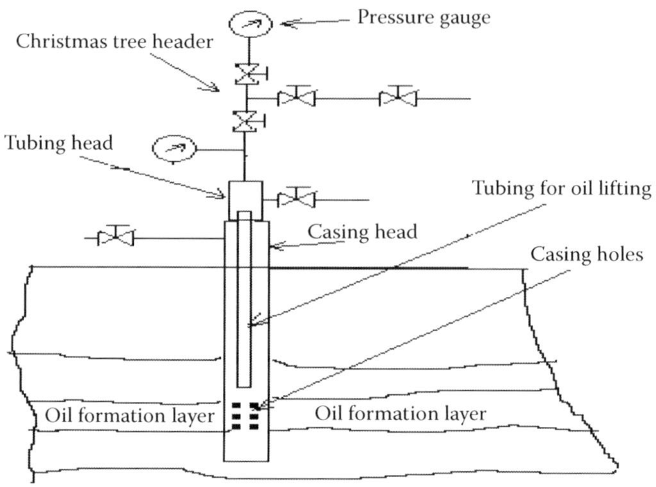

FIGURE 1.12 A completed well ready for production.

1.7 WELL LOGGING

Well logging is a continuous recording process of the activities during drilling, well development, and production until the closure of the well. Thus, the record identifies the history of the well. Well logging is carried out during the drilling operation using special probes (electrical resistivity, inductance, or magnetic resonance), physical sampling of the drilled soils and rocks, core samples, monitoring drilling fluid, etc. Various parameters, such as porosity, permeability, and water saturation in oil, of the formation are also obtained by the resistivity probes. During the drilling operation, information about the drill bit, its movement, and direction are determined by these probes. The direction of drilling is ascertained by the dipole sharing investigation tool (DSI). Information is also gathered to release drill bits stuck in the well, monitoring the perforation operation of the casing to communicate with the formation, the properties of oil and gas in the formation, etc. At various stages of production, well probing is used to inspect the casing, the wall of the uncased well, etc., for necessary maintenance operation of the well.

1.8 OIL PRODUCTION PROCESSES

The *gas lift* method employs high pressure gas, usually air or carbon dioxide, which is introduced into the well through the annulus and oil is carried through the inner tubing, leading to the well-head piping. Initially, the well is filled with the mud fluid and the oil cannot move up owing to the hydrostatic head of the mud fluid. As the gas enters the annulus and piping, the density of the mud column decreases and the hydrostatic head decreases, and as a result, the mud fluid is lifted by the oil pressure. A mud–oil mixture is collected and separated on the surface tanks. When complete displacement of mud takes place from the well and from the pores of the layer near

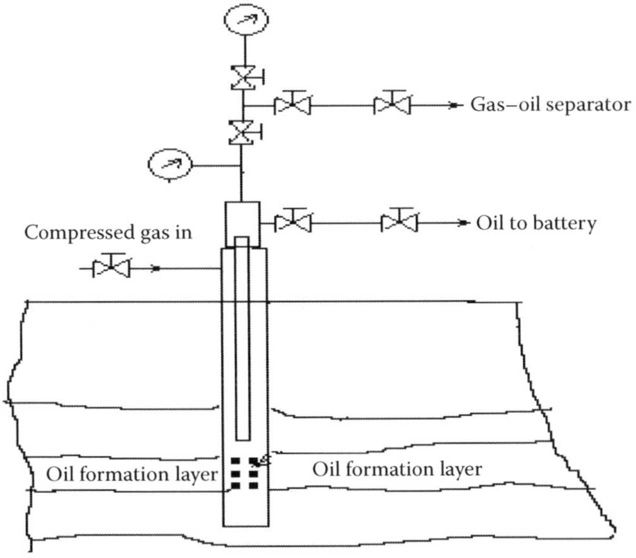

FIGURE 1.13 Scheme of a gas lift method of production.

the borehole by the oil pressure, oil production starts increasing. Such a schematic gas lift operation is shown in Figure 1.13.

A *sucker rod lift* well contains a piston (or a plunger) pump lowered into the inner tubing. The piston is operated by a metallic wire or rod leading through the tubing and above the well head and connected to a wire rope from a hanger attached with a reciprocating driving system at the base of the well head. The piston is contained in a cylinder with non-return valves fitted at both ends. During the upstroke of the piston, the bottom valve opens, keeping the top valve closed and, as a result, the cylinder pressure falls below the reservoir pressure, forcing oil to enter the cylinder. While during the down stroke of the piston, the upper valve opens and the bottom valve closes and oil in thc cylinder is pushed up to the tubing through the upper valve. Thus, the volume of oil displaced upward in the tubing is proportional to the stroke length of the piston. When the tubing is filled with oil after repeating the reciprocating operation, oil starts flowing upward and is collected. A schematic sucker rod lift arrangement is shown in Figure 1.14. A *submersible pump well* contains a centrifugal or screw pump installed in the tubing lowered into the borehole. Both the electric motor and the pump are submersed in the well bottom. Electric cable sealed in a flameproof arrangement is lowered into the well hole through the tubing. The motor is usually kept below the pump in the tubing. Pumps are small in diameter (3–6 in), multistage centrifugal or screw pumps. Since entrainment of sand particles and gas may cause problems to the centrifugal pumps, modern wells are using high capacity multistage screw pumps that can carry slurries, viscous oil, and even gas.

In fact, future wells will deliver more viscous oil contaminated with sand and clay materials, therefore, increasing use of submersible screw pumps will take place in modern and existing wells. Modern screw pumps with a diameter as small as 6 in with a capacity of 100 m^3 or more per day and with a head of 1000 m are being used

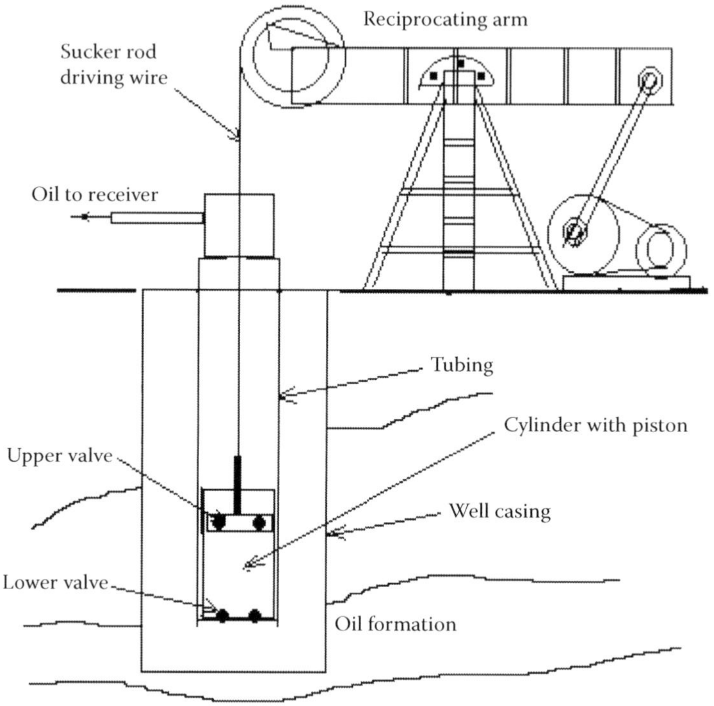

FIGURE 1.14 Schematic arrangement of a sucker lift production well.

in wells. The number of stages of a pump may be more than 100 tightly fitted in a tubing. A submersible pumping well is shown in Figure 1.15.

The *hydraulic pumping* method employs a special type of tubing that consists of two tubes. The inner tube is of a larger diameter in which the plunger or the diaphragm pump is lowered into the borehole. The plunger or the rod of the diaphragm is forced by pumping a liquid over it in a reciprocating manner. Oil is discharged through the outer pipe through its annular space and is delivered to the surface tank. This method does not require lowering any electrical cable and no wire for actuating the plunger. A high-pressure reciprocating surface pump delivers the liquid forced up and down the plunger of the pump in the borehole in a reciprocating manner. The plunger pump can be withdrawn on the surface from the inner pipe by forcing liquid through the annular outer pipe. Such a pumping arrangement is shown in Figure 1.16. The rate of production from a single well may not be large. Hence, a good number of wells, varying from 100 to 1,000 wells depending on the rate of production, are drilled in the area where the formation is spread. Excitation (stimulation) of the wells by gas or water injection from the surrounding injection wells (judiciously located) is extremely necessary to increase reservoir pressure to the flowing wells. Modern methods also employ combustion of oil in the surrounding wells to push the oil in the formation by heat effect on reducing viscosity in the porous channels of the formation. A proper temperature gradient is essential from the channels of the combustion zone to the target well. Crude oil from all these wells are collectively routed to storage and conditioning.

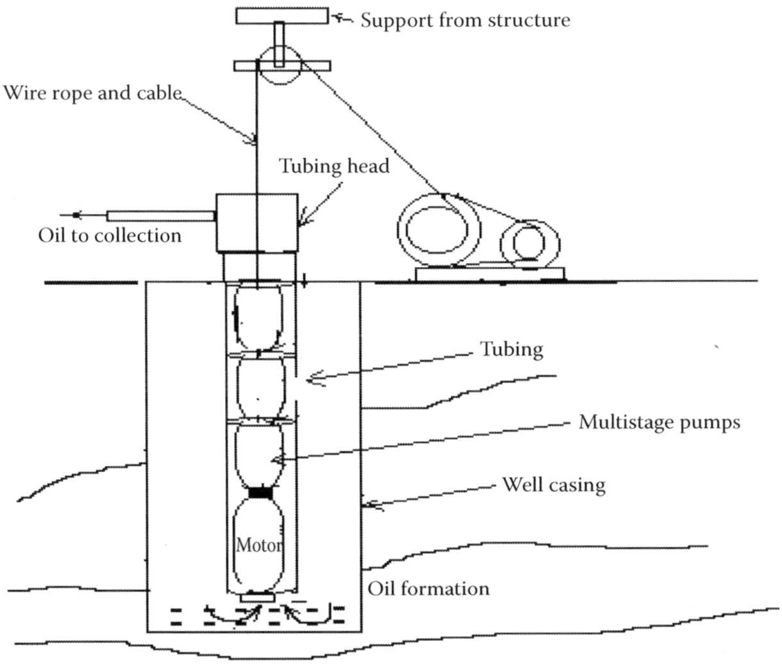

FIGURE 1.15 Schematic submersible pumping arrangement for oil lifting.

1.9 CRUDE CONDITIONING AND STORAGE

Hydrocarbon gases and light and heavy hydrocarbon liquids are all present in a single homogeneous phase under pressure in the formation before it is drilled. After this is released from the formation, components of crude oil separate into layers of light and heavy hydrocarbons. Raw crude oil collected from the wells also contains sand, mud, and water as impurities which may vary from 20% to 30% by volume. Hence, raw crude is collected in a battery of treatment tanks where both storage and treatment of crude oil are carried out. Treatment steps involve gravity settling and removal of sand and water followed by chemical treatment to remove emulsified water and, finally, to a crude conditioning unit. Gases lighter than propane have a tendency to escape, whereas propane and heavier gases are found dissolved in crude oil at atmospheric pressure. Proper mixing and repeated heating above room temperature (usually 45°C–50°C at low pressure and up to 90°C at a pressure of 2–3 atm) followed by cooling to storage temperature can increase the dissolution of gases and homogenise the layers of light and heavy liquid hydrocarbons.

Segregated wax in crude oil may choke the pipeline and pumping equipment due to deposition. Heating of crude with a mixing facility reduces segregation of wax by making it uniformly distributed in the bulk and thus it can be stored and transported without risk of deposition for many hours at room temperature or lower temperature above the *pour point*. Asphaltic and heavy hydrocarbons with high viscosity are also mixed up with wax and other hydrocarbon components during the heating and cooling cycle, making them less viscous. Heating can be

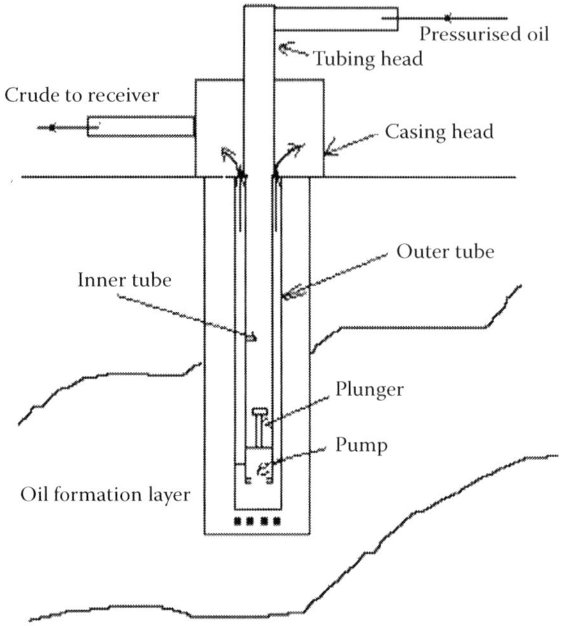

FIGURE 1.16 A schematic hydraulic pumping arrangement for lifting oil.

provided by a steam coil in the storage vessel with mechanical mixers, and cooling is done in another vessel with cooling coils in which refrigerants may be the vaporizing hydrocarbon gas or other liquids. Heating and cooling vessels are connected in sequence. Conditioning of crude oil can also be done by heating and cooling in sequence in pipe coils at the required temperatures. Mixing is enhanced for high velocity in the pipe coils. This type of conditioning method reduces the cost of storage and loss of gases. Conditioning may also be used for mixing oil and gases from various wells in the desired proportion to meet the desired quality. Water and slop (oil and water mixture) from the treatment unit is also treated in the battery before recycling to wells.

1.10 TRANSPORTATION AND METERING OF CRUDE OIL

Treated crude oil is received in large storage tanks usually under pressure to avoid loss of hydrocarbon vapours and is despatched by tankers (ships), trailers (large tank cars), and most conveniently through pipelines. Pipelines as long as 1000 m or more from the oil field tanks to the refineries or to the shipping ports are most common in any oil-producing country.

Booster pumping stations are placed at the required positions to maintain delivery pressure to the receiving ends. High pressure centrifugal or screw pumps are employed for pumping through pipelines. The horse power of such pumps may vary from 500 to 2,000 hp with a capacity to transport 500–1,000 m^3/h with a discharge pressure of 100 atm or more. Power consumption depends on the pressure loss and rate of delivery, which is ultimately determined by the viscosity of the oil, the roughness of the surface, and the diameter and length of the pipe, flow rate, etc. Injection of methane in the pipeline may be useful for boosting pressure and this is separated from crude only at the refinery. However, this is not suggested for loading in a tanker or while crude contains water, which may form abrasive methane hydrate crystals at low temperature. Quantity loaded in the vessels of the tanker or trailer or to refinery tankages are measured directly by physical dipping using calibrated dip rods or tapes as required. Flow rate through a pipeline is measured by low-pressure drop flow meters to reduce power loss. Ultrasonic, electromagnetic, or tracer type flow meters with signal-generating devices are suggested as these do not come in contact with the oil and hence no pressure loss occurs. Pressure sensing gauges are also installed at vital distances of the pipeline to monitor any failure of pumps or oil leakage in the line. Electrical signals from the flow and pressure sensors are transmitted continuously to control rooms at the originating and booster pumping stations.

Nowadays, verbal and digital communication and recording between these stations are maintained by computers. Private microwave towers are also built for this purpose. Since pipeline routes are on the surface, underground, or under river or sea bed, it may be necessary to have proper vehicles to check the pipelines for maintenance. Helicopters are commonly employed to access remote places. Pipeline washing is carried out using a leather or polymer ball (known as pig) pushed by pump pressure at one end and collected at the pig-trapping-pit end before reaching the suction of the booster pump. Pressure safety valves are located at places to release oil or gas to prevent damage to the pipeline during excessive pressure rise owing to blockage of the line for any reason.

1.11 GAS HYDRATES

Methane occurs naturally due to the biological decomposition of organic matter available on the earth's surface and is frequently manifested as marsh gas from wastelands, ponds, etc. Large gas reserves under the earth have been detected where methane exists as crystals (gas hydrates) formed due to the interaction with water and methane molecules as a clathrate compound at high pressure and low temperature, usually below 20°C. Because of the abundance of methane gas formed in the geological period under the earth's surface, usually in the Arctic region below the sea bed which is enough to saturate water at a very low temperature, methane hydrates are formed. Gas hydrates have been found at depths of 200 m to 1,000 m in the sea. Below this depth, ice layers, permafrost (permanent frost), of water is prevalent. The temperature of sea water decreases as depth increases and the rate of decreases in temperature with respect to the depth is known as hydrothermal gradient. While the geothermal temperature below the sea bed increases at a rate of 34°C per km. Thus, gas hydrates exist only a few hundred meters below the sea bed where the temperature is below 20°C and after this depth it cannot exist because of higher temperature due to geothermal conditions. Gas hydrates are lighter than water once dislodged from the bulk it floats on the sea water and gradually releases methane from the sea surface. Hydrates are stable up to a maximum temperature of 15°C. Hence, if the temperature increases above 15°C, gas will be released from the hydrate. In fact, other hydrocarbon gases, such as ethane, propane, butane, carbon dioxide, and helium, may also be present in the hydrates along with methane. The presence of these gases also contributes to the stability of gas hydrates in a range of temperatures. Hydrate reserves are identified by the seismic exploration method. Approximately, 164 Nm^3 of methane is available from 1 m^3 of gas hydrate containing 0.2 m^3 of gas and 0.8 m^3 of water. Exploration is carried out during winter while the sub-sea level is favourable for stable gas hydrates in the reserve.

1.11.1 Production Method

A well is drilled and hot water is introduced to release the gas. But propagation of heat through the well to the reserve will cause the release of gases from the surroundings of the well, as a result a special type of dome-shaped collecting device is used. Methane is traced by infrared (IR) sensor. A schematic production method is presented in Figure 1.17. The economics of the production is dependent on the cost of energy supply for generating hot water by burning a part of the recovered gas and cost of collection. The ratio of heating value of methane to the heat required to release methane per cubic meter of hydrate is a good indication of the economic viability of such a production method. Compressed carbon dioxide as the product of combustion in this process is injected back into the reserve to replenish the gas collected. Since gas hydrates are available in the deep sea area, exploration and production require huge investments and operating expenses. Storage of methane and its transportation are also troublesome due to gas hydrate formation. In a pipe transfer under the sea or low temperature area, methanol is injected to avoid hydrate formation at the prevailing temperature. Though large amounts of gas hydrate reserve is available in the permafrost zone, it is difficult to produce with the existing technology.

1.12 COAL BED METHANE

Methane is strongly adsorbed by coal and hence coal mines contain methane in the coal bed. Methane obtained from the coal bed is called coal bed methane (CBM).

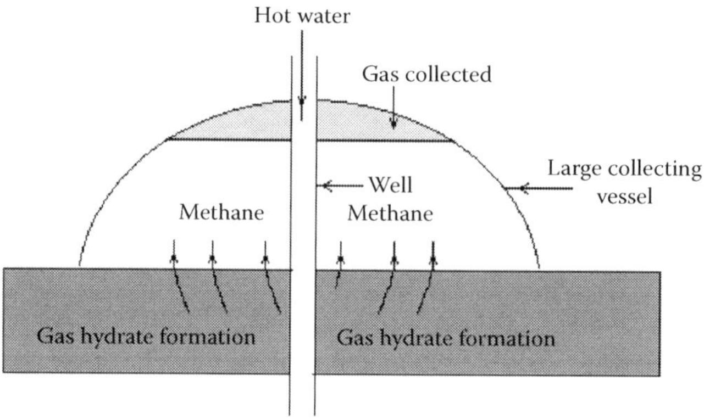

FIGURE 1.17 A typical gas production well from a gas hydrate reserve.

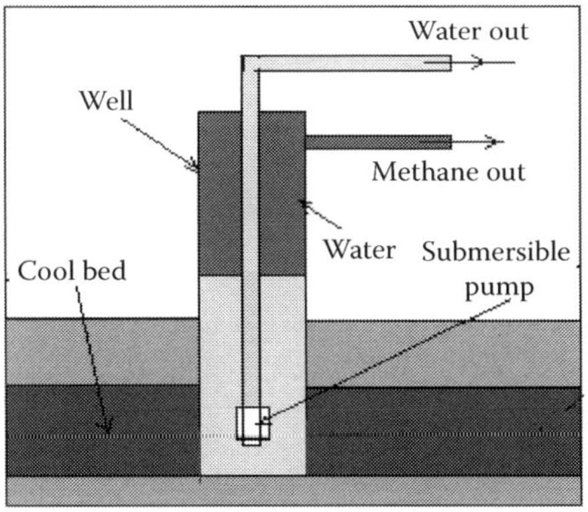

FIGURE 1.18 A coal bed methane well scheme.

Hazards in coal mining are commonly encountered owing to the presence of methane, which is poisonous and explosive. Extraction of methane from the coal bed is a recent activity by the oil and gas companies. In fact, methane extraction can reduce the risk of coal mining and generate methane as a valuable fuel. However, production of methane from the coal bed is based on some fundamental rule of adsorption. Methane and carbon dioxide are usually adsorbed in the coal bed. Both gases are strongly adsorbed, in fact, carbon dioxide is more strongly adsorbed than methane. Coal bed well is usually full with water, which exerts a hydrostatic pressure over the bed resisting desorption of gases from the bed. The presence of water causes carbon dioxide to be present in the water phase due to high solubility, whereas methane remains adsorbed in the coal. According to the adsorption isotherm, the quantity of adsorbed methane increases with pressure. Thus, if pressure is reduced by dewatering, methane will be desorbed from the bed. With this theory, methane is extracted from the coal bed by dewatering through a well. Thus, methane comes out from the annular portion of the well piping. This is schematically explained in Figure 1.18.

QUESTIONS

1. What are the types of hydrocarbon compounds present in crude oil?
2. What does potential of crude reservoir mean?
3. What are the scientific methods for exploration of crude reserve?
4. What are the various methods of production of crude oil from reservoirs?
5. Why is crude oil conditioning required?
6. What are the physical properties of drilling mud considered important during drilling of a potential reservoir?
7. How is crude oil stored after production in a field?
8. What are the various modes of transportation of crude oil?

2 Petroleum Products and Test Methods

2.1 CRUDE OIL ANALYSIS

Crude petroleum oil is a mixture of hydrocarbons. The hydrocarbon gases, methane, ethane, propane, and butane, are present in crude oil in its dissolved state. Methane has a high vapour pressure and it escapes from crude oil unless pressure above the vapour pressure is maintained. Usually it separates out from crude oil itself from the well and is collected separately as *natural gas* and a trace of it may be found in crude oil under atmospheric pressure. Though ethane has a higher vapour pressure than propane and butane, traces of it are usually found in crude oil right from the well. Propane and butane are present in the liquid state at slightly above atmospheric pressure owing to their low vapour pressure and high solubility in crude oil. The propane–butane mixture is separated from crude oil and is used as liquified petroleum gas (LPG). The remaining liquid hydrocarbons can be separated as boiling fractions such as naphtha cut (boiling up to 140°C), kerosene cut (boiling between 140°C and 270°C), gas oil or diesel cut (boiling between 270°C and 350°C), by heating and vaporising the crude oil by a gradual increase in temperature followed by collection after condensation. When crude oil is heated in a distilling flask, vapours start emanating as the temperature rises and these vapours are collected after condensation using ice cold water. The temperature of vapour giving the first drop of condensate is reported as the initial boiling point (IBP), which may be above or below 0°C depending on the presence of the lowest boiling hydrocarbon in crude. This vaporizing phenomenon is so fast at the beginning that temperature measurement is quite uncertain as the vapour of the first drop is immediately followed by the mixture of vapours with increasing boiling points, hence this must be noted in the shortest possible period (within 5–10 min) after the charge is heated. As heating is continued, more and more hydrocarbon vapours with increasing boiling temperature are separated from crude and collected as condensates. This process is continued with gradual heating until no further vaporisation takes place. The vapour temperature and the volume of liquid condensates (boiling fractions) collected are measured and reported as the distillation analysis of crude oil. In the laboratory, such a batch distillation is carried out for a specified amount of crude oil (500 cc, 1 litre or more). During distillation at atmospheric pressure, the rate of vaporisation decreases gradually after 40–45% of crude is distilled and, finally, negligible or no vaporisation takes place. If attempts are made to increase the temperature by further heating, the residual crude in the flask may undergo thermal cracking (break down of hydrocarbon compounds present originally in the charge), which is a reaction phenomena other than distillation. Alternatively, when heating of the residual oil is continued under vacuum, vaporisation is restored and condensates are collected. Distillation, in fact, is done at atmospheric pressure and then followed under vacuum. The separation of hydrocarbons can be improved if the vapour and its condensates (*reflux*) are in intimate contact for some time during distillation and a reproducible distillation analysis is possible. Such a method of distillation is known as true boiling point (TBP) distillation. A reflux ratio of 5:1 is usually sought as the standard separation. Separation of fraction by 1% volume can be obtained at a very close temperature difference and such a close separation is comparable with a distillation tower of 10–12 plates. A routine crude distillation test is carried out in a standard TBP distillation apparatus in a packed column of specific size and packing material with a certain amount of reflux ratio (reflux/vapour ratio). A TBP apparatus is shown in Figure 2.1 and a typical TBP analysis curve is shown in Figure 2.2.

Usually, the composition of crude oil varies from well to well and it is essential to have a separate TBP analysis for each batch of oil purchased before refining. Hence, the yields of light to heavy fractions will vary from crude to crude. A small variation in TBP analysis may have a wide variation in the chemical constituents of the boiling fractions. This analysis indicates the maximum possible yields of raw cuts of products that could be obtained by distillation. In a refinery, a distillation column is used to separate these cuts in large rates *continuously*, where the yields of the raw cuts are slightly different from the TBP analysis. In fact, a distillation column is designed on the basis of the TBP analysis of a crude oil. Crude oils having wide difference in TBP analysis cannot be distilled in the same column without sacrificing the yields

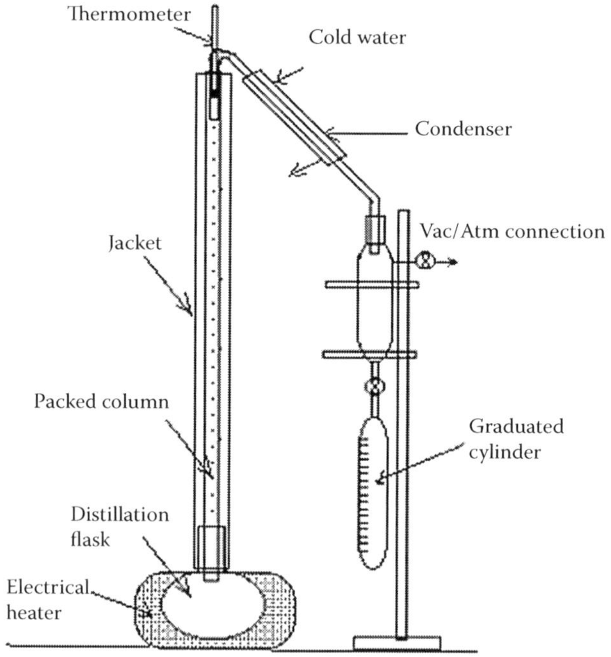

FIGURE 2.1 A TBP distillation apparatus.

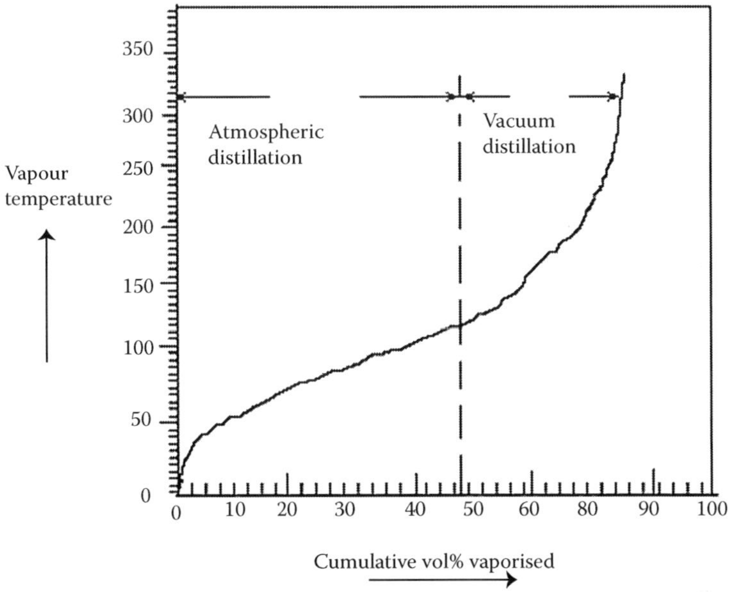

FIGURE 2.2 TBP distillation analysis of a crude oil.

and quality of the products. It is, therefore, inevitable that the design of a distillation column of a refinery must be done judiciously, depending on the availability of the crude oils, which may have little difference in TBP analysis.

2.1.1 API Gravity

API is a gravity measuring parameter for all petroleum oil. This is related as

$$API = 141.5/s - 131.5, \quad (2.1)$$

where "s" is the specific gravity of oil at 15.5°C (60°F) with respect to water at the same temperature. The greater the specific gravity, the lower the API gravity. For water, API gravity becomes 10 and for oil it is greater than 10. Crude oil having an API gravity as low as 9 has been found, which is heavier than water but most commonly it is always greater than 10. Thus, high API gravity crude oil is rich with lighter fractions and thus costlier. The price of oil is fixed depending on the API gravity as it is an easily measured entity and is directly related to the presence of lighter hydrocarbons.

2.1.2 Characterisation Factor

The next entity is the characterisation factor (CF), which is most commonly used with API gravity to judge the quality and many physical properties of crude oil and its products. This is defined as the ratio of cubic root of the molal average boiling point (T_b, in Rankine) of oil to its specific gravity (s) at 15.5°C. Thus, it is expressed as

$$CF = (T_b)^{1/3}/s. \tag{2.2}$$

2.1.3 Bottom Sediment and Water

Bottom sediment and water (BSW) is a measure of the quantity of residual sediment mostly settleable from the crude oil (if sufficient time is allowed for settling) and water. This may contain both heavy asphaltic hydrocarbon oil and non-hydrocarbons, such as inorganic salts. An amount of BSW is routinely tested for every batch of crude oil received in a refinery. Water, salt, and sediment are removed from the storage tanks, followed by electrical desalting.

Throughout the world, petroleum products are tested according to the methods and equipment specified by the American Standard for Testing Materials (ASTM). Detailed testing procedures are available from the ASTM handbooks for testing petroleum products. Some of the common properties and an outline of the testing methods for petroleum products are presented in the following sections.

2.2 DOMESTIC FUELS

2.2.1 Liquified Petroleum Gas

Liquified petroleum gas (LPG) is a mixture of propane and butane in liquified form at a pressure of about 10–15 atm, depending on the proportion of the components present. This is the first product of distillation from crude oil. Raw hydrocarbon gases contain sulfurous hydrocarbons, which may be odorous and corrosive. Hence, treatment of these gases is carried out to remove unwanted impurities, such as mercaptans, hydrogen sulfide associated with the gases, and moisture. LPG is tested for its quality as required by its use as a domestic cooking gas. One of the simplest tests of LPG quality is the *weathering test* (Figure 2.3).

In this test, 100 mL of liquid LPG is kept in a test tube with a mercury-in-glass thermometer dipped into it. Liquid LPG starts vaporising through the open mouth

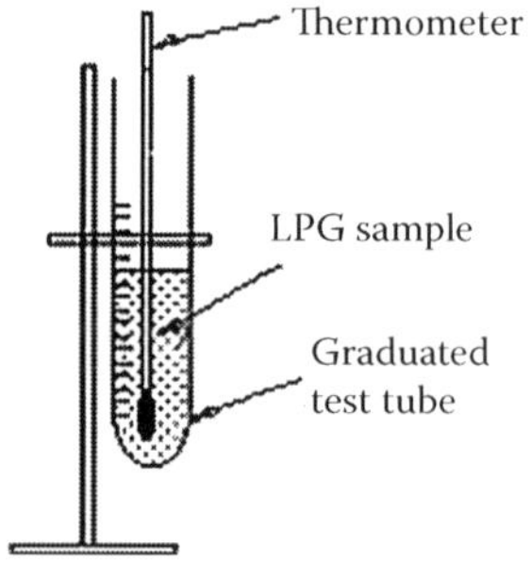

FIGURE 2.3 Weathering test of LPG.

of the test tube to the atmosphere, causing the temperature to fall rapidly. After 95 mL of liquid is vaporised, the residual liquid temperature is noted. If the temperature is very low (below 0°C), then the proportion of propane in the mixture is more and if it is high (above 2°C), then the proportion of butane will be more. For domestic use, the maximum weathering temperature should be 2°C. Slightly lower than this temperature may be useful, but the carrying capacity in the cylinder will be reduced. Whereas, higher than this temperature should not be acceptable because incomplete vaporisation may occur at a low ambient temperature (e.g., in winter or in hilly places like Darjeeling). In order to avoid excessive pressure in the filled cylinder for the required mass (usually 14.2 kg for domestic cylinders) it is required to test the vapour pressure of LPG at a temperature slightly above room temperature. This is measured in a high-pressure vessel (known as a Reid vapour pressure vessel) fitted with a pressure gauge and a constant temperature bath with a thermometer. The experiment is known as Reid's vapour pressure test and the pressure measured is reported as Reid vapour pressure (RVP). Domestic LPG has a maximum RVP of 8 kg/cm^2 at a temperature of 37.8°C. A RVP measuring arrangement is shown in Figure 2.4.

In fact, commercial LPG may not have a uniform composition of only propane and butane, but may contain an amount of lighter gases in traces and also contain some olefinic gases obtained from petrochemical plants. The presence of mercaptans and hydrogen sulfide causes corrosion to vessels, pipelines, and joints, which may be further aggravated by the presence of moisture. LPG is tested for its corrosive properties by the *copper strip corrosion* test. In this test, a standard polished copper strip is dipped into liquid LPG in a thick-walled small vessel, known as a *bomb*, kept at a constant temperature, usually 30°C. After 1 h (or maybe more depending on its end use), the bomb is opened and the copper strip is removed and compared its brightness of surface against a set of standard copper strips indicating different degrees of corrosion as ASTM 1, 1a,1b, 2, 2a, etc. The desired specification is that the surface brightness of the strip under study should not be any worse than ASTM 1. Hydrogen sulfide in LPG is usually not present in the domestic cylinder. As, in the refinery hydrogen sulfide is removed from hydrocarbon gases by absorption in a solvent (usually in diethyl amine, DEA). Traces of hydrogen sulfide and mercaptan can

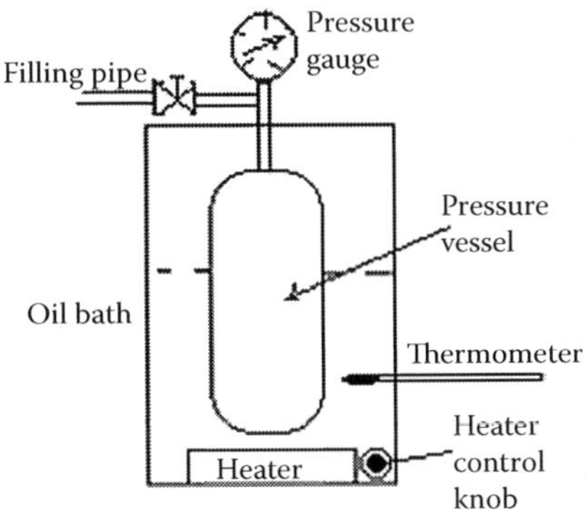

FIGURE 2.4 RVP measuring experiment.

be measured by Doctor's test, which uses a solution of sodium plumbite and traces of elemental sulfur for the black precipitation of lead sulfide owing to the following reactions.

$$H_2S + Na_2PbO_2 = PbS \downarrow + 2NaOH, \tag{2.3}$$

$$2RSH + Na_2PbO_2 + S = PbS \downarrow + 2NaOH + R_2S_2. \tag{2.4}$$

Removal of sulfur compounds from LPG is discussed in detail in Chapter 3. The presence of *moisture* in LPG is troublesome as it may be converted to ice during vaporisation (due to the latent heat of vaporisation), which will ultimately choke the valve and may damage it. Pipelines that transfer LPG and the pumping equipment may be damaged due to the formation of ice. Hence, LPG should be dry and no free water should be present.

Important parameters as per the Indian Bureau of Standard (BIS) of LPG

Property	Specification
Copper strip corrosion, 1 h at 38°C	Not worse than ASTM 1
Dryness	No free water
Odour	Easily detectable
Total volatile sulfur	0.02 max by weight
Vapour pressure at 37.8°C, max	8 kg/cm^2
Volatility, maximum temperature for 95% by volume evaporation at 760 mm Hg pressure	2°C

2.2.2 Kerosene

Kerosene or superior kerosene oil (SKO) is another domestic fuel mainly used for lighting or lamp oil. It is also used as a domestic stove oil. It is a petroleum product that boils in the range of 140°C–280°C and is available from crude petroleum oil. It is heavier than naphtha or petrol (motor spirit, MS), but lighter than diesel oil. The major properties of kerosene that determine its burning quality are the smoke point and the flash point.

2.2.2.1 Smoke Point

The smoke point is determined as the height of the flame (in millimetres) produced by this oil in the wick of a stove or a lamp without forming any smoke. The greater the smoke point, the better the burning quality. Domestic kerosene should have a smoke point of 20 mm (minimum). This is measured in a standard testing apparatus consisting of a standard lamp with a wick of specified dimension and mass. In this apparatus, a mirror is provided to position the flame in the centre such that the straight height of the flame can be measured in a graduated scale. Smoke is produced mainly due to the presence of carbon and heavy hydrocarbon particles in the flue gas. The presence of aromatic hydrocarbons contributes carbon atoms when burnt. In the refinery, aromatic hydrocarbons are removed by extraction to a desirable extent so that the smoke point becomes greater or near 20 mm. A smoke point testing apparatus is shown in Figure 2.5.

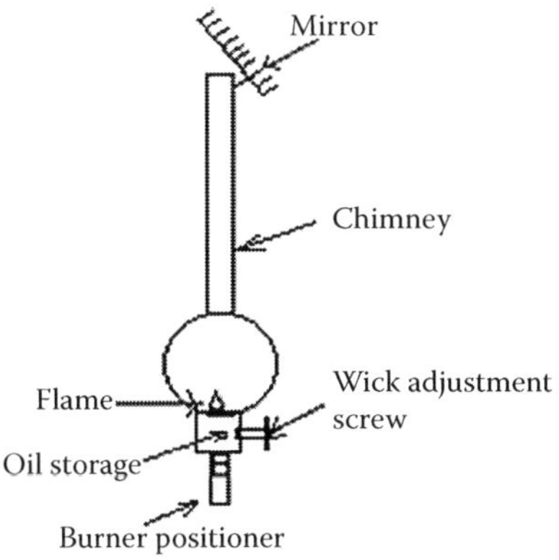

FIGURE 2.5 A smoke point testing apparatus.

2.2.2.2 Flash Point

The flash point is defined as the temperature of the oil at which it momentarily flashes in the presence of air and the igniting source. For domestic kerosene, it should not be below the ambient temperature. In India, this value is 35°C (minimum). This is determined in the laboratory using a standard cup, standard flame, and under standard atmospheric conditions. There are two standard test methods for measurement—the open cup method (Abel's method) and the closed cup method (Pensky–Mertinus method). In either test, oil is taken in a closed container provided with a stirrer and a heater as shown in Figure 2.6.

The temperature of the oil in the bath is measured by a thermometer. The oil is heated at the rate of 5°C–6°C per minute and a standard flame is introduced every 30 sec. The temperature at which a momentary flash occurs is noted as the flash point.

2.2.2.3 Char Point and Bloom

Other burning qualities of kerosene are char point and bloom formation. Char point is defined as the coke and ash left on the wick after complete burning of the oil. For domestic kerosene, the maximum amount of char allowed is 20 mg/kg of oil burnt. If the oil contains more aromatics, more organic char may be found. Bloom is the

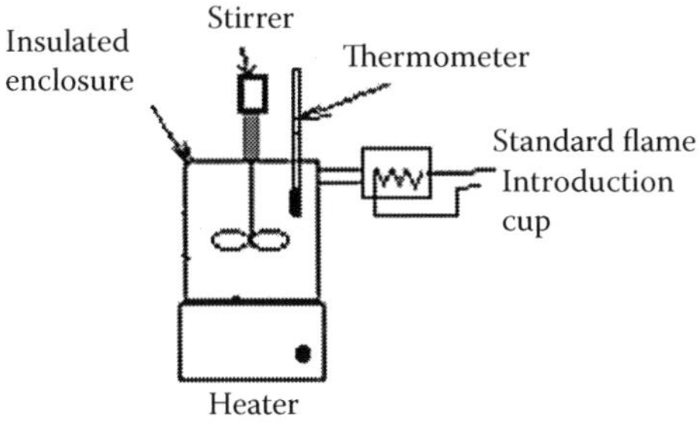

FIGURE 2.6 A flash point testing apparatus.

darkness produced by the flame of the oil while burning in a standard glass. The bloom should not be darker than a standard brightness desired in a lamp.

2.2.2.4 Distillation Test

ASTM distillation of kerosene is carried out under atmospheric conditions. Finished kerosene must be at least 20% distilled at 200°C or lower and the final boiling point (FBP) should be no more than 300°C. Kerosene that is too light is dangerous for use in a domestic kerosene stove as explosion may occur owing to the presence of a very light boiling fraction especially in the range of naphtha. The presence of too heavy boiling fractions in kerosene will make it a poorly burning fuel.

2.2.2.5 Sulfur Content and Corrosion

Since sulfur produces hazardous sulfur dioxide, kerosene must be free from it, although a trace of it is always present. Domestic kerosene must not have a sulfur content of more than 0.25% by weight. A corrosion test of kerosene is also carried out using the copper strip corrosion method. Corrosiveness of oil will damage the storage tanks, stove containers, pipelines for transportation, etc. For domestic kerosene, it should not be worse than ASTM 1 in the copper strip test for 3 h at 50°C in a standard bomb.

Important parameters as per the Indian BIS of kerosene

Property	Specification
Copper strip corrosion, 3 h at 50°C	Not worse than ASTM 1
Char value, max	20 mg/kg of oil
Bloom on glass chimney	Not darker than grey
Smoke point, min	20 mm
Flash point (Abel), min	35°C
ASTM distillation:	
Recovery at 200°C, min	20% vol
FBP, max	300°C
Sulfur total, max	0.25 wt%

2.3 AUTOMOTIVE FUELS

2.3.1 Motor Spirit

MS is known as petrol in India and Europe but as gasoline in the U.S. It is an automotive fuel for running motor cars. Diesel is the other automotive fuel.

There are two types of vehicles in India, classified as petrol and diesel vehicles according to the type of automotive fuels they use. The engine of a petrol car is different from the engine of a diesel car. Petrol or MS is suitable for spark ignition (SI) type engines and diesel for compression ignition (CI) engines. In the SI engines, fuel is burnt directly by introducing an electrical spark into the mixture of air and vapour of MS through the carburettor (a mixing device of air and vapour of fuel) of the engine. Power is developed due to the volumetric expansion of burnt gases (flue gas) in the engine cylinder during the expansion stroke (forward stroke) and exhausted

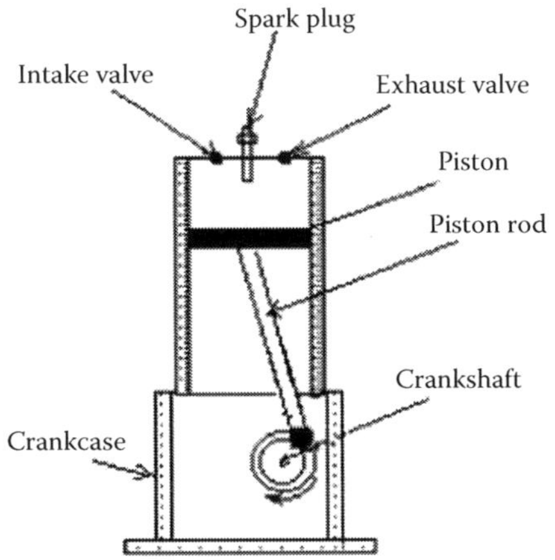

FIGURE 2.7 Schematic representation of an internal combustion engine.

to the atmosphere in the compression stroke (return stroke) in a repeated cycle. The power developed is dependent on the quality of the fuel and the design of the engine. The working principle of an internal combustion engine is shown in Figure 2.7.

2.3.1.1 American Standard for Testing Material Distillation

MS contains a mixture of hydrocarbons boiling from 30°C to 40°C and up to a temperature slightly above 200°C. In the laboratory, MS is distilled at atmospheric pressure according to the standard ASTM method of distillation. A sample of 100 mL is placed in a standard distilling flask and the vapour is condensed through a condenser. Liquid is collected in a graduated cylinder. At the beginning of distillation, the vapour temperature is reported as the IBP while the first drop is collected (Figure 2.8). Then, distillation is continued and the temperature of the vapour and the cumulative volume percent collected are simultaneously reported. The maximum vapour temperature at which the distillate collection is negligible is reported as the FBP. In SI engines, light

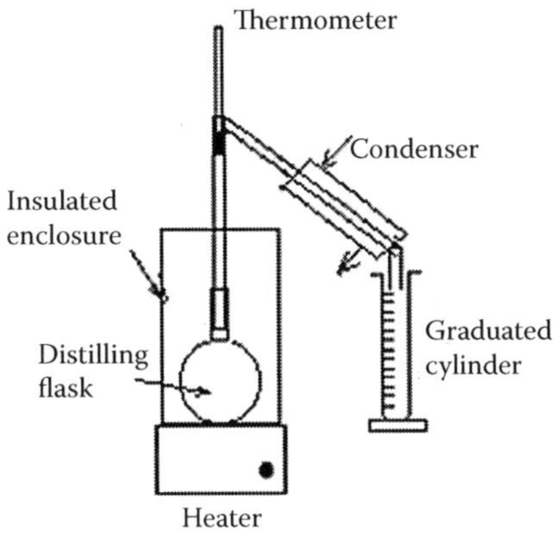

FIGURE 2.8 ASTM distillation apparatus.

hydrocarbon liquid is needed as the fuel because it is readily vaporised, which is the most essential requirement for combustion in an SI engine. In the ASTM distillation, no internal reflux is maintained, like TBP, and as a result the separation is not as good as it could be in TBP distillation. Correlation for the conversion from ASTM to TBP and vice versa is available and hence ASTM distillation data can be readily converted to TBP. ASTM distillation is the easiest test for the boiling ranges of hydrocarbons in MS. The required ASTM distillation analysis of MS is that 10%, 50%, and 90% vaporisation should occur at the maximum vapour temperatures of 70°C, 125°C, and 180°C, respectively. Though the IBP may vary, the FBP will not exceed 215°C. These vaporisation temperatures have significance in engine performance. The 10% point of distillation is important for starting an engine. In winter, when ambient temperature is low and if the 10% point is high, vaporisation of MS will be difficult unless heated. Again, if it is very low it may cause vapour locking due to excessive vaporisation. The 50% point is an indication of uniform vaporisation in the engine during the warming-up period of driving and the 90% point is important as far as the crankcase dilution is concerned. If the 90% point is high, a residual part of the fuel will be unburnt and will mix-up with the motor oil used for lubricating the piston of the engine, causing dilution of crankcase oil. This is also the reason for increased fuel consumption per kilometre. Figure 2.9 depicts a ASTM distillation analysis of MS.

2.3.1.2 Octane Number

Engine performance is measured by the maximum power development and the rate at which it is developed at different engine speeds. If the rate of power development

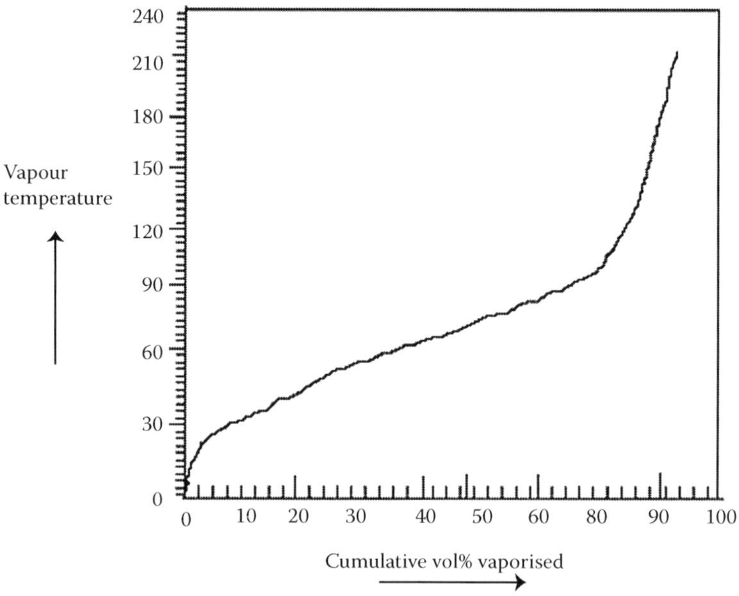

FIGURE 2.9 ASTM distillation of MS.

is not uniform with speed, rather it fluctuates, then this type of situation is called knocking or hammering of the engine. Several experiments found that the composition of MS plays the main role in knocking. If MS contains more propane, butane (short chain hydrocarbons), iso-octane (branched chain hydrocarbons), benzene, toluene, xylenes, and aromatics (ring compounds) in the boiling range of MS then the rate of power development is smooth without knocking, whereas the long chain hydrocarbons give rise to severe knocking if present in the fuel in substantial amounts. Since MS is a mixture of all the short or long chain, branched, and ring hydrocarbons, it is common practice to use iso-octane as the fingerprint for measuring the relative engine performance of the mixture. Specification of MS as far as engine performance is concerned is denoted by octane number. Octane number is defined, for any MS fuel, as the percentage of iso-octane in a mixture of iso-octane and n-heptane, which will give the same engine performance as could be achieved by the actual fuel sample.

Thus, if an engine runs with 100% pure iso-octane, the power rating is 100% (knock free) and is defined as 100 octane number. If the engine is run with n-heptane, a straight chain hydrocarbon, there will be tremendous knocking in the engine and the octane number is taken as zero. If the engine is run with a mixture of iso-octane and n-heptane with different proportions, the engine power rating will fall with the decrease in the proportion iso-octane in the fuel. The octane number of the MS sample, therefore, falls within 0 and 100. This is measured in a standard engine, known as the co-operative fuel research (CFR) engine.

However, the octane number may be greater than 100, which is called the performance number. An engine test is carried out at two different speeds and the octane numbers are defined as the research octane number (RON) and the motor octane number (MON), which are conducted at two different speeds, 600 and 900 rpm, respectively. However, tested octanes may not correspond to real performance when the road conditions differ. RON is more applicable for city driving whereas MON is applicable for highway driving. The octane number falls with the increase in speed, which is why MON is always less than RON for the same MS fuel. Aircrafts using gasoline require very high RON fuel as necessary with the cruising speed. It is also found that the octane number falls with the altitude. The octane number of a MS fuel can be increased by adding tetra-ethyl lead (TEL), but this has been prohibited by environment protection laws. Hence, organic substances like methyl- or ethyl tertiary butyl ethers (MTBE or ETBE) are used instead of TEL for boosting the octane number.

2.3.1.3 Corrosion

The presence of mercaptan sulfur may cause corrosion in the fuel pipes and the engine cylinder and produce sulfur dioxide during combustion. In the past, merox (a catalytic mercaptan oxidation method) treatment was done to convert corrosive mercaptans to non-corrosive disulfides, but this did not remove the sulfur originally present in the fuel. However, it did give rise to the formation of sulfur dioxide during combustion. Since emission of sulfur dioxide is prohibited by environmental protection laws, nowadays mercaptans and other sulfur compounds are mostly removed by a catalytic hydrodesulfurisation unit in a refinery. The corrosive effects of other organic compounds along with traces of sulfur-bearing compounds and additives must be tested in the laboratory. The copper corrosion test similar to that described in the testing of LPG is also carried out in the laboratory at standard temperature (50°C) for 3 h.

2.3.1.4 Reid Vapour Pressure

RVP of MS should not exceed 0.7 kg/cm^2 (gauge) otherwise there will be the tendency for vapour locking in the engine and vapour loss as well during its storage. Too low vapour pressure will give rise to problems starting the engine. RVP is measured in a similar manner as described for the testing of LPG.

2.3.1.5 Oxidation Stability

MS is, in fact, a mixture of hydrocarbons obtained from various units of a refinery, such as reformate (obtained from a naphtha reforming plant), thermally (viscosity breaking unit, coking unit, etc.), and catalytically cracked gasoline components (from a fluidised bed catalytic cracking unit). By-products from petrochemical plants, like aromatics and pyrolysis gasoline components, are also blended. All these components may contain much unsaturated hydrocarbons especially the di-olefins, which are susceptible to mild polymerisation when in contact with oxygen or air. These polymerised products make layers of films or gum like substances and reduce the effectiveness of the fuel while used. This gum formation is a slow process but may cause severe damage to fuel if stored for a long time. Therefore, the storageability of MS is measured in terms of the period during which the fuel does not form any gum. This is tested in the laboratory by contacting fuel and oxygen (or air) in a pressure vessel at a constant temperature for a specified period (usually 6 h or longer) and a fall in pressure, if any, is observed.

2.3.1.6 Additives

Various additives, like colour (to distinguish octane numbers), anti-icing (to reduce ice formation), anti-static (to disperse the generation of statical electric charges), anti-oxidant, anti-corrosive, and octane boosting agents are added to the fuel.

2.3.2 High Speed Diesel

High speed diesel (HSD) is another liquid fuel that is used most abundantly in our country in a variety of vehicles and appliances, such as cars, buses, tractors, lorries, barges, speed boats, railway engines, irrigation pumps, and generator sets. Like MS, diesel is also a mixture of hydrocarbon compounds, boiling in the range of 250°C–360°C. Unlike MS, diesel oil is not vaporisable at ambient condition but requires heating. Diesel is burnt in a CI engine where the fuel is atomised and sprayed in the hot compressed air. Since the temperature for autoignition (the method of ignition without the aid of fire or spark) for diesel oil is much lower than MS, the temperature during compression of air causes this ignition.

Important Parameters as per the Indian BIS of MS

Property	Specification
Copper strip corrosion, 3 h at 50°C	Not worse than ASTM 1
Octane No, min, RON	87/93
ASTM distillation:	
IBP	To be reported
Recovery up to 70°C, min	10% vol
Recovery up to 125°C, min	50% vol
Recovery up to 180°C, min	90% vol
FBP, max	215°C
Sulfur total, max	0.25 wt%
RVP at 38°C, max	0.70 kg/cm^2

2.3.2.1 Cetane Number

It has been found that paraffinic hydrocarbon has a lower autoignition temperature than that of aromatics. Therefore, if diesel oil is rich with aromatic hydrocarbons than the paraffinic counterparts, most of the oil will be unburnt at the beginning and will simultaneously burn suddenly when the temperature rises due to the combustion of paraffinic hydrocarbons. This type of situation will give rise to non-uniform pressure (shock wave or knocking) in the engine.

Important Parameters as per the Indian BIS of diesel

Property	Specification
Copper strip corrosion, 3 h at 100°C	Not worse than ASTM 1
Distillation	
Recovery up to 366°C, min	90% vol
Final boiling	To be reported
Diesel index, min	45
Flash point (Abel), min	33/38 depending on users
Flame height, max	18 mm
Viscosity, kinematic, at 37.8°C	2.5–7 cSt
Sulfur content, total, max	0.25% wt
Carbon residue, Ramsbottom, max	0.20% wt
Pour point, max	6°C
Ash content, max	0.01% wt

Knocking quality of diesel oil is measured by cetane number, which is defined as the percentage of normal cetane (a straight chain hydrocarbon) in a mixture of n-cetane and α-methyl naphthalene (an aromatic hydrocarbon), which gives the same performance as that of the diesel sample. This is tested in a standard diesel engine of CFR.

2.3.2.2 Diesel Index

The presence of paraffinic hydrocarbons in diesel may be related by the aniline point, which is the temperature at which aniline solubilises the fuel in equal amounts and a homogeneous mixture results. The greater the paraffin content, the higher the aniline point. Also, the API gravity of oil increases as the paraffin content rises. With the help of these properties, the diesel index (DI) is given as

$$DI = API \times \text{aniline point in } {}^\circ F/100. \tag{2.5}$$

DI has been correlated with the cetane number and it has been found that DI is directly proportional to the cetane number. The cetane number is thus obtained from the DI. The value of the DI should be at least 45.

2.3.2.3 Sulfur

In order to reduce pollution, the sulfur content of diesel should not be large. In refineries, catalytic desulfurisation is carried out to remove sulfur from diesel oil. The sulfur content should not be more than 0.25% by weight of the oil.

2.3.2.4 Corrosion

A corrosion test of diesel is also carried out using the copper strip method at 100°C for 3 h.

2.3.2.5 Flash Point

The flash point of diesel is a minimum of 33°C for automobiles. However, depending on the ambient temperature, it may be higher. The lower the flash point, the greater the chance of autoignition. During winter a low flash point is preferable while in summer it should be more (around 35°C). Too high a flash point may cause knocking in the engine.

2.3.2.6 Flame Length

In the combustion chamber of an engine, a long flame length may damage the chamber. Hence, a short flame length is desirable. Diesel oil should not produce a flame length of more than 18 mm. In the refinery, sometimes kerosene and other hydrocarbons, like heavy naphtha, are also blended in diesel with the result that the flame length may increase. Hence, reduced crude oil (RCO) is injected to adjust the flame length to the desired value.

2.3.2.7 Pour Point

The pour point is defined as the temperature at which oil will cease to flow due to the formation of wax crystals. In India, the pour point of diesel has been fixed at a value less than 6°C. However, in colder places, a lower pour point is advisable.

2.3.2.8 Viscosity

In a flow process, fluid experiences a kind of friction opposing the flow. This friction is known as fluid friction, which is defined as the resistive force (shear) exerted between two parallel sliding layers of fluid moving in the direction of flow. This shear force is proportional to the rate of change of velocity (shear rate) of the sliding layers. Viscosity is the proportionality constant between the shear force and the shear rate. The higher the viscosity, the greater the friction and the fluid is termed more viscous. Most hydrocarbon liquids follow this rule (Newton's law of viscosity). Low viscosity is preferred for diesel oil at ambient temperature. It should be between 2.5 and 7 centi stoke (cSt) at 38°C.

2.4 AVIATION FUELS

The fuels used in aeroplanes are called aviation fuels. Depending on the type of aircraft, like jet planes or turbine planes, different types of aviation fuels are used. They are either gasoline based for jet planes or kerosene based for turbine planes. Aviation gasoline is usually polymer gasoline or alkylated gasoline having an octane number greater than 100, usually expressed as the performance number. Kerosene-based aviation fuel is known as aviation turbine fuel (ATF) and is mostly consumed by passenger aeroplanes. This fuel is the hydrocarbon fraction boiling in the range of 150–250°C and is similar to the kerosene fraction. Though it resembles kerosene, tests are carried out under stringent conditions for the safety of the airborne people in the flying machines. A corrosion test is carried out using the copper strip test for 2 h at 100°C and a silver strip test is carried out for 16 h at 45°C. Distillation tests are conducted as for kerosene while the 20% recovery should be at 200°C and the FBP should not be more than 300°C. Besides freezing point is to be below –50°C as the sky temperature may be very low at high altitude.

Specifications of ATF

Property	Specification
Copper strip corrosion, 2 h at 100°C	Not worse than ASTM 1
Silver strip corrosion, 16 h at 45°C	Zero
Distillation:	
Recovery up to 200°C, min	20% vol
Final boiling	300°C
API × aniline Point (°F), min	5,250
Flash point (Abel), min	38°C
Freezing point, max	–50°C
Smoke point, minimum	20 mm
Viscosity, kinematic, at –34.4°C, max	6 cst
Sulfur content, total, max	0.20% wt
Carbon residue, Ramsbottom, max	0.20% wt
Pour point, max	6°C
Ash content, max	0.01% wt
Aromatic percent vol, max	20
Olefin percent vol, max	5

The product of the aniline point and the API gravity of oil should be at least 5,250. The sulfur content should be below 0.20% wt. In addition to these tests, aromatic and olefinic hydrocarbons must be analysed and these hydrocarbons, if present, should be no more than 20% and 5%, respectively.

2.5 FURNACE FUELS

2.5.1 Gaseous Fuels

Industrial furnaces are mostly fuel-fired furnaces and use gaseous and liquid hydrocarbon fuels. Off-gases and by-product gases commonly generated in refineries and petrochemical plants consist of methane, ethane, propane, butane, and their olefinic homologues. Gaseous fuel can generate a higher temperature than liquid petroleum owing to its higher heating value.

2.5.2 Liquid Fuels

Light diesel oil or LDO (much heavier than HSD) is a liquid fuel. It is usually a blend of vacuum gas oil, coker gas oil, deasphalted oil, waxy distillates, etc., and falls into the category of black oil. A few properties measured as given in the table are the flash point, viscosity, total sulfur, pour point, and carbon residue. It is a cheap liquid fuel as compared to HSD and is commonly used in furnaces.

Property of LDO	Specification
Flash point (PMC), min	66°C
Flame height, max	10 mm
Viscosity, kinematic, at 37.8°C	2.5–7 cSt
Sulfur content, total, max	1.8% wt
Carbon residue, Ramsbottom, max	0.25–1.5% wt
	−18°C (summer grade)

Furnace oil or fuel oil is similar to LDO but with higher viscosity. It is obtained from viscosity broken crude residue and vacuum distillates and is commonly used as the cheapest furnace oil in industrial furnaces. It is a black petroleum oil, which is a viscous fluid and requires heating up to a temperature of 50°C or higher to maintain liquidity. Its properties are presented in the following table.

Property of furnace oil	Specification
Flash point, (PMC), min	66°C
Flame height, max	10 mm
Viscosity, kinematic, at 50°C	125 cSt (winter grade)
	175 cSt (summer grade)
Sulfur content, total, max	4.5% wt

Internal fuel oil (IFO) is the common furnace oil used in refineries. It is predominantly a mixture of asphalt, short residue, and wax. Sometimes visbroken fuel oils are also injected to adjust the viscosity.

2.6 LUBRICATING OILS

A lubricant is a solid, semi-solid, or liquid material that is used to reduce friction between two solid surfaces. Examples of lubricants are graphite, sulfur, wax, soap, and mineral oil. Lubricating oils are liquid lubricants mainly made from petroleum oils (lube oil base stocks) blended with soaps of fatty acids and other additives. A lubricating oil present between two solid surfaces reduces the friction while in motion and also helps in cooling down the heat of friction. The main ingredients are petroleum base stocks that are obtained from vacuum distillates of crude oil. These distillates are heavy petroleum fraction and cannot be vaporised like petrol or diesel. The quality of a lube oil is determined by properties like viscosity, viscosity index, pour point, and some additional properties determined by their end uses.

2.6.1 Viscosity

Viscosity is a parameter that measures resistance to flowability of liquid or gas. It is defined as the force exerted between two sliding layers of liquid or gas moving at different velocities. Visosity (or coefficient of viscosity) is the proportional constant between this force (shear) and the velocity gradient (rate of deformation due to shear or rate of shear). In the centimetre–gram–second system (cgs) unit, it is expressed in terms of gm cm/sec or poise. Usually it is expressed in centipoise or cP, where 1 poise = 100 cP. In SI, unit viscosity is expressed as pascal. Sec. However, in the hydrocarbon industry, the term "centistoke" is more popularly used as the measure of viscosity, which, in fact, is the kinematic viscosity. This is the ratio of viscosity (cP) to the density (g/cc) of the fluid. There are various methods of viscosity measurement, e.g., capillary, orifice, rotary, float or falling sphere, and vibrational methods, depending on the ranges of viscosity from light to heavy. A pressure drop in a tube of definite length and diameter can be used to measure viscosity by Poisulle's equation of flow. Stoke's law may be used to measure viscosity in a falling sphere method where time of fall at a definite height is reported. Viscous shear or drag in terms of power consumption for a typical rotating disk on a fluid film over a static one is a measure of viscosity.

2.6.2 Saybolt Method

This is a capillary method. For light and non-sticky hydrocarbon liquids, the Saybolt method is used where time (t) of efflux of a definite quantity (60 cc) of liquid through a typically designed capillary tube kept in a constant temperature bath is measured. This is reported as Saybolt seconds. Viscosity in centistoke is related with Saybolt second (t) is given as

$$\text{Viscosity in cst} = 0.219t - 149.7/t. \tag{2.6}$$

However, ASTM standard conversion tables are available for exact measurements.

2.6.3 Redwood Method

This is an orifice or jet method. Time taken by a definite volume of liquid flowing through an orifice of a definite dimension is measured. Viscosity is reported as Redwood seconds. A conversion table or an equation is used to evaluate the kinematic viscosity in centistoke from Redwood second.

2.6.4 Brookfield Method

This method uses a rotary type viscometer consisting of two disks containing the liquid under test. The shear force experienced by the rotating disk is measured against the liquid over the static one. Standard liquids are used for calibration.

2.6.5 Viscosity Index

The viscosity index (VI) is a parameter that indicates the rate of change of the viscosity of the oil due to a variation in temperature. This index is defined as the ratio of the difference of viscosity (U) of the lube oil to be used with respect to the viscosity (L) of petroleum (aromatic) oil having zero VI to the difference of viscosity (H) of high VI oil (paraffinic; 100 VI) to the viscosity (L) of zero VI oil for a temperature change from 37.8°C to 98.9°C. This can be written in the following way,

$$\mathrm{VI} = (\mathrm{L} - \mathrm{U}) / (\mathrm{L} - \mathrm{H}) \times 100, \tag{2.7}$$

where L, U, and H are the viscosities of the low VI reference oil (VI = 0), the sample oil, and the high VI reference oil (VI = 100), respectively, all at a temperature of 37.8°C. This can be explained graphically as shown in Figure 2.10. Note that the viscosity of the sample and reference oils must be so selected that they have the same viscosity at 98.9°C. In the case when the VI, calculated as above, is greater than 100 the following relation is applicable for calculating the VI as

$$\mathrm{VI} = (10^{a} - 1) / 0.00715 + 100, \tag{2.8}$$

where

$$a = \{\log(\mathrm{H}) - \log(\mathrm{U})\} / \log(\mathrm{V}). \tag{2.9}$$

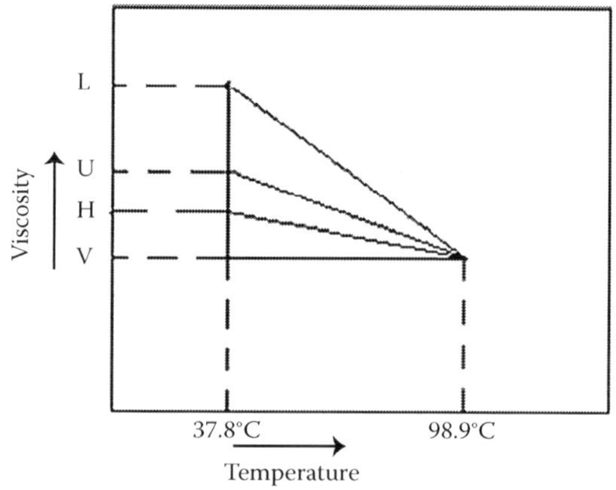

FIGURE 2.10 Graphical representation of the viscosity index.

Viscosity terms L, H, V, and U are shown in Figure 2.10.

2.6.6 Cloud Point

The cloud point is the temperature at which oil becomes hazy or cloudy due to the onset of wax crystallisation or solidification. This is tested in a standard tube in an ice bath. The tube fitted with a thermometer is placed in the bath. At intervals, the test tube is removed (without disturbing the oil) to observe cloudiness. The temperature is reported to be the cloud point.

2.6.7 Pour Point

The pour point is the temperature at which a liquid hydrocarbon ceases to flow or pour. This is measured by a standard method where a definite quantity of an oil sample is taken in a test jar or tube (with a thermometer properly stoppered), heated to 115°F (46°C) to make all the wax dissolve in oil, and cooled to 90°F (32°C) before testing. An ice bath containing ice and salt is made ready at a temperature of 15°F–30°F (–9°C to –1°C) below the estimated pour point based on cloud point and the test tube containing the sample is placed with the thermometer. At intervals of 5°F, the test tube is removed from the ice bath and tilted to see if the oil is mobile or static. If it is found that at a certain temperature the oil shows no movement even when the test tube is kept horizontal for 5 sec, this temperature is reported to be the solid point. The pour point is taken as 5°F above this solid point. Lube oils used in engines, gears, bearings, etc., vary with the properties. For example, in reducing the bearing friction, lube oil should be of high viscosity and high VI so that it is not squeezed out during use and is not thinned out or thickened with frictional heat and also should be capable of bearing load. Heavy semi-solid lubricants containing metallic soaps with base oil classified as grease which are the common load-bearing lubricant. Lubricants used for lubricating the surface of an engine cylinder and piston should have a very high VI but low viscosity. Crankcase oil should also have the same property and it also serves as a cooling medium for engine cylinders. For aviation services, lube oil should have a high VI with a very low pour point. It has been found that the VI of paraffinic oils is greater than napthenic and aromatic oils, whereas viscosity is low for paraffinic oils as compared to naphthenic and aromatic oils. Contribution toward the viscosity, VI, and pour point from the paraffin (P), naphthene (N), and aromatic (A) hydrocarbon groups are listed below.

Hydrocarbon groups and properties of lube oils

Hydrocarbon	Viscosity	Viscosity index	Point point
P	Low	High	High
N	Medium	Low	Low
A	High	Low	Low

Besides the above three properties, additional properties are required for a lube oil specific for uses. These are flash point, carbon residue, saponification value, acidity, foaming characteristics, sulfur content and corrosiveness, drain out interval, etc.

During use, the lubricant should be free from abrasive particles, free from water, free from carbon particles, etc. When it is used for cooling purposes, it should be of low viscosity to reduce fluid friction during pumping. Power transformers are cooled by lubricating oil, known as transformer oil, which should have a high dielectric constant to avoid breakdown in the high electrical voltage. Varieties of lubricating oils and greases are produced by blending lube base stocks and other additives to meet different uses; presentation of all these are out of the scope of this book. However, a few lube base stocks and their properties according to the BIS are listed in Table 2.1.

2.7 MISCELLANEOUS PRODUCTS

2.7.1 Jute Batching Oil

Jute batching oil (JBO) is a straight run product obtained from the atmospheric crude distillation unit. This has a boiling fraction slightly different from that of diesel cut and it is 340°C–365°C. There are two grades of this oil—JBO(p) and JBO(c). Both of these oils are used in the jute industry for processing and bailing jute fibres. Jute packaging for food products and drugs are commonly employed and precautions must be taken to avoid any type of contamination. Hence, oils used for jute processing must conform to the specifications laid down by the Food and Drug Administration (FDA). The BIS specification of JBO is listed below.

Properties	JBO(p)	JBO(c)
FBP, max	–	371°C
Recovery at 371°C	50% min	–
Recovery at 413°C	95% min	–
Pyrene content, max	–	25 ppm
UV absorbance	–	Must be reported
Flash point, min	100°C	100°C

The quality "c" stands for confirmed FDA and "p" for pre-FDA. The smell of kerosene and pyrene content, a carcinogen, are the most important parameters for selecting JBO to be used for jutes used in food packaging. JBO(c) has a FBP of 371°C and is lighter than JBO(p) whose FBP is near 420°C. Hence, JBO(C) is not free from the smell of kerosene and is usually not suitable for packaging food, whereas JBO(p) is used for processing jute for food packaging. JBO is also used as a quench oil in a coke oven gas cleaning tower.

2.7.2 Mineral Turpentine Oil

Mineral turpentine oil (MTO) has a boiling fraction similar to kerosene cut in the boiling range of 125°C–240°C. This cut may be obtained directly from the crude distillation unit as a straight run product or by splitting kerosene cut. It is mainly used in the paint industry as a solvent or thinner and hence desirable properties are guided by the requirements of the paint industries. Typical properties are given in Table 2.2.

TABLE 2.1
Properties of Certain Lube Base Stocks

Name of the Lube Base Stock	Spindle Oil	Intermediate Oil	Heavy Oil	Bright Stock
Property	**Specification**	**Specification**	**Specification**	**Specification**
Flash point (closed cup), min	150°C	243–260°C	266–288°C	288–300°C
Viscosity index, min	95	95	90	90
Viscosity, kinematic at 98.9°C, max	2.6–2.75 cst	8–9.9	13–17	35–45
CCR % wt, max	0.01	0.05	0.3	0.6–2.5
Pour point, max	−12°C	−12°C	−6°C	−6°C
Saponification value, max	1	1	1	1
Neutralisation number, max	1	1	1	1

TABLE 2.2
Important Properties for a Typical MTO

Property	Specification
Copper strip corrosion, 3 h at 50°C	Not worse than ASTM 1
Aromatic content, max	40% vol
Flash point (Abel), min	30°C
ASTM distillation:	
IBP, min	125°C
FBP, max	240°C

2.7.3 Carbon Black Feed Stock

Carbon black feed stock (CBFS) is a petroleum product rich in aromatics suitable for yielding carbon particles or coke for battery and metallurgical industries. It is a by-product from a lube extraction plant where aromatic hydrocarbons from distillate oil are removed as an extract. The resulting aromatic rich oil after recovery of the solvent is obtained as the CBFS. As aromatics yield good carbon black. Carbon yield is determined by the Bureau of Mines Co-relation Index (BMCI) number and Conradson carbon residue (CCR), where the BMCI is given as

$$\text{BMCI} = 48{,}640/\text{K} + 473.7\,G - 456.8,$$

where K is the boiling point in kelvin and G is the specific gravity of the oil. The value of the BMCI number should be above 80 for the CBFS. The CCR is the amount of carbon generated while heated at high temperature in the absence of oxygen. This is represented as the weight of carbon formed per unit weight of oil sample. The higher the percentage, the greater the yield of coke and carbon, while the oil is cracked at high temperature. In addition to the above, carbon required for electrodes or for the metallurgical industry must be free from metal contents. Specification of this product is dictated by the consuming industries.

2.7.4 Bitumen

Asphalt or bitumen is obtained from short residue (residual mass from the bottom of the vacuum distillation unit) after extraction of valuable oil (known as bright stock) by propane. Asphalt is a very sticky, semi-solid, viscous petroleum, containing various hetero-atomic compounds of hydrocarbons enriched with metallic, sulfurous, nitrogenous, and oxygenated compounds. At a temperature above 90°C, it flows like liquid and solidifies at room temperature. It is mainly used as paving material, paint, water-proofing agent, etc. As a paving grade material, it should have a desirable penetration index, which is defined as the depth of submergence or penetration of a standard weight through a needle penetrator. The greater the penetration, the greater the softness of the asphalt. Typical paving grade bitumen has a penetration of 60/70 or 80/100 using a 100-g cone (1/100 cm)

TABLE 2.3
Typical Specifications of Bitumen

Property	Specification	
Penetration index	60/70	80/100
Flash point (PMC), min	175°C	175°C
Softening point, min	40–55°C	35–50°C
Matter soluble in carbon disulfide, min	99% wt	99% wt
Density at 15°C	0.99	0.99
Ductility at 27°C in centimetres, min	75	75

at 25°C for 5 sec during the test. The other property is the flash point. As this has to be applied on the open space surface, the flash point should be above 175°C and the softening point should be above the ambient temperature depending on requirements. In addition to these, mechanical properties like ductility must also be measured. A list of the important properties of a typical paving grade bitumen is given in Table 2.3.

2.7.5 Petroleum Coke

Petroleum residue contains heavy hydrocarbons that crack at high temperature in the absence of air yielding gases, light hydrocarbon liquids, and black solid carbon rich residue or coke. The greater the amount of heavy aromatic hydrocarbons in the feed stock, the greater the yield of coke. Thermal cracking of petroleum involves various reactions, like decomposition, dehydrogenation, isomerisation, polymerisation, etc. Decomposition reaction takes the lead in the reactions and the yield of coke increases with the increase in time to cracking. In the old processes, cracking was carried out in a coking vessel (drum) until the drum was filled with coke. The modern method involves cracking in a tube-still heater followed by quenching of the product mass to avoid choking of the furnace tubes. By-products like gases, hydrocarbon liquids known as coker gasoline, and gas oil are also formed. The yield and quality of coke depend on the temperature, pressure, and reaction time. Petroleum coke is available in various forms, such as graphitic soft coke, needle coke, and fluid coke. The quality of coke is determined by the highest content of carbon (usually from 88% to 95% by weight) and with minimum impurities, such as metallic components, moisture, volatile matters (hydrocarbons), and low sulfur, to make them amenable to manufacturing electrodes, for use in metallurgical furnaces as a carbon-reducing element and a clean fuel as well. There are a variety of uses for coke in the paint and petrochemical industries. Residues from most of the Indian crude oil are rich with heavy aromatic hydrocarbons and low sulfur and metallic components and, therefore, are suitable for making good quality coke. However, residues from Middle East crudes are not suitable for making coke due to their high content of sulfur and metallic contents and are converted to paving grade asphalts or bitumen or used as a feed for catalytic cracking or as a fuel.

2.7.6 Wax

Wax is obtained from the refinery as a by-product from a de-waxing unit. It is roughly defined as petroleum hydrocarbons that solidify at a temperature above –20°C. High molecular weight normal paraffinic hydrocarbons are the major constituent of wax, which freeze even at atmospheric temperature.

Vacuum distillates usually contain much wax and are removed by freezing at low temperature. Paraffinic wax is usually available as a slack wax, whereas naphthenic wax is available as microcrystalline wax. Slack wax is soft and has a low melting point suitable for candle making, paper coating, etc. Microcrystalline wax is hard with a high melting point and has a variety of applications, such as cosmetics, paints, and as an additive in rubber and plastics. Raw wax is obtained as a milky white to brown-coloured solid product and coloured wax is made out of this by adding additives. The quality of wax is guided by the user industries.

QUESTIONS

1. What is a TBP analysis of crude oil? What is meant by reflux and theoretical plates in such an apparatus?
2. What are the parameters used to identify the quality of crude oil?
3. What are the various fuel products available from crude oil?
4. How does the ASTM distillation test method differ from TBP distillation?
5. Why and when is it necessary to carry out distillation in vacuum?
6. What are the various non-fuel products available from crude oil?
7. What is a "weathering test" and for which product is it applicable?
8. Define octane and cetane numbers.
9. Distinguish between petroleum coke and bitumen with respect to their chemical properties.
10. Define viscosity and the viscosity index.
11. What are the various methods of measurements of the viscosity of a liquid?
12. Define cloud and pour points and their importance.

3 Processing Operations in a Petroleum Refinery

3.1 CRUDE OIL RECEIVING

In a refinery, crude oil is received and stored in a floating roof tank. The roof is made of compartmented deck or pontoon floating over the oil to avoid loss of hydrocarbon vapours of dissolved hydrocarbon gases and low boiling fractions present in crude oil. The floating roof also floats up and down to compensate for the breathing operations during pumping in and out of the tank thereby safeguarding the loss of vapours and ingress of atmospheric air into the tank. A floating roof storage tank of crude oil is shown in Figure 3.1, where a floating roof (1), made in the form of a ship's deck, floats over the crude oil stored. As the vapour is formed, floating roof does not allow the vapour to leak into the atmosphere, thereby avoiding loss. During low level in the tank the roof is supported by legs (2) while it rests on the tank bottom. A dip pipe (4) is provided to measure the level of crude stored, through a dip hatch (5) with a cover (6). The level of liquid in the tank or dip can be manually measured with a graduated dip tape by climbing onto the tank from the operating platform (7) via a ladder (8). A manual inspection using a swing ladder (3) onto the surface of the floating roof can be done for deck compartments, the rolling seals (16), and the roof drain (14) connected with a flexible steel pipe, which is mobile with the roof. Crude oil received from tank cars, pipelines, or from tankers (ship) may contain much water either carried with the cargo or due to flushing of the receiving line. This water must settle in the receiving tank before oil accounting is started. Temperature, oil cut, water cut, and the density of the sample crude are the most important parameters in determining the quantity of oil received. After the quantity and valuation of oil are established, the water in the tank must be drained. Steam coils and side mixers are usually provided near the bottom of the receiving tank to homogenise the crude layers before processing. However, water and salt, which do not settle in the tank, can only be removed by desalting located within the battery limit of the crude distillation unit.

3.2 DESALTING OF CRUDE OIL

Crude oil received in a refinery contains much water, salts, clay, and sand, which do not settle in the tank and are desalter at the beginning of the refining operations. The amount of water and settleable solids present in crude oil is indicated by bottom sediment and water (BSW) analysis. The presence of salts of magnesium, sodium, alkaline earth metals in crude oil varies from crude to crude. These salts are dissolved

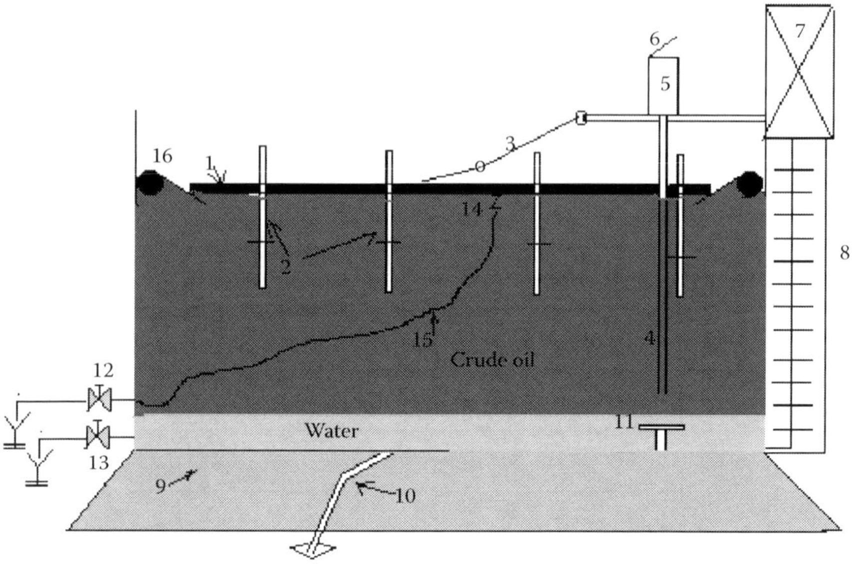

FIGURE 3.1 Crude storage in a floating roof tank.

in water associated and entrained with crude oil from the drilling fluids or wash water from the crude-carrying vessels. Salts in the presence of water are responsible for corrosion. For example, chloride salts of magnesium generate hydrochloric acid at a temperature above 150°C and may cause severe corrosion both in the liquid and vapour phases.

The modern method of electrical desalting removes water and the dissolved salts simultaneously from crude oil. A modern electric desalter is shown in Figure 3.2, where crude oil enters the bottom of a horizontal vessel provided with two flat electrodes supplied with a very high voltage (20–33 kV AC). Crude containing salt and water is mixed with fresh demineralised water (DM water) through a mixing valve at the bottom portion of the vessel in order to dissolve salts to the aqueous phase. The optimum temperature for solubilising the salts to the aqueous phase is 120°C–130°C, hence crude must be preheated to this temperature before it enters the desalter. Crude, water, and salt travel upward as an emulsified mixture and are brought in contact with the flat electrodes. Drops of water are ionised by electric charge and coalesce to form bigger drops of water, which then fall by gravity toward the bottom of the vessel. Thus, emulsion is broken as long as the crude–water mixture is in contact with the electrodes. De-emulsifying agents are also injected in small amounts to break the unbroken emulsion. The practice of electric desalting can reduce the salt content by 90%–98%. Single or multiple desalters in a series may be employed depending on the concentration of salt in the incoming crude. At the desalting temperature, low boiling hydrocarbons may vaporise and the desalting operation is affected. This also causes cavitation to the discharging pump at the exit of the desalter. To avoid vaporisation, the drum pressure is monitored and controlled by a pressure controller, which senses the pressure through a pressure transmitter and manipulates the feed flow rate to and from the desalter. Pressure in the drum is usually maintained at a pressure

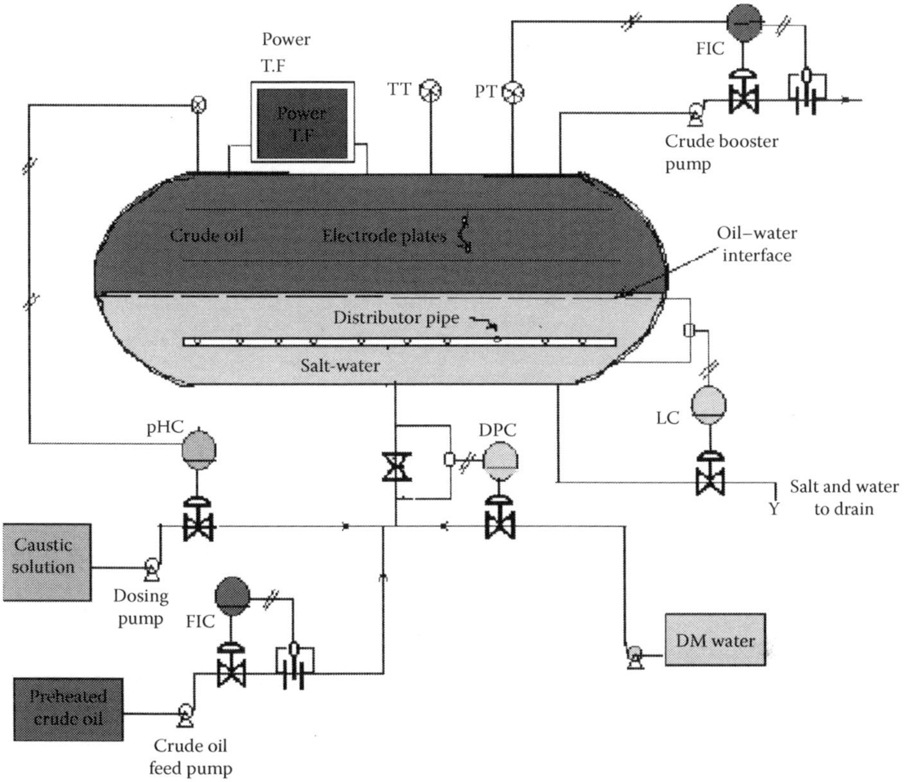

FIGURE 3.2 A typical crude oil desalting unit

slightly above the vapour pressure of mixed hydrocarbon gases, usually 8–10 kg/cm^2, to maintain the liquid phase, if the pressure increases above this due to vaporisation, vapours will pop through a pressure safety valve leading to the crude distillation column and the power supply to the transformer is automatically switched off to avoid an accident. Formation of hydrochloric acid from chloride salts and the presence of organic acids in crude oil may cause severe corrosion in the desalter drum. Hence, a caustic solution is injected with the incoming crude and the pH of the aqueous phase is monitored and controlled to maintain a pH near 7. The water level in the desalter is controlled such that the water level should not reach the electrodes, which may cause a short circuit between them as water is a good conductor of electricity. An interface level transmitter senses the water–oil interface level and sends an alarm signal if the level is high or low and it automatically switches off the power supply in case the water rises above a high level. A low level of water may cause poor desalting and may allow carryover of crude oil with discharged brine to drain.

3.3 DISTILLATION AND STRIPPING

Distillation of crude oil is carried out at atmospheric pressure and under vacuum. Low boiling fractions vaporise at atmospheric pressure up to a temperature below

400°C without cracking the hydrocarbon compounds. High boiling fractions are vaporised under vacuum at a temperature much lower than 400°C and cracking is avoided. Thus, all the low and high boiling fractions of crude oil are separated by atmospheric distillation followed by vacuum distillation without cracking or degradation of the original hydrocarbons present in crude oil. After the removal of oil fractions, the vacuum residue left requires solvent extraction (known as deasphalting) to recover the remaining valuable petroleum fraction. Atmospheric distillates are the major source of petroleum fuels and vacuum distillates are the source of fuels, wax, lubricating oil base stocks, bitumen, petroleum cokes, etc., mainly by secondary treatments like extraction, thermal, and catalytic treatments.

3.3.1 Atmospheric Distillation

Crude oil from the desalter is heated in a train of heat exchangers up to a temperature of 250°C–260°C and further by a tube-still heater (also known as a pipe-still furnace) to a temperature of 350°C–360°C. Hot crude is then flashed into a distillation column, which is a tall, multiplated cylindrical vessel that separates the petroleum fractions according to the differences in volatility. Top pressure is maintained at 1.2–1.5 atm such that the distillation operation can be carried out at a pressure close to atmospheric pressure and for this reason this column is also known as the atmospheric distillation column. Vapours from the top of the column, consisting of hydrocarbon gases and naphtha, emerge at a temperature of 120°C–130°C. The vapour stream associated with steam used in the column bottom and the side strippers are condensed by a water cooler and the liquid is collected in a vessel (known as a reflux drum) situated at the top of the column. Part of this overhead liquid (oil phase only) is returned to the top plate of the column as overhead reflux and the rest of the liquid is drawn and sent to a stabilizer column which separates gases from liquid naphtha. A few plates below the top plate, the kerosene fraction is drawn at a draw temperature of 190°C–200°C. A part of this fraction is returned to the column after it is cooled by a heat exchanger. This cooled returned stream is known as circulating reflux and is essential to control the heat load in the column. The rest of the draw is passed to a side stripper where steam is used to strip out the undesirable low boiling fractions to maintain the desirable flash point of kerosene fraction, which is then further cooled and sent to a storage tank as raw kerosene, also known as straight run kerosene, which boils in the range of 140°C–270°C. A few plates below the kerosene draw plate, the diesel fraction (also known as atmospheric gas oil or straight run diesel or gas oil) is drawn at a temperature of 280°C–300°C. After cooling, part of this is returned to the column as circulating reflux and the rest is steam stripped through a side stripper, like kerosene, to adjust the flash point. This diesel fraction, with a boiling range of 270°C–340°C, from the stripper is then cooled and sent to storage. In some refineries, jute batching oil (JBO), which is an intermediate cut between diesel and the bottom residue, is also drawn at a draw temperature of 300°C–330°C followed by steam stripping. Residual oil drawn from the bottom of the column is known as reduced crude oil (RCO), topped crude, or long residue. The temperature of the stream at the bottom reaches around 340°C–350°C, well below the cracking temperature of oil. RCO is a brown to deep chocolate-coloured stream containing high boiling fractions that cannot be vaporised at the prevailing temperature and atmospheric pressure of the column without cracking. Since depression of the boiling point occurs as the system pressure is reduced, boiling fractions remaining in the atmospheric residue are, therefore, distilled under vacuum in a separate column (known as the vacuum distillation column) where high boiling fractions are separated and collected as vacuum distillates.

A typical atmospheric distillation column (containing 41 plates), where the hot feed crude oil is flashed in the fifth plate, producing top distillate (IBP–140°C cut), straight run kerosene (140°C–270°C cut), straight run diesel (270°C–340°C cut), JBO (340°C–365°C cut), and the bottom residue (RCO, 365°C + cut), is shown in Figure 3.3. The top product goes to the gas and naphtha separation plant directly from the column. Kerosene, diesel, and JBO from the strippers are sent to fixed roof storage tanks. By introducing the liquid from the seventh plate back to the fifth plate, 4%–5% of the feed is over flashed. Overflash helps in further removal of heavier components carried over with the lighter stream. Part of the RCO from the bottom of the column goes to a vacuum distillation unit and the rest goes to a steam-heated fixed roof storage tank. The pipeline carrying the RCO to a vacuum distillation unit is well insulated and steam traced to avoid congealing of the product in the line. A temperature between 180°C and 200°C in the transfer line is usually maintained. Typical draw temperatures of the straight run distillates from an atmospheric distillation column are listed in Table 3.1.

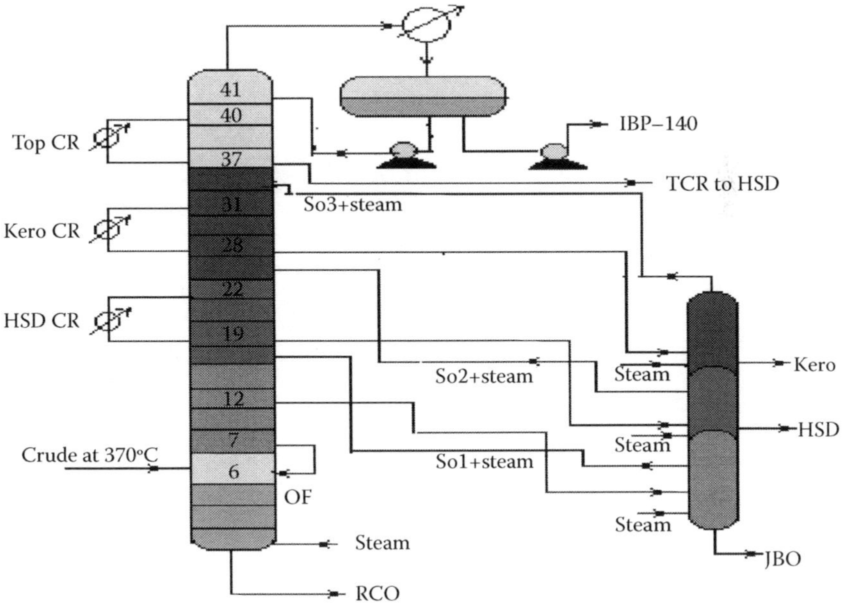

FIGURE 3.3 A typical crude distillation unit.

TABLE 3.1
Typical Draw Temperatures of Straight Run Products

Cuts	Boiling range, °C at 1 atm	Draw temperature, °C
Crude oil		Feed enters at 370
Overhead	IBP–140	120
Top reflux	IBP–140	50
SR.Kero	140–270	223
SR.HSD	270–340	307
JBO	340–365	332
RCO	365+	362

3.4 STABILISATION

The top product from the atmospheric distillation column (IBP–140°C cut) is a mixture of hydrocarbon gases, e.g., methane, ethane, propane, butane, and naphtha vapours. Methane, ethane, propane, and butane exist in gaseous form, whereas naphtha vapours liquify at ambient temperature and atmospheric pressure.

Hence, if the overhead product is kept at atmospheric pressure, these valuable liquifiable fuel gases will be lost to the atmosphere. The initial boiling point (IBP) of this mixture will vary due to the relative presence of these gases. These gases are easily separated from naphtha by distillation in a multiplated column where the top product is the mixture of gases containing methane to butanes and the bottom product is stabilised naphtha boiling in the range of C_5–140°C. A typical stabiliser column is shown in Figure 3.4.

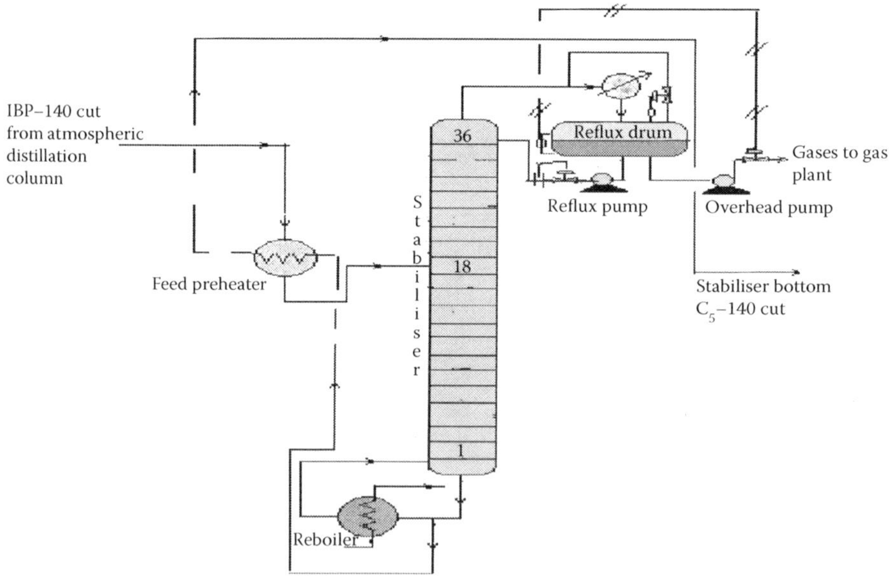

FIGURE 3.4 A typical stabilisation column.

3.5 AMINE ABSORPTION

Gases from the stabiliser go to a gas plant where hydrogen sulfide associated with hydrocarbons is removed by solvent extraction, usually by diethyl amine (DEA) solution. In a plated tower, the solvent flows down and the gases flow up the column countercurrently. Used amine solvent containing hydrogen sulfide emerges from the bottom of the tower and the top of the column produces hydrocarbon gases free of hydrogen sulfide. The amine solution containing hydrogen sulfide is regenerated in a separate steam stripper. A typical amine absorber and a regenerator are shown in Figure 3.5.

3.6 DE-ETHANISER

A gas mixture from the amine wash column enters a de-ethaniser column where methane and ethane are separated from the propane and butanes by distillation. This is a distillation column where a preheated feed gas mixture is introduced in the tower, reboiled by a heater at the bottom of the column, and refluxed by the top product from the reflux drum located at the top of the tower to separate the methane and ethane gases from the rest. The top product from this column is sent to a fuel gas system for the furnaces used in the refinery. However, top gases may also be used as a source of petrochemicals and hydrogen. The bottom product is a mixture of propane and butanes that are sent to a meroxing unit for the removal of mercaptans. A typical de-ethaniser is shown in Figure 3.6.

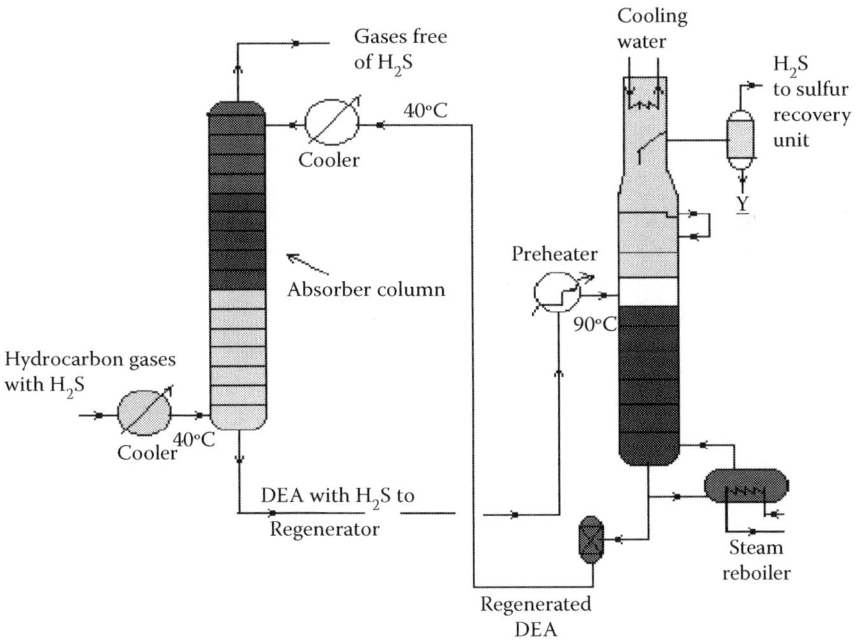

FIGURE 3.5 Diethyl amine absorption and regeneration unit for H_2S removal.

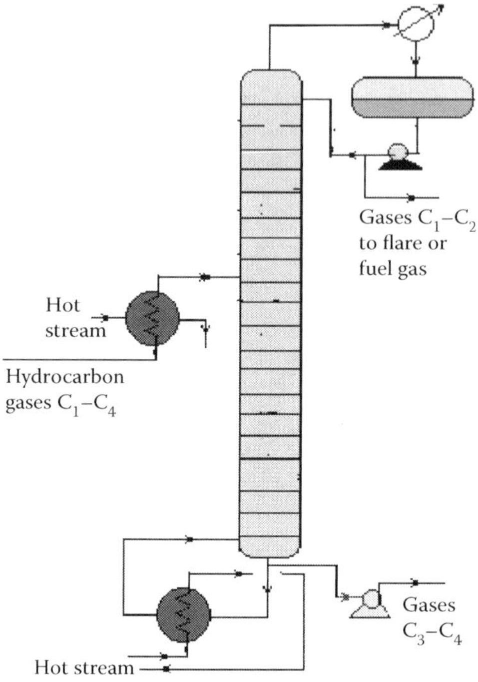

FIGURE 3.6 De-ethaniser unit.

3.7 MEROXING AND CAUSTIC WASH

Merox is an abbreviation of the words "mercaptan oxidation," which is done catalytically in a packed bed reactor. Mercaptans are a sulfur compound of hydrocarbons and are designated as RSH, where "R" is the alkyl radical, "S" is the sulfur atom, and "H" is the hydrogen atom. Mercaptans are highly corrosive, hence their presence in petroleum products is not desirable. Mercaptans can be extracted by a caustic solution, a process known as extractive merox (as shown in Figure 3.7).

Mercaptans can also be catalytically converted to disulfides (cobalt salt in the form of a chelate compound as the catalyst containing in a charcoal packed bed). Disulfide $(RS)_2$ are not corrosive as mercaptans and remain with the product. This type of treatment is called sweetening merox (as shown in Figure 3.8). The following reaction takes place in a caustic solution.

$$NaOH + RSH = NaSR + H_2O.$$

Sodium mercaptide "NaSR" formed is further oxidised to disulfide in the presence of air and the catalyst as

$$NaSR + RSH + 1/2O_2 = NaOH + (RS)_2.$$

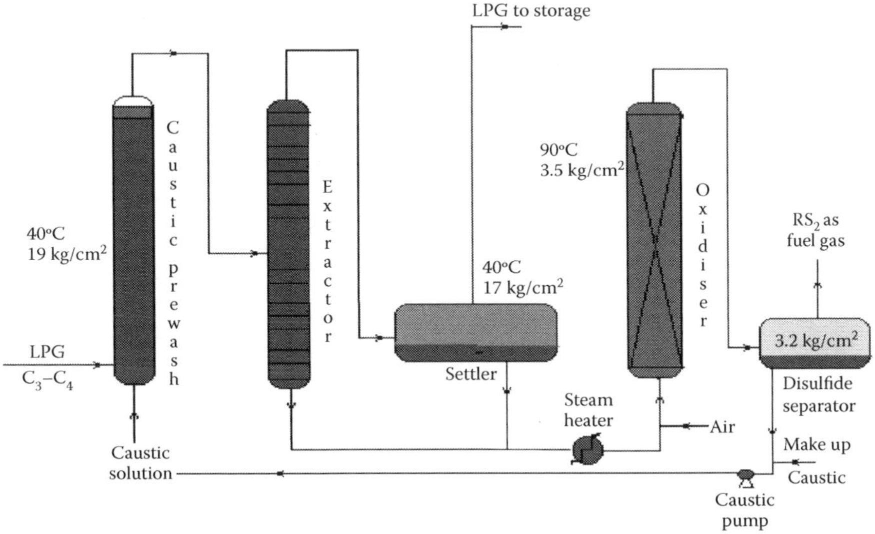

FIGURE 3.7 An extractive merox unit.

This step also regenerates sodium hydroxide. Meroxing is usually done for liquified petroleum gas (LPG), kerosene, and gasoline. Sodium mercaptides of the low molecular weight mercaptans present in LPG separates out from the gas mixture and is extracted from the product, a process known as extractive merox. For gasoline and kerosene, heavier sodium mercaptans remain in the oil phase and are converted to disulfide in reaction, i.e., sweetening merox is carried out.

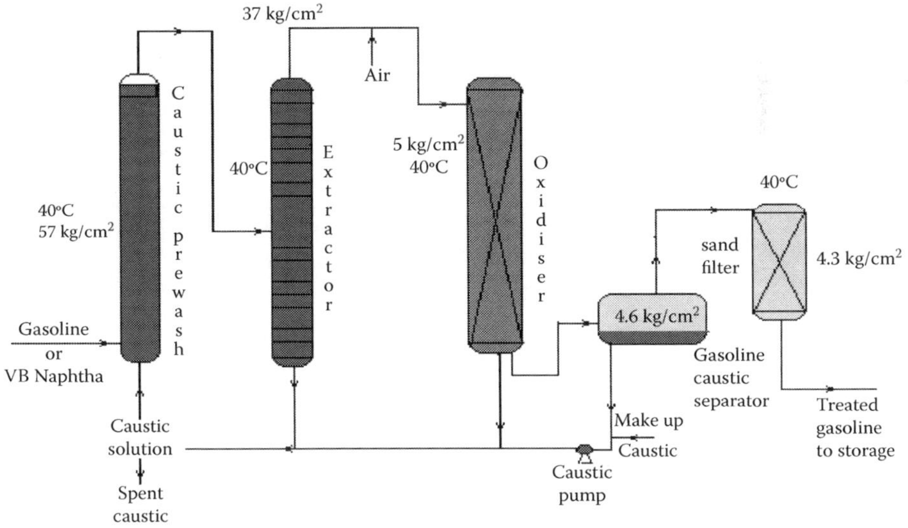

FIGURE 3.8 Sweetening merox treatment for gasoline.

Though meroxing is a convenient method for reducing the corrosiveness of the product, both mercaptans and disulfides produce sulfur compounds of oxygen (SO_x), which is a major pollutant to the environment, therefore, meroxing is replaced by catalytic hydrodesulfurisation in modern refineries.

3.8 LIQUIFIED PETROLEUM GAS SPLITTER

Liquified petroleum gas (LPG) mainly contains a mixture of propane and butane with traces of methane and ethane left after de-ethanisation. Propane and butane can be separated from LPG by distillation in a plated tower, the top product is enriched propane and the bottom product is butane. In refineries that use propane as the solvent for the propane deasphalting unit or as a refrigerant, they can obtain their propane requirement from such a distillation column, commonly known as an LPG splitter. Butane is either recycled to LPG or blended in gasoline to adjust the Reid vapour pressure (RVP). A typical LPG splitter column is shown in Figure 3.9.

3.9 NAPHTHA REDISTILLATION

As already mentioned in Section 3.3, a naphtha stream is obtained from the stabiliser bottom, which is normally liquid boiled in the range of C_5–140°C, which means hydrocarbons having carbon number five and hydrocarbons boiling up to 140°C are present in the mixture. This is further separated into two fractions in a column where the top is the lighter fraction, C_5–90°C, and the bottom is 90°C–140°C. This lighter fraction is suitable for petrochemical plants in the production of olefins and hydrogen because of the presence of paraffinic hydrocarbons, whereas the heavier naphtha is suitable for the production of high octane gasoline and valuable petrochemicals like benzene (B), toluene (T), and xylenes (X). A typical naphtha redistillation column is shown in Figure 3.10.

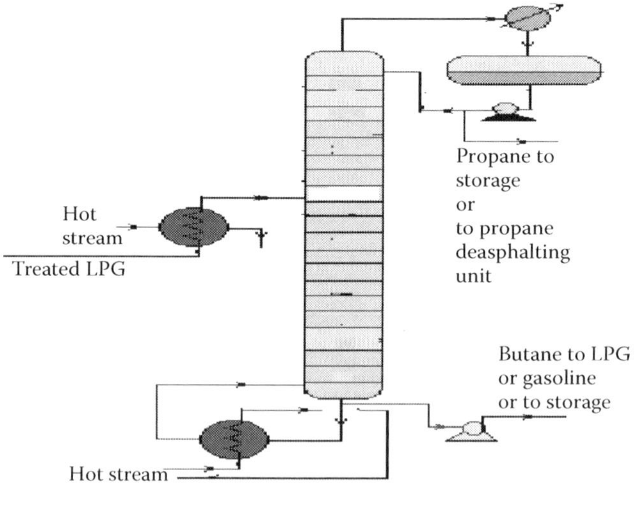

FIGURE 3.9 LPG splitter to separate propane and butane.

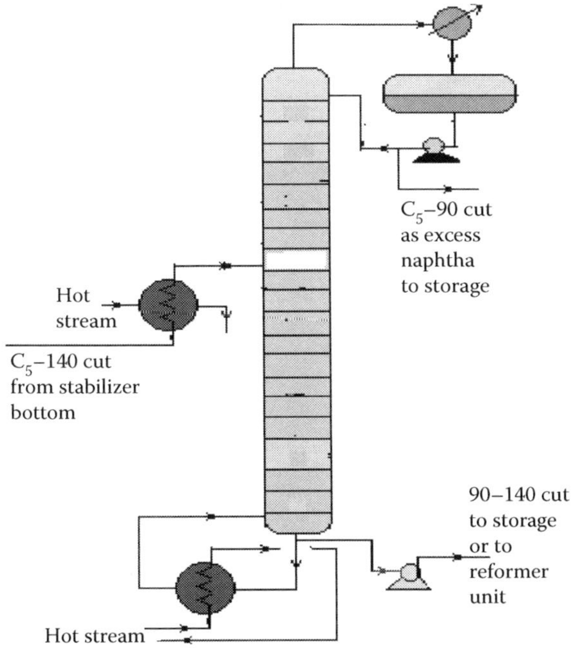

FIGURE 3.10 A naphtha redistillation unit.

3.10 NAPHTHA PRETREATMENT

A catalytic reforming unit uses platinum as the catalyst to convert naphtha into high octane gasoline or aromatics. As a platinum catalyst is easily poisoned by the presence of sulfur, oxygen, nitrogen, and metallic components, it is essential to pretreat naphtha before reforming. This, in fact, is done simultaneously in a catalytic hydrogen pretreatment reactor.

Naphtha stream, i.e., 90°C–140°C fraction, from the naphtha redistillation column is usually desulfurised in a catalytic hydrogenation unit and is then sent to a platinum reforming unit where high octane gasoline or aromatics (BTX) are produced. In this unit, naphtha is preheated by a train of heat exchangers and further heated in a pipe-still heater to a temperature of 350°C in the presence of hydrogen under a pressure of 20–25 kg/cm^2 over a catalyst (Co-Mo sulfides supported on alumina) packed bed reactor.

The following types of reactions take place in the reactor.

Desulfurisation reaction:

$$RSH + H_2 = RH + H_2S \uparrow.$$

Deoxygenation reaction:

$$ROH + H_2 = RH + H_2O \uparrow.$$

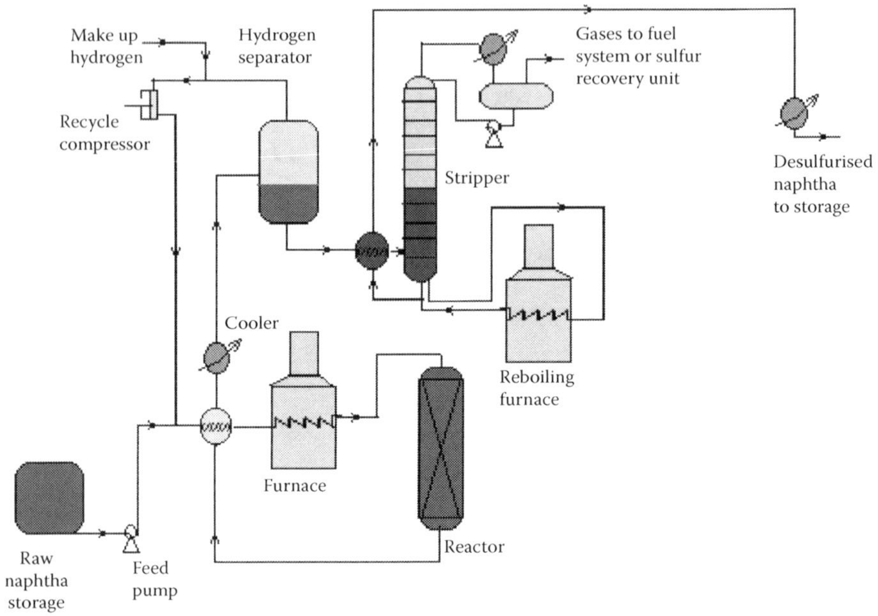

FIGURE 3.11 A naphtha pretreatment unit.

Denitrogenation reaction:

$$RNH + H_2 = RH + NH_3 \uparrow.$$

Demetallation reaction:

$$RM + H_2 = RH + {}^*MH.$$

*MH indicates metals adsorbed on the surface of catalyst.

The feed naphtha is premixed with hydrogen and heated by preheaters and a tube-still heater to raise the temperature up to 350°C at a pressure of 20–25 kg/cm^2. The product mixture from the reactor is cooled to 50°C and flashed in a separator vessel to release unreacted hydrogen, which is then recycled back to the reactor. The product mixture containing hydrocarbon gases (C_1–C_4), hydrogen sulfide, ammonia, moisture, etc., are steam stripped in a plated column. Desulfurised naphtha comes out as the bottom product which is further cooled before it is sent to storage. A flow sheet of naphtha pretreatment is shown in Figure 3.11.

3.11 NAPHTHA PLATINUM REFORMING (PLATFORMING)

Desulfurised naphtha (90°C–140°C) is catalytically converted to high octane gasoline by the reaction of a platinum (now platinum–rhenium) catalyst in a reactor in a hydrogen environment. This type of treatment is also known as platforming. The major reactions involved during platforming are dehydrogenation, isomerisation, and dehydrocyclisation along with a small amount of hydrocracking and hydrogenation reactions. The following are examples.

1. Dehydrogenation of paraffins and naphthenes:
   n-paraffins = alkenes + hydrogen
   naphthenes = aromatics + hydrogen
   These reactions are endothermic.
2. Dehydrocyclisation of paraffins:
   n-paraffins = naphthenes + hydrogen
   This reaction is also endothermic.
3. Isomerisation of paraffins:
   n-paraffins = i-paraffins
   This reaction is nearly thermo-neutral.
4. Hydrocracking:
   n-paraffins + hydrogen = hydrocarbon gases + lighter paraffins
   naphthenes + hydrogen = hydrocarbon gases + lighter naphthenes
   These reactions are exothermic.
5. Hydrogenation
   alkenes + hydrogen = alkanes
   aromatics + hydrogen = naphthenes
   These reactions are exothermic.
6. Coking reactions:
   hydrocarbons = carbon or cokes

Desulfurised naphtha and hydrogen are mixed and preheated by heat exchangers followed by heating in a tube-still furnace to raise the temperature of the vapour mixture to around 500°C. The reaction temperature is 470°C at the start of the run (SOR), i.e., when the catalyst is fresh or regenerated. A traditional reformer catalyst contains 0.3–0.35% wt of platinum. Since, during the reaction, coke formation takes place over the surface of the catalyst, the reactivity of the catalyst comes down and the temperature has to be raised to maintain the uniform reactivity. When the catalyst is used for about a year, the temperature has to be maintained near 520°C. The temperature is not raised further to avoid permanent damage to the catalyst. However, deactivation due to coke laydown is temporary and can be removed by burning the coke in the presence of air during regeneration. However, damage due to high temperatures above 600°C is permanent because of the sintering of the catalyst; hence the temperature at the end of the run (EOR) is never allowed to increase above 520°C. Traces of sulfur, nitrogen, and oxygen present in the feed naphtha permanently deactivate the catalyst and that is why naphtha pretreatment (as discussed earlier) is carried out before reforming. In order to reduce coke formation, it is essential to maintain hydrogen circulation at a desired partial pressure. Usually a minimum of 5 mol of hydrogen per mole of feed naphtha must be maintained during the reaction. The total pressure of the system is about 25–30 kg/cm^2. Usually, three reactors in a series are employed while regeneration is carried out for all the reactors after complete shutdown of the unit. Alternatively, four reactors may be employed while three reactors, in fact, are in service and the fourth reactor (swing reactor) is

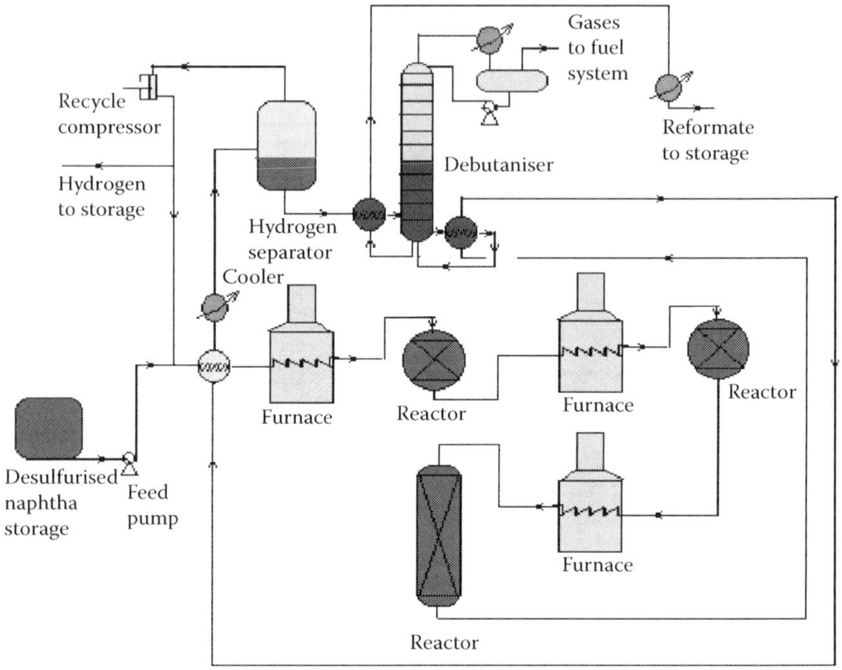

FIGURE 3.12 A naphtha reforming unit for gasoline.

taken out for regeneration. This unit does not require complete shut down and hence production is not affected. A typical reforming unit is shown in Figure 3.12. Three reactors are used in this flow sheet, the first two are spherical in shape while the last one is cylindrical. The catalyst is loaded and distributed in these reactors in the ratio of 15:35:60 by percent weight of the total weight of the catalyst loaded, respectively. Since the overall reactions are endothermic in nature, the product temperature falls while exiting from a reactor and is reheated by the intermediate furnace before it enters the second and third reactor. Finally, the product mixture is cooled and separated from hydrogen by flashing in a vessel at a temperature of 50°C. Hydrogen from this vessel is partly recycled back to the reactors and a part is sent to other hydrogen-consuming units in the refinery. The product from the bottom of the separator vessel is sent to a plated column to separate the butane and lighter hydrocarbons from the final product, known as the debutanised reformate, which is cooled before storage.

3.12 KEROSENE HYDRODESULFURISATION

Straight run kerosene (SKO, ATF, MTO, RTF) fractions from the atmospheric unit contain sulfur as compounds of hydrocarbons. This sulfur is catalytically hydrogenated to hydrogen sulfide and is thus removed from the kerosene fractions. This process is similar to the naphtha pretreatment process. Here, kerosene feed is mixed with hydrogen and preheated to 250°C–260°C followed by heating in a tube-still furnace to raise the temperature further to around 350°C. This hot mixture then

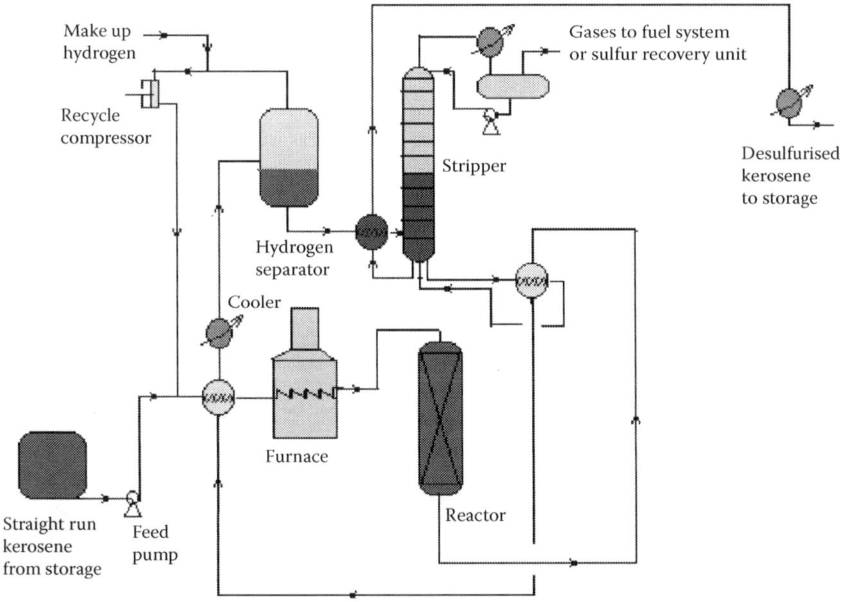

FIGURE 3.13 A kerosene hydrodesulfurisation unit.

enters a fixed bed reactor containing a bed of cobalt or nickel-molybdenum catalysts. The products are hydrogen sulfide, ammonia, and moisture, which are formed due to reactions with hydrogen and sulfur, nitrogen, and oxygen-containing hydrocarbons. A small amount of C_1–C_5 hydrocarbon gases may also be formed due to the cracking reactions. Excess hydrogen is always used to maintain sufficient partial pressure of hydrogen for the reactions and also to suppress coke formation. Excess hydrogen is separated from the cooled products in a high-pressure separator drum and recycled. Desulfurised liquid kerosene, containing dissolved hydrogen sulfide, moisture, ammonia, and hydrocarbon gases, is then stripped off in a stabiliser column. Gases from the top of this stabiliser may be treated in an amine absorption unit before it is used as a refinery fuel. A flowsheet diagram of a kerosene hydrodesulfurisation unit (KHDS) is shown in Figure 3.13. Usually, hydrogen consumption in a KHDS unit and the naphtha pretreatment unit is not much, however, hydrogen generated in a reforming unit is more than sufficient and, in fact, excess is used elsewhere, as in the lube hydrofinishing unit in a refinery.

3.13 DIESEL HYDRODESULFURISATION

The main ingredient of diesel oil is the straight run atmospheric gas oil, i.e., the straight run diesel fraction. However, many other fractions, such as distillates from a vacuum distillation unit (vacuum gas oil (VGO), spindle oil (SO) fraction), by-products from a vis-breaking unit (VB gas oil), light cycle oil from a fluid catalytic cracking (FCC) unit, and heavy naphtha fractions, join in a diesel pool. Thus, diesel contains a lot of sulfur that is removed in a catalytic hydrogenation unit similar to the KHDS unit

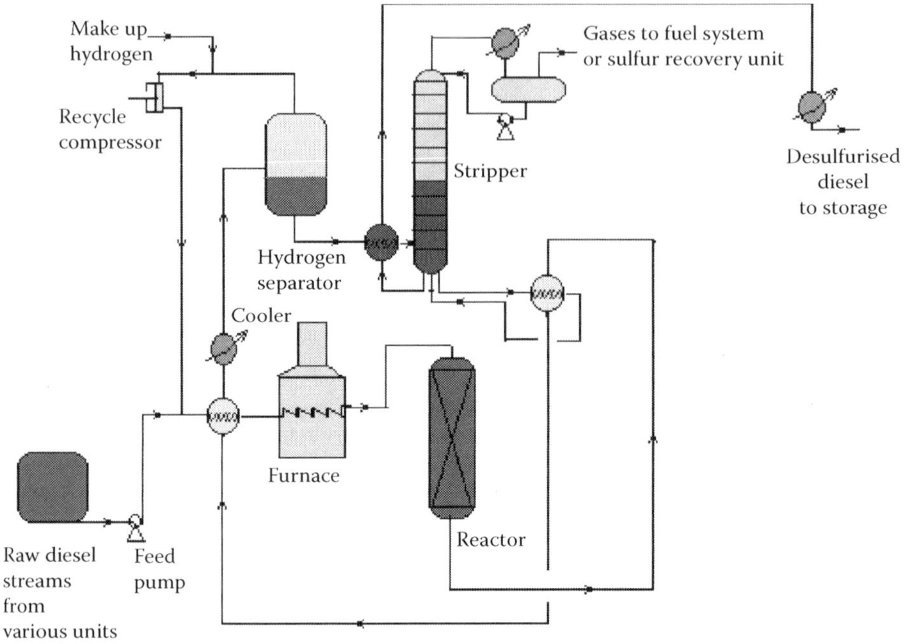

FIGURE 3.14 A typical DHDS unit.

described. The only difference is that higher hydrogen pressure (about 40–50 kg/cm^2) is required to remove sulfur from the heavier sulfur-containing hydrocarbons present in the diesel oil. Because of this, mild hydrocracking reactions may occur and produce cracked C_1–C_5 hydrocarbon gases. High-pressure separation of hydrogen and stabilisation of the final product is also similar to the KHDS unit. A typical diesel hydrodesulfurisation (DHDS) unit is shown in Figure 3.14. Since hydrogen consumption is larger and variable with the rate of production of diesel fractions, a separate hydrogen generation unit (which is discussed in detail elsewhere in this book) is required.

3.14 VACUUM DISTILLATION

As already mentioned in Section 3.1, vacuum distillation of the atmospheric residue yields additional and valuable distillates, which could otherwise be thermally destroyed if further distillation was attempted at atmospheric pressure and above. A typical vacuum distillation unit is shown in Figure 3.15. Hot RCO either from the steam-heated storage tank or from the bottom of the atmospheric distillation column is pumped through a series of preheaters (heat exchanger train) followed by heating in a pipe-still heater to raise the temperature to 360°C–370°C. This hot stream is then flashed in a multiplated distillation column where vacuum (below atmospheric pressure) is maintained by steam ejectors by medium pressure superheated steam as the motive fluid that entrains the top hydrocarbon vapours, which are condensed by water coolers. Usually, three numbers of ejectors are used; the first stage sends the uncondensed vapour to the second stage followed by condensation. The uncondensed vapour then enters the third stage followed

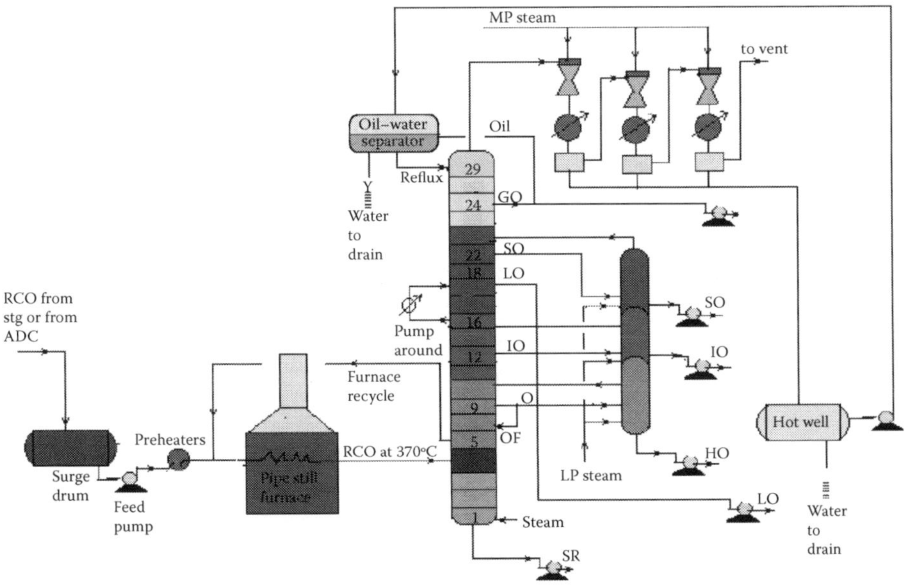

FIGURE 3.15 A vacuum distillation unit.

by condensation and the uncondensed vapour from the third stage is vented out through a flare or stack. A vacuum of 30–40 mm of mercury is maintained at the top of the column and 100–120 mm at the bottom. Condensates from these ejectors are collected in a drum, known as a hot well. The oily layer is then sent to an oil–water separator vessel from which oil is drawn as vacuum gas oil (VGO), part of which is sent to the column as the reflux and the rest to storage. Condensates from the hot well and the separator drum are drained to the sewer or water collection system. A few plates below the top plate of the column, an additional VGO is drawn and sent to the diesel/gas oil pool. SO is the next vacuum distillate, which is drawn a few plates below the gas oil draw plate. Similarly, the other vacuum distillates drawn from the plates below are light oil (LO), intermediate oil (IO), and heavy oil (HO) in this sequence. LO is sent to the vis-breaking unit for the production of low viscous fuel oils, whereas SO, IO, and HO distillates are further stripped in a side stripper by steam to remove the lighter components to adjust the flash points. The bottom residue from the tower is called the short residue (SR), which is stripped by the bottom steam followed by cooling through a steam generator and sent for storage in the deasphalting unit. A portion of the hot vacuum distillate is drawn from the column and returned back after cooling to control the heat load of the column. This stream is called the pump around. Unlike circulating refluxes in the atmospheric column, pump around is only one stream in the vacuum column. To aid washing of the HO fraction, 4%–5% of the feed RCO is overflashed. It is to be mentioned that SO, IO, and HO are the lube oil base stocks (LOBS), which are the main ingredients of lubricating oils in the market. Valuable lube oil stock, known as bright stock, is obtained from the SR by solvent deasphalting. However, the quality of vacuum distillates varies from crude to crude and, as a result, many crudes may not yield LOBS, but are converted to fuel products. It is commonly found that most of the Middle East or Gulf crudes are suitable for the manufacture of LOBS from vacuum distillates. Typical boiling ranges at atmospheric pressure and the draw temperature under vacuum are listed below.

Typical Draw Temperatures of the Distillates from a Vacuum Distillation Column

Cuts	Equivalent boiling range (°C) at 1 atm	Draw temperature (°C) under vacuum
RCO	356+	Feed enters at 370
Overhead		80
Reflux		65
GO	265–362	180
SO	362–385	232
LO	385–462	276
IO	462–504	336
HO	504–542	356
SR	542+	360

3.15 SOLVENT EXTRACTION

Vacuum distillates (SO, IO, HO) and also the deasphalted oil from the residue (bright stock) are suitable for the production of LOBS, which are the main ingredients of finished lubes (which are, in fact, obtained by blending metallic soaps and additives along with the LOBS). In order to meet the desired quality of lubes, certain properties, such as viscosity, viscosity index, pour point, flash point, carbon residue, foaming tendencies, sulfur content, and colour, must be adjusted by secondary processing after vacuum distillation. Solvent extraction is the traditional process for adjusting the viscosity index of the oil by partially removing the aromatic hydrocarbons, which are high viscosity and low viscosity index components. Solvents, such as phenol, furfural, and nitro methyl pyrolodine (NMP), are commonly used in refineries. These solvents have a high affinity to aromatic hydrocarbons, which are solubilised by these and form a different layer or phase, called extract.

The rest of the oil lean with aromatics are separated as the raffinate. The amount of separation of aromatic hydrocarbons by the solvent depends mainly on the operating parameters, such as solvent-to-oil ratio, temperature, partial solubility with the non-aromatic compounds present in the oil, or selectivity. Solvent and oil are contacted in a plated column usually in a rotating disc contactor (RDC) type gentle agitator provided over the plates through a common rotating shaft. The down coming stream is gradually enriched with aromatics and finally drawn from the bottom of the column as the extract. The uprising stream is gradually lean with aromatics and rich with paraffins and other non-aromatics and finally emerges as the raffinate stream. In fact, complete removal of aromatics from oil is not desirable as aromatics are the main contributors to the viscosity of the oil and hence removal of aromatics is carefully selected based on the required adjustment of the viscosity and viscosity index of the final product. Solvent from both the extract and raffinate phases is recovered by steam strippers and is recycled back to the extraction column. A typical furfural extraction unit is shown in Figure 3.16.

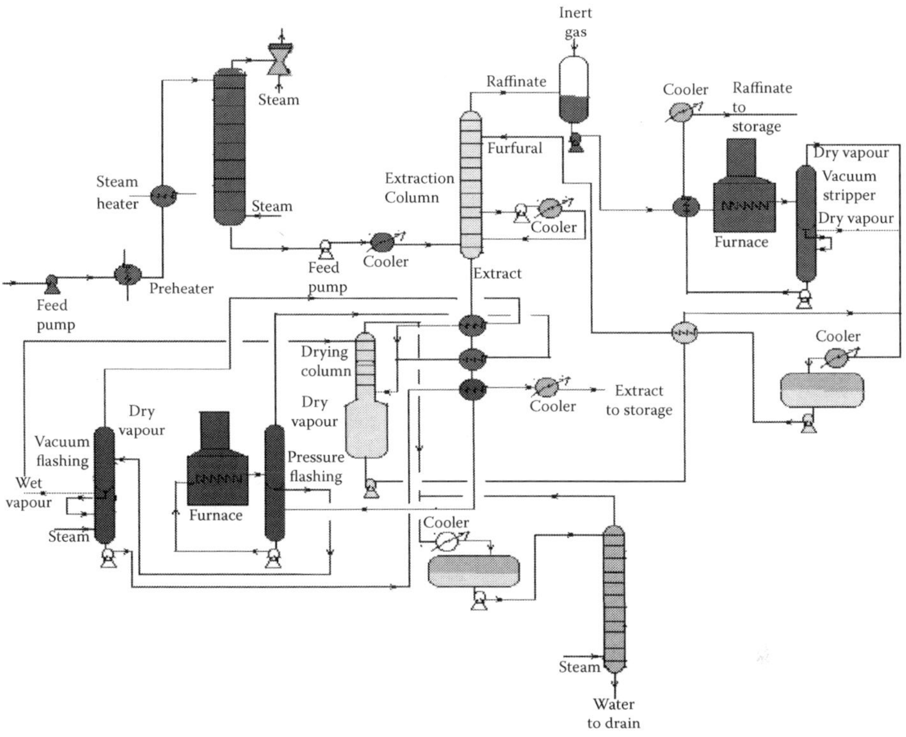

FIGURE 3.16 A furfural extraction unit.

3.16 PROPANE DEASPHALTING

Liquid propane is a good solvent for hydrocarbon oil and mainly rejects heavy hydrocarbons (asphaltenes) and non-hydrocarbons present as asphalt in the vacuum residue. Hence, propane is used as a solvent for extracting oil from vacuum residue or SR. The extracted oil is known as deasphalted oil, which is suitable LOBS (Lube oil base stock) for making high viscous lubricants, the viscosity index is then corrected by selective solvent extraction of aromatics, as discussed in the previous section. A deasphalting column is also a plated tower where the down coming stream becomes rich with asphalt and emerges from the bottom of the tower as the raffinate and uprising stream becomes rich with the deasphalted oil and finally emerges as the extract. Propane is separated by vaporising both the extract and raffinate streams in separate steam strippers. A typical propane deasphalting unit is shown in Figure 3.17.

3.17 SOLVENT DEWAXING

Solvent-treated vacuum distillates and deasphalted oil, as discussed in the previous sections, are further treated to adjust the pour point of the oil. Normal paraffinic hydrocarbons present in the oil have a high solidification temperature and form wax as the solid or semi-solid mass of hydrocarbons. There are paraffinic molecules

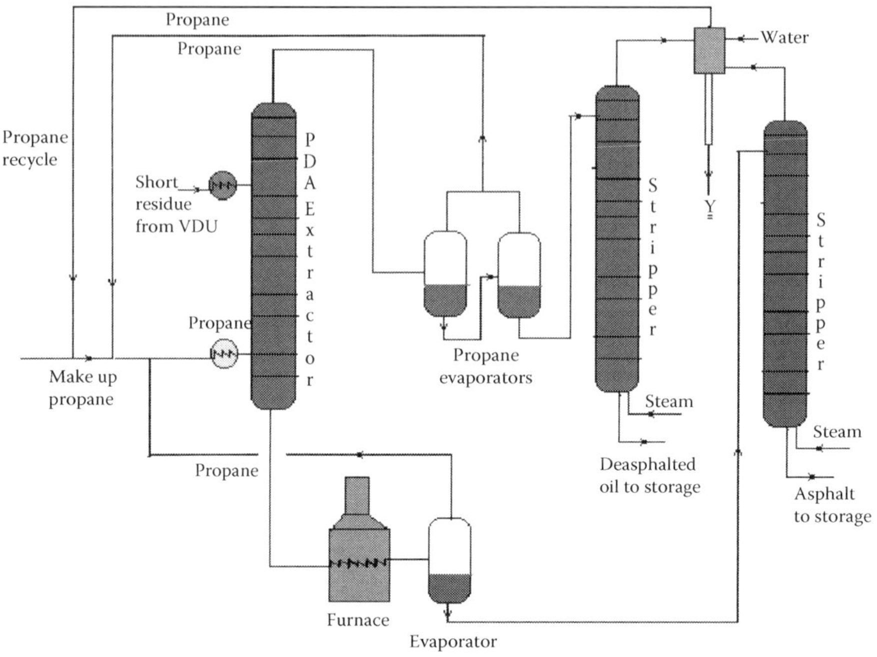

FIGURE 3.17 A propane deasphalting unit.

that can solidify even at room temperature. Since lubricating oil is used at various temperatures, ranging from very low to very high temperatures, it is essential that lubricating oil does not solidify in this range of temperatures. Hence, the pour point (a few degrees above the solidification temperature) of lube oil must be low. Though paraffins (mainly normal varieties) are responsible for wax formation, wax is defined as the hydrocarbon matter that solidifies at a temperature below –20°C (or the lowest temperature of the environment determined by the usage). Wax-forming hydrocarbons are selectively removed by low temperature freezing and separation. This is usually carried out by mixing the feed oil with a solvent, such as methyl ethyl ketone, benzene, or toluene, mixed in various proportions to prevent crystallisation of the wax-forming molecules in the precoolers and chillers, until they are separated from the solvents along with the desired oil components through a filter. A typical dewaxing plant, as shown in Figure 3.18, uses a vacuum drum filter where the solvent–oil mixture is sprayed over a filter cloth on a rotary drum enclosed in an inert atmosphere. The wax crystallises over the filter cloth and is scraped out with a scraping blade (also known as doctor's knife) in a continuous manner. The drum is divided into four chambers, for filtering, washing, purging, and scraping. Each section is connected separately with the vacuum, inert gas or steam, solvent, and filtrate oil collection systems through a rotary valve arrangement. The drum is partially submerged in a vessel containing the oil–solvent mixture and, as it rotates, the various chambers come in contact in succession. When the filtering zone comes in contact with the oil–solvent mixture, separation of the wax from oil starts and a wax layer is developed on the surface of the filter cloth. As this section is about

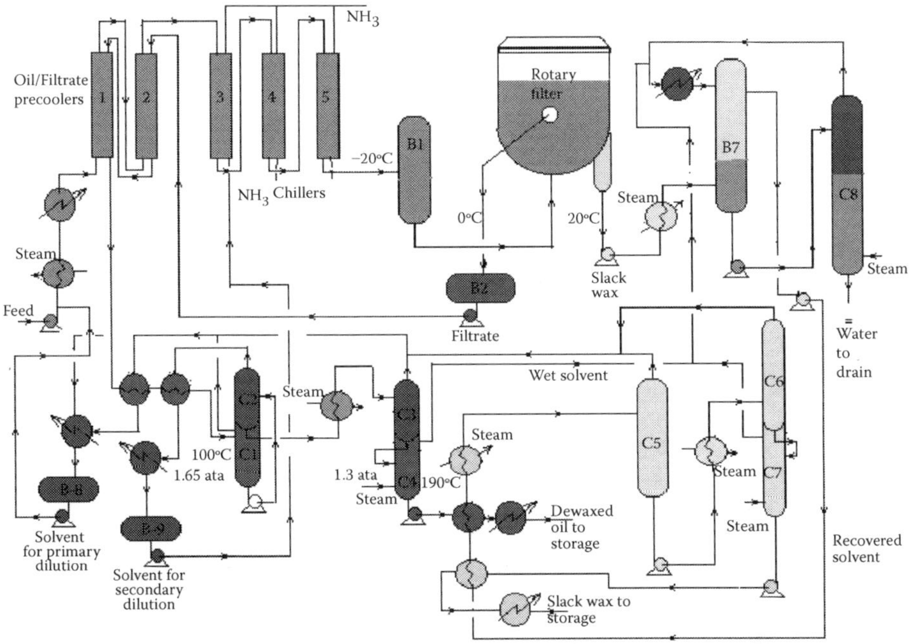

FIGURE 3.18 A solvent dewaxing unit.

to leave the oil–solvent level, the wax cake is completely developed over the filter cloth and the majority of oil is dewaxed. As the drum enters the washing section, an additional solvent is sprayed over the wax cake, which removes the entrapped oil, if any, and is removed by the vacuum connection of that section. As the cake reaches the purging section, inert gas purges the cake adhering to the cloth surface in order to loosen the wax cake to aid easy removal of the wax in the next section. This helps prevent the pores of the filter cloth from being plugged by wax. As the scraping section is entered, loose wax cake travels under a blade and is scraped from the cloth. Sometimes, steam is also purged to free the pores of the cloth. The scraped wax is then passed over a steam-heated screw conveyor to a wax collection pit. Figure 3.19 depicts a vacuum drum filter with various facilities.

3.18 HYDROFINISHING

Dewaxed oil from the dewaxing unit contains sulfur, oxygen, nitrogen, and organometallic compounds, which must be removed to improve the quality of LOBS. Sulfur is present as mercaptans and as heteroatomic hydrocarbons in oil and is responsible for the corrosive effects. Nitrogen compounds are responsible for the colour of the lube. Metallic compounds are responsible for unwanted deposition and may cause degradation of metallic surfaces in contact with the lube. Corrosiveness due to the presence of oxygen compounds for the formation of acid may occur. High molecular weight olefinic or unsaturated compounds, especially diolefins, may give rise to poor oxidation stability.

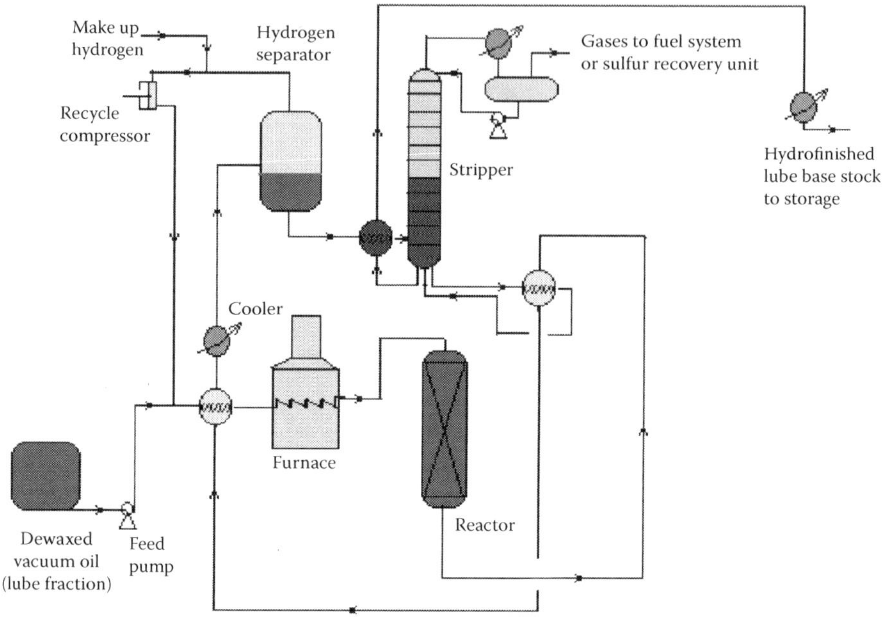

FIGURE 3.19 A catalytic lube hydrogen finishing unit.

In the hydrofinishing unit, catalytic hydrogenation is carried out in a fixed bed reactor. A nickel–molybdenum or tungsten catalyst is used at a high partial pressure of hydrogen. Temperature and pressure are maintained at around 350°C and 50–80 kg/cm^2, depending on the feedstock to be treated.

The reactions involved over the catalyst are similar to those already discussed in Section 3.10 in the naphtha pretreatment unit. A typical hydrofinishing unit for dewaxed oil is shown in Figure 3.19. It is to be noted that in a lube processing plant extraction, dewaxing and hydrofinishing operations are carried in a chain for each distillate separately, e.g., IO cycle, HO cycle, and BS cycle. The capacity of the units in a chain must be designed to handle the maximum production of the desired vacuum distillates.

3.19 CATALYTIC PROCESSES FOR LUBE OIL BASE STOCK MANUFACTURE

Selective dearomatisation and dewaxing of vacuum distillates using solvent extraction processes, as described in the previous sections, are being replaced by catalytic hydrogen processes. Hydrogenation of unsaturates and aromatics can be carried out selectively to adjust the viscosity and viscosity index, followed by catalytic dewaxing and hydrofinishing processes. All these processes are carried out using different catalysts and different operating conditions in the presence of hydrogen. In these processes, neither any removal of any aromatic hydrocarbon molecule takes place nor any paraffin wax is removed, but are converted to saturates and iso-paraffins as part of the lube oil components.

However, unwanted nitrogen, sulfur, oxygen, and metals are removed. A small amount of gases and lighter fractions may be formed as by-products. Hence, the yield of lube stocks per ton of feed distillate processed is much higher than that from solvent processes. In addition, the cost of recovery and loss of solvent and the huge cost in chilling and the refrigeration unit for dewaxing are avoided. The only disadvantage is that the by-products from solvent processes, such as aromatic rich extract (carbon black feedstock) and wax, are not produced in the catalytic processes and only the deasphalted oil can be handled. All these catalytic processes are divided into the following stages.

1. Deasphalting of residual feed by a solvent, e.g., propane, to avoid deactivation of the catalysts in the successive catalytic processes.
2. Hydrotreatment for removal of sulfur, nitrogen, oxygen, and metallic impurities from feedstock.
3. Hydrogenation to saturate aromatics in the dearomatising unit.
4. Hydroisomerisation of normal paraffins to iso-paraffins with or without selective mild hydrocracking of paraffinic hydrocarbons as per requirement to reduce the pour point.
5. Hydrofinishing reactions to adjust minor colour, sulfur, etc. A typical catalytic dearomatisation and dewaxing plant is shown in Figure 3.20.

3.20 HYDROCRACKING

Commercial hydrocracking processes

Name of hydrocracking process	Feedstock	Products	Operating conditions	Hydrogen to oil ratio
H-Oil	atmospheric and vacuum residues, asphalts to tar sands	Gases, light and middle distillates	Ebulliated bed reactor, 90%, 450°C–500°C, 10–30 MPa conversion	200–250 Nm^3 per cubic metre of feed
H-H	Vacuum distillates and deasphalted oils	Light to middle distillates, lubricating oils	Fixed bed process conversion 90%, 400°C–500°C, 10–30 MPa	200–250
CANMET	Heavy distillates and residues	Light and middle distillates	Fixed bed reactor, ferrous sulfate catalyst conversion 90%	200–400
ISOMAX	Vacuum distillates	Light and lube distillates	Fixed bed reactor, 400°C–450°C	200–400
BASF-IFP	Vacuum gas oil	Light and middle distillates	Fixed bed reactor, 400°C–450°C	200–400

Catalytic cracking in the presence of hydrogen is known as hydrocracking. This process involves high pressure and high temperature cracking of heavy petroleum stocks, such as vacuum distillates, and atmospheric and vacuum residues.

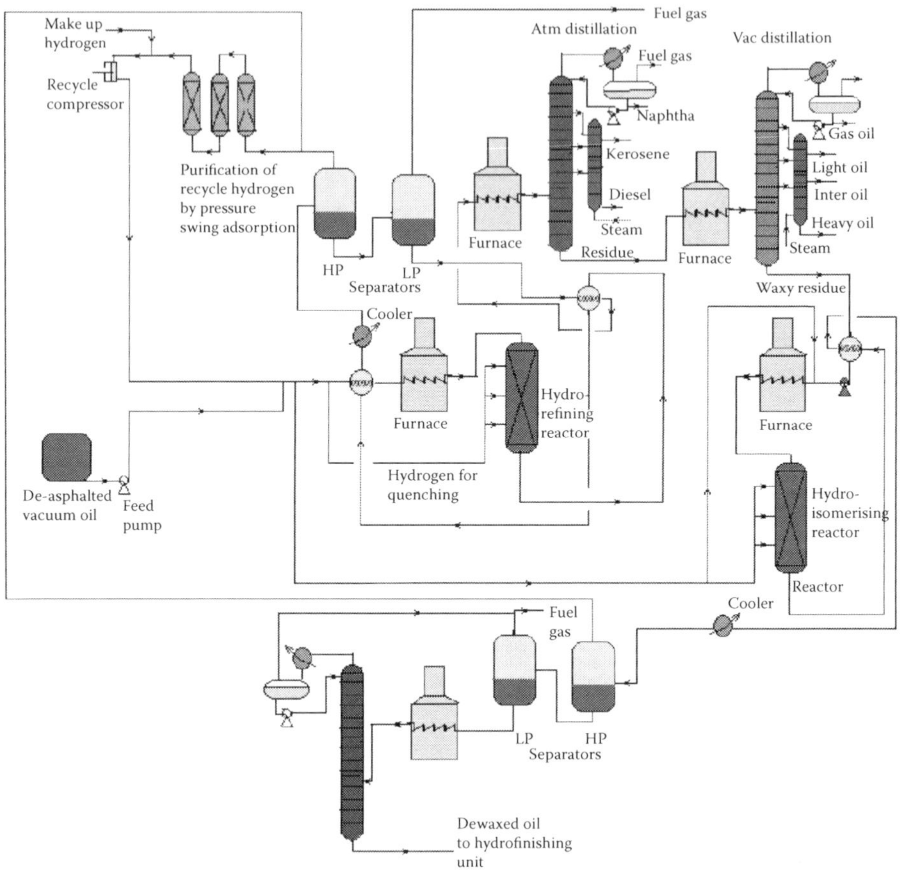

FIGURE 3.20 A catalytic hydrorefining and hydroisomerising unit for simultaneous dearomatisation and dewaxing of deasphalted lube oil.

The reactions involved are dissociation of hydrocarbon molecules, hydrogenation of unsaturates, along with desulfurisation, denitrification, deoxygenation, and demetallation, yielding good quality light and middle distillates, good lube distillates, fuel oil, etc. Various hydrocracking reactors and operating conditions are in use, depending on the type of feedstocks to be treated. For instance, VGO may be cracked in a fixed bed reactor to yield light and middle distillates, whereas a residue hydrocracker is an ebulliated bed reactor yielding light distillates to heavy fuel oils. Since a large excess of hydrogen is used, the final products have excellent qualities, i.e., low sulfur (almost free) kerosene and diesel oil, good quality aviation turbine fuel, good gasoline with high oxidation stability, good quality lube distillates with low sulfur, nitrogen, oxygen, and metallic impurities, and with a reduced coke yield. The table lists various well-known hydrocracking processes for various feedstocks. Though there are many hydrocracking processes, they differ mainly by the type of feedstock, product pattern, and catalyst, but the operating temperature varies from 400°C to 500°C and pressure above 10 MPa using in excess of

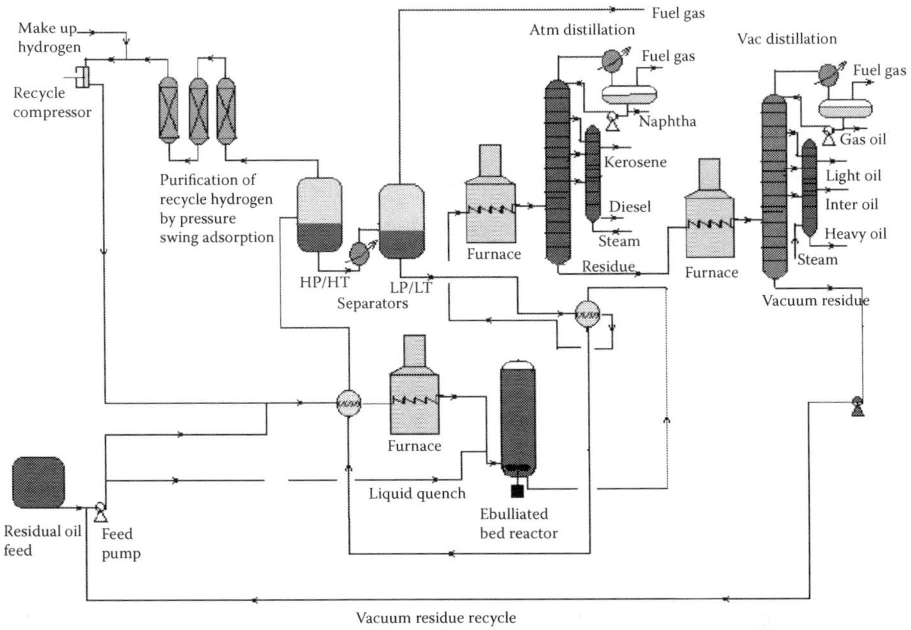

FIGURE 3.21 A residue hydrocracking plant using ebulliated bed reactors.

hydrogen circulation rate. The presence of nitrogen and metals in the feedstock are a major concern for the hydrocracking catalyst. Nitrogen yields ammonia, which destroys the acidic cracking sites of the catalyst, and metals deposited on the catalyst, poison the metal sites and also plug the porous catalyst. Hence, for such feedstocks, guard reactors are used to reduce the nitrogen and metal content of the feedstocks. Deasphalting is suggested for vacuum residues. Heavy aromatics present in the feedstock potential for yielding polynuclear aromatics (PNA) during hydrocracking reactions may need to be separated by vacuum distillation. It has been found that even with severe hydrocracking, PNA molecules cannot be destroyed which are found to be main ingredients of cokes. Most of the catalyst are made from transitional metals supported on alumina or silica. New catalysts are being prepared from molecular sieves. A typical fixed bed hydrocracking unit is presented in Figure 3.21.

3.21 MILD HYDROCRACKING

The term "mild hydrocracking" implies hydrocracking at mild operating conditions. It is carried out mostly for producing middle distillates and fuel oils from vacuum distillates, catalytic dewaxing for lube base stock at a pressure below 10 MPa and at a temperature range varying from 350°C to 450°C. Since pressure is low compared to hydrocracking reactors, mild hydrocracking can be carried out in the traditional hydrotreatment reactors. Conversion of vacuum distillates may not exceed 50%.

3.22 HYDROGEN GENERATION

Hydrogen is essential for the various processing units in any modern refinery, e.g., catalytic hydrogen-aided processes, such as desulfurisation, denitrogenation, demetallation, hydrogenation of aromatic and unsaturates, hydroisomerisation, and hydrocracking. Although hydrogen is obtained as the by-product from platforming reaction from naphtha, it is not sufficient for all the hydrogen-aided processes. Hence, a separate hydrogen plant is essential to produce hydrogen of the required quantity and purity. Electrolysis of water produces purest hydrogen but at an enormous cost, hence commercial hydrogen is obtained from petroleum or coal. Depending on the type of feed used, commercial hydrogen processes are gas-, naphtha-, or coal-based plants. Coal-based hydrogen plants are the oldest and are mainly used for the production of water gas, producer gas, and synthesis gas. Gas-based plants use methane as the raw material, which reacts with steam in the presence of a catalyst to generate hydrogen. Naphtha-based plants use naphtha with a low aromatic content as the raw material, which generates hydrogen in the presence of steam and a catalyst. These processes were mainly developed for the production of synthesis gas (a mixture of hydrogen and nitrogen) required for ammonia production in fertiliser plants. Hydrogen yield is maximum from methane and it decreases with heavier feedstocks (C:H ratio increases). Since heavier petroleum produces coke, catalysts are quickly deactivated, therefore methane is preferred in most commercial processes. Alternatively, carbon monoxide from partial oxidation of heavy petroleum is used to produce hydrogen in the presence of steam. Hence, the selection of the routes of hydrogen manufacturing processes depends on the availability of the raw material, the purity and yield of the hydrogen, and the cost of separation of the by-products and their value. Catalyst are mainly nickel based, usually supported on alumina. Since sulfur, arsenic, and some other elements are found to be permanent poisons to this catalyst, raw material must be desulfurised and treated to remove the poisons before reforming. A steam reforming plant consists of a feed desulfurisation unit, a primary reformer, a secondary reformer, shift converters, a carbon dioxide absorber, a methanator, and a hydrogen purification unit. The purpose of each unit is described in brief in the following subsections.

3.22.1 Feed Desulfurisation

When the feed is methane or hydrocarbon gases, sulfur compounds are mainly present in the form of hydrogen sulfide and mercaptans, hence the gases must be treated with amine solution to absorb the hydrogen sulfide, followed by catalytic desulfurisation using a cobalt–molybdenum oxide catalyst in the presence of hydrogen. The reactions are similar to those described earlier in the catalytic hydrodesulfurisation processes. For naphtha or heavier petroleum feedstocks, the second step of desulfurisation is carried out.

3.22.2 Primary Reforming

Desulfurised gas or naphtha preheated and mixed with steam is passed through a catalyst-packed (nickel) tube-still furnace where hydrogen is obtained due to the following reactions.

Dehydrogenation reactions:

$$CH_4 + H_2O = CO + 3H_2,$$

$$C_nH_{2n+2} + nH_2O = nCO + (2n + 1)H_2,$$

$$C_nH_{2n} + nH_2O = nCO + 2nH_2.$$

Cracking reactions:

$$CH_4 = C + 2H_2,$$

$$C_nH_{2n+2} = CH_4 + C_2H_6 + C_2H_4 + C_3H_8 + C_3H_6 + \ldots + H_2,$$

$$C_nH_{2n} = CH_4 + C_2H_6 + C_2H_4 + C_3H_8 + C_3H_6 + \ldots + H_2,$$

where $n < 5$ for gases other than methane and $n > 4$ for naphtha and heavier hydrocarbons. Each dehydrogenation reaction is endothermic and proceeds at a temperature of around 850°C. The above reactions are also associated with coke formation and condensation of aromatics and unsaturates. Suppression of the generation of coke and unsaturates is done by high partial pressure of hydrogen. By-product aromatics and cracked hydrocarbons are always formed and further reformed in the secondary reformer.

3.22.3 Secondary Reforming

At this stage, unconverted methane or hydrocarbons are partially oxidised to carbon monoxide and hydrogen. In a fertilizer plant, ammonia is obtained from nitrogen and hydrogen (1:3 ratio) as the feed gas mixture, known as synthesis gas. Nitrogen, hydrogen, carbon monoxide and carbon dioxide are obtained from steam reforming reactions with hydrocarbons. While oxygen is consumed to oxidise the hydrocarbons to carbon monoxides and nitrogen as the inert. Reactions are simply incomplete combustion reactions and are highly exothermic in the presence of excess steam.

$$C_nH_{2n+2} + (3n + 1)/2O_2 = nCO_2 + (n + 1)H_2O \qquad \text{highly exothermic,}$$

$$C_nH_{2n+2} + nCO_2 = 2nCO + (n + 1)H_2 \qquad \text{mildly endothermic,}$$

$$C_nH_{2n+2} + nH_2O = nCO + (2n + 1)H_2 \qquad \text{mildly endothermic.}$$

The overall reaction is, therefore,

$$C_nH_{2n+2} + (3n + 1)/6O_2 = nCO + (3n + 2)/3H_2 + 1/3\ H_2O \qquad \text{exothermic.}$$

The temperature of the reaction is maintained above 1000°C in the absence of a catalyst.

3.22.4 Shift Reactors

Carbon monoxide formed in the secondary reformer is converted to an additional hydrogen at this stage by reacting with steam.

$$CO + H_2O = CO_2 + H_2.$$

This is an exothermic reaction and in order to dissipate the heat, this reaction is carried out at two temperature ranges, one at high temperature (HT) between 400°C and 450°C and the other at low temperature (LT) between 200°C and 300°C. In the HT reactor, an iron oxide–chromium oxide catalyst is used and in the LT reactor, a copper-zinc catalyst is used. About 65%–80% of the carbon monoxide is converted to hydrogen in the HT reactor and the balance takes place in the LT reactor.

3.22.5 Hydrogen Purification

Gas mixture coming from the LT shift reactor contains much steam, carbon monoxide, carbon dioxide, unconverted hydrocarbons, nitrogen, and hydrogen. Cooling of the gas mixture separates the steam as water and the dehydrated gas mixture is then passed through a series of adsorbers cyclically operated to adsorb the gases, except hydrogen, in a pressure swing adsorption unit. Hydrogen purified by this method produces 99.99% pure hydrogen. A naphtha steam reforming plant for the production of hydrogen is shown in Figure 3.22.

3.23 FLUID CATALYTIC CRACKING

Fluidised catalytic cracking (FCC) is a process in which lighter boiling fractions can be generated from heavy petroleum stocks ranging from VGO to residues. Zeolite, silica or alumina acts as the cracking catalyst. The temperature of cracking reactions

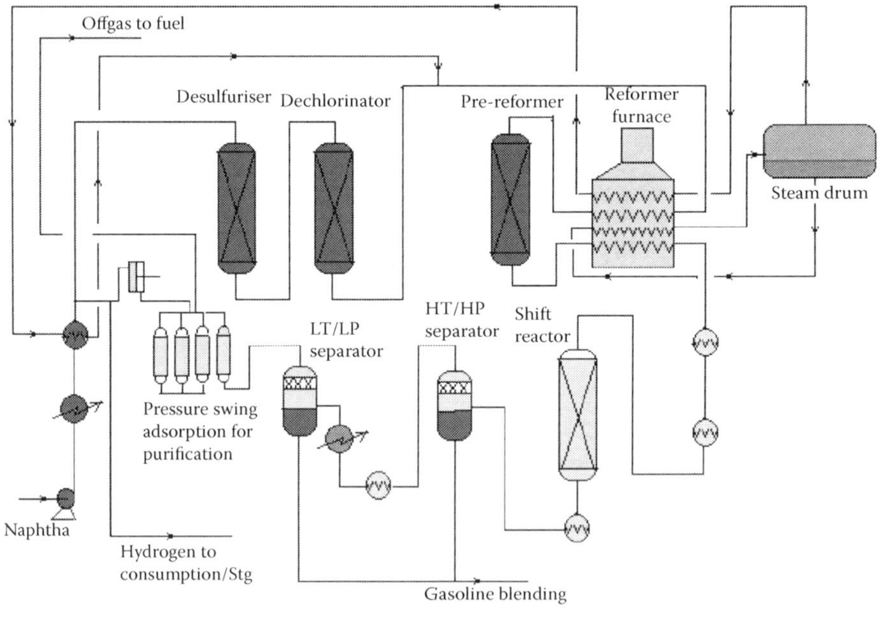

FIGURE 3.22 A modern plant for hydrogen manufacture by steam reforming of naphtha.

is in the range of 500°C–600°C. Feedstock rich in paraffinic hydrocarbons are preferred, however cracking of naphthenes and aromatic rings also occurs. Simultaneous reactions, such as dehydrogenation of saturated hydrocarbons, cyclisation of straight chain compounds, isomerisation, decomposition of heavy hydrocarbons, and polycondensation of aromatics to form polynuclear aromatics (PNA), may occur during the reactions. Much coke is generated due to the breakdown of aromatic and heavier hydrocarbons and the formation and condensation of polynuclear aromatics (PNA), thereby reducing the activity of the catalyst. Most of the reactions are initiated by the acid sites of the catalyst that donates H^+ ions.

Fine catalyst powder is fluidised in the tall tubular part (known as riser) of the reactor with the help of steam and light vaporisable hydrocarbons, like naphtha, which reduces the viscosity of the feed oil. Heavy feed oil is atomised by steam in the riser. Products are carried to the wider disengagement section of the reactor where products are disengaged from the catalyst through multistage cyclones. Hydrocarbon vapours from the reactor then enters a distillation column with side strippers for the recovery of various fractions. Spent catalyst laden with coke is transported to a fluidised bed regenerator either below or at the side of the reactor where coke on the catalyst is burnt out to regenerate the catalyst from its temporary deactivation. The regenerated and hot catalyst is then returned to the riser reactor to continue the process of cracking. Air is used as the fluidisation medium in the regenerator and the heat of combustion is used to maintain the reaction temperature. A small amount of coke (called equilibrium coke) is always retained on the surface and additional fresh catalyst must be maintained to compensate for the loss of activity. The organometallic hydrocarbons in the feed deposits metals on the surface of the catalyst, causing permanent deactivation of the catalyst. Usually, vanadium, nickel, and sodium are found in the feedstock and cause the deactivation. The presence of nitrogen in the feed also destroys the acid sites of the catalyst. Gases, gasoline, light cycle oil, heavy cycle oil, and residual oil contaminated with the catalyst are obtained from the distillation column. The light cycle oil goes to the diesel pool and the heavy cycle oil goes to the fuel oil pool. Residual oil goes to a catalyst decantation unit before it is recycled to the reactor. Decanted oil also acts as a quenching medium in the reactor. Gases and gasoline from the distillation column must be treated for sulfur removal separately. Because of the presence of olefinic hydrocarbons in the gasoline, hydrogenation may be required to improve its oxidation stability. A typical FCC unit is shown in Figure 3.23. The success of an FCC unit depends on the quality of the feedstock. Paraffin-rich feedstock is good for cracking, whereas aromatic-rich feed produces more coke than light fractions. The carbon residue of the feed analysis is an indication of potential coke that could be generated during cracking. Coking is also desirable, to some extent, for heat generation in the regenerator for maintenance of the reaction temperature. There are also some disadvantages owing to the presence of cracked products, such as the high content of unsaturated hydrocarbons and aromatics, which give rise to poor burning quality and oxidation stability. Cracked gasoline has a high octane number but it has poor oxidation stability due to the presence of unsaturated hydrocarbons. It is also corrosive due to the presence of mercaptans originally present in the feed stocks. Blending of LCO in diesel may reduce the cetane number due to the increased amount of aromatics and branched hydrocarbons present in LCO.

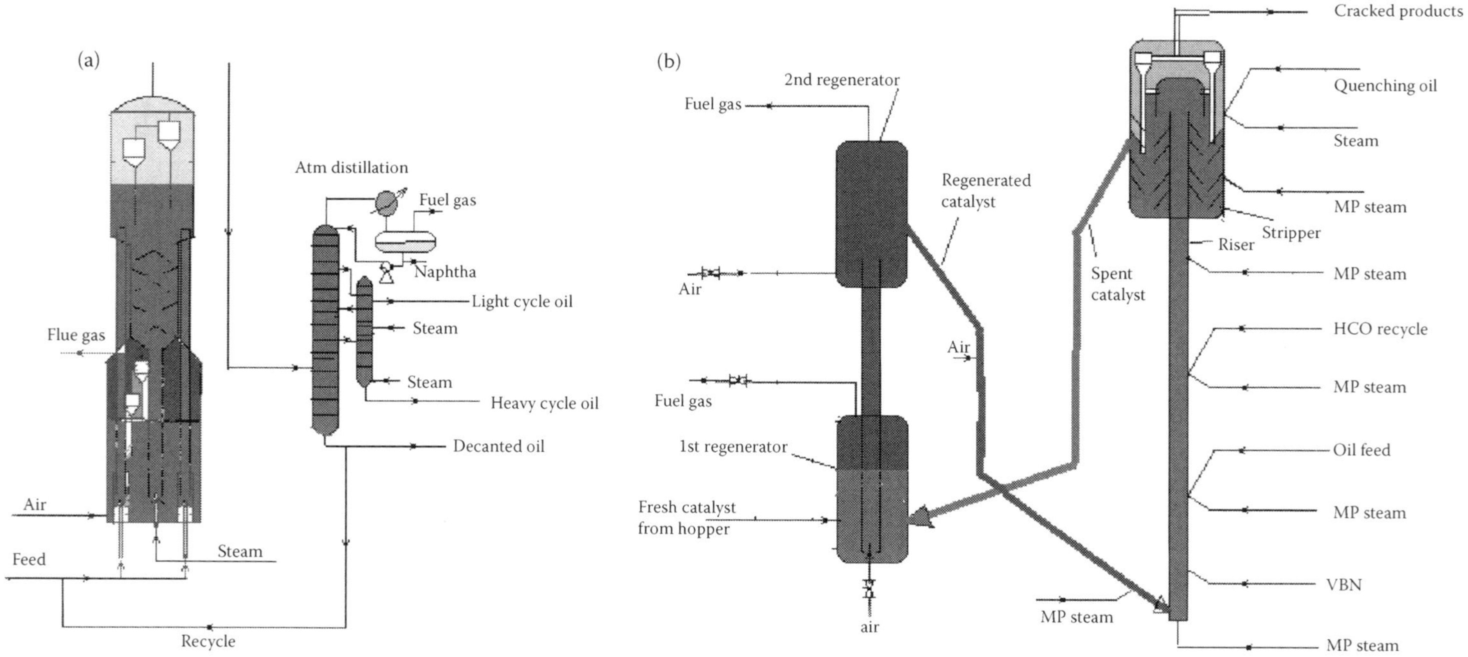

FIGURE 3.23 (a) A modern catalytic cracking unit— reactor and regenerator combined, (b) A modern catalytic cracking reactor unit with separate reactor and regenerator.

3.24 BITUMEN BLOWING

Bitumen is another name for asphalt. Raw asphalt from the deasphalting unit is blown with hot air in a furnace to adjust the softening point and penetration index for the production of paving grade bitumen. Depending on the surface temperature and the environment of application, the softening temperature and penetration index are adjusted by varying the air/feed ratio, temperature, and blowing time in the furnace. A typical bitumen blowing unit is shown in Figure 3.24. If the asphalt contains lower amounts of metals, these can be routed to the coking unit for production of metallurgical coke. It is a matter of fact that lube bearing crude oil yields asphalt of high metal content whereas non-lube bearing crudes yield asphalts with low metallic contents. As a result, asphalts from lube bearing crude is suitable for the production of bitumen, whereas asphalts and SRs and even heavy vacuum distillates from non-lube bearing crude are suitable for coke production.

3.25 VIS-BREAKING

Vis-breaking or viscosity breaking is a mild thermal cracking unit that produces low viscosity fuel oil from a high viscosity oil stock. In this method, the feedstock is usually a mixture of high vacuum distillates and residues, even asphalt, heated in a furnace at a cracking temperature (slightly above 400°C) at a pressure above atmosphere for a short time and quickly quenched and flashed in a plated column. Sufficient steam is used to separate the cracked light hydrocarbons. Products include gases, gasoline (VB gasoline or naphtha), gas oil (VB gas oil), and low viscous fuel oil or furnace oil as the major product. A typical vis-breaking unit is shown in Figure 3.25. As shown in the figure, heavy viscous vacuum oils, residue from the

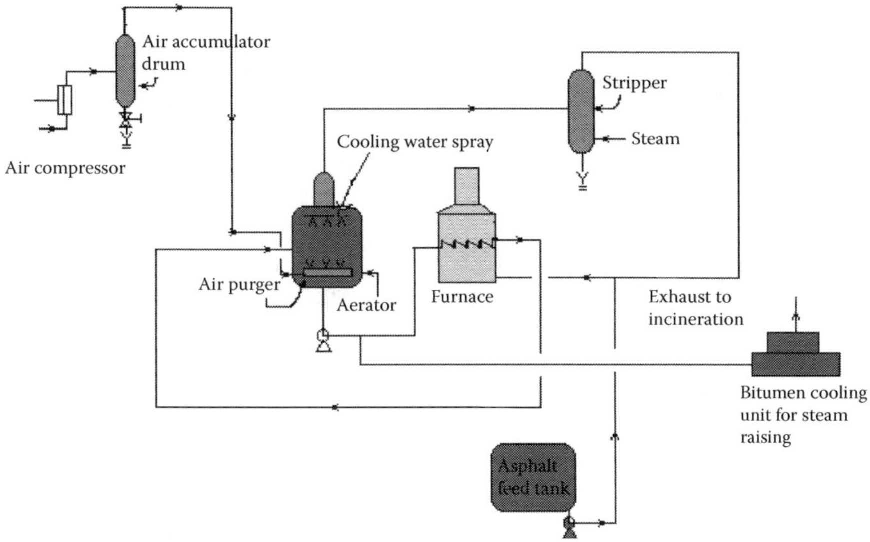

FIGURE 3.24 A typical bitumen blowing unit for the production of paving grade bitumen.

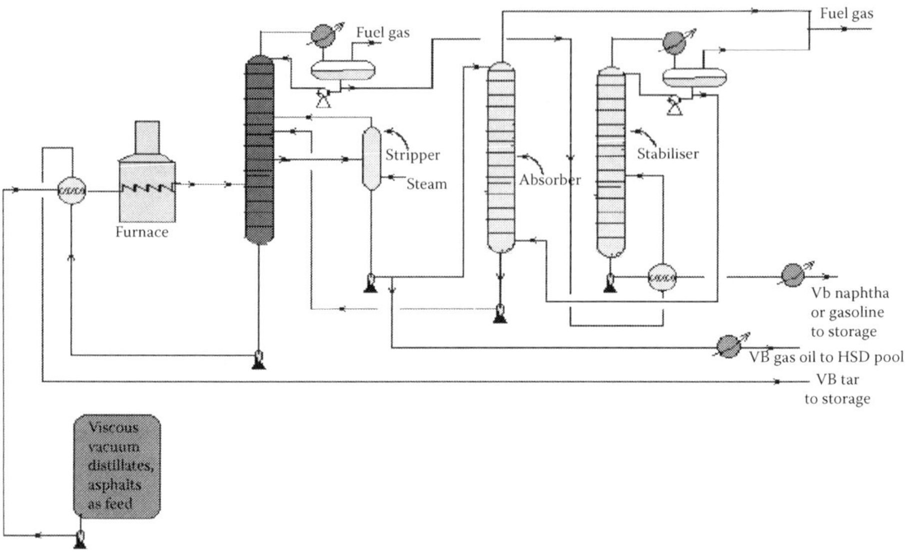

FIGURE 3.25 A vis-breaking unit for the production of fuel oil from vacuum distillates and residues.

vacuum distillation unit, and asphalt from a propane deasphalting unit are the feedstocks, which are preheated by hot products followed by heating in a tube-still furnace and then flashed in a distillation column. Top vapours enter a stabiliser column to separate the gases and VB naphtha components. Gases are further scrubbed in an absorber tower by VB gas oil from the main distillation column. Gases leaving the absorber tower are used as fuel gas and the rich gas oil stream containing scrubbed hydrocarbons from the gases is recycled back to the main fractionator column. The bottom of the main fractionator column is the VB tar (black oil) which is the cheapest and major fuel for the industrial furnaces.

The by-products are VB gas oil and VB naphtha. VB gas oil is mixed with the straight run, vacuum, light cycle gas oils, etc., for HSD as the final product. VB naphtha has medium octane number and is blended with high octane components like cat cracked gasoline and reformate. Since the products from this unit contain much mercaptans and unsaturated hydrocarbons, meroxing or desulfurising is essential. Catalytic hydrodesulfurisation may be beneficial as far as the removal of sulfur and olefins/di-olifins is concerned, but at the cost of an octane number.

3.26 COKING

The coking unit of a refinery yields petroleum coke, which is heavily condensed hydrocarbon with more than 90% carbon. This high carbon stock is used in the metallurgical and graphite industries for extraction of metals from ores and also as a clean fuel. As already mentioned, non-lube bearing crude oil yields large residual masses and asphalts with minimum metal and sulfur content, which are suitable for the production of petroleum coke. The coking unit uses various methods depending

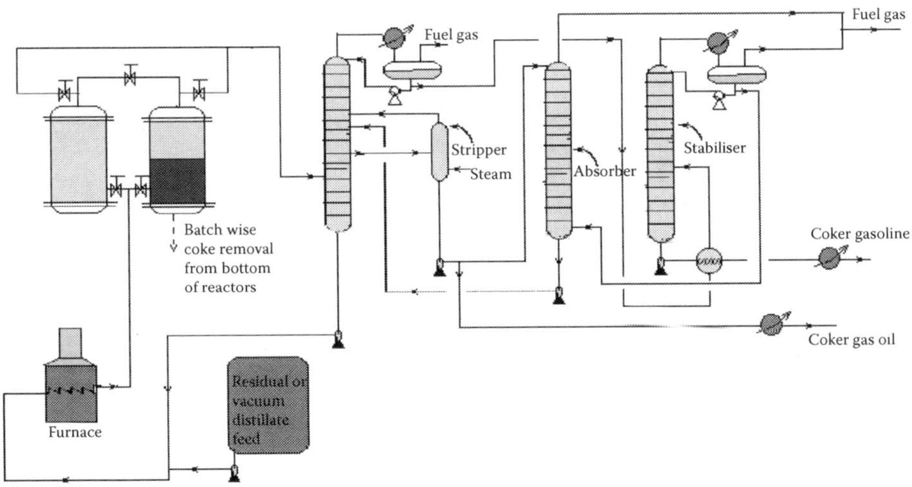

FIGURE 3.26 A delayed coking unit.

on the properties required by the users. Delayed coking and fluid coking plants are commonly employed in refineries.

In the delayed coking unit, feedstock is heated in a furnace to a temperature of around 480°C–500°C at high velocity to before sending to a coking drum where a long residence time allows the coking reactions to go to completion, thereby maximising coke formation. Once the coke drum is filled, another empty coking drum is pressed into service, the filled drum is isolated and coke is cut off by a high velocity water jet. In any such unit, a minimum of two coking drums or chambers are required, however more drums or high volume chambers may be used. A typical delayed coking unit is shown in Figure 3.26. In the fluid coking method, a fluidised bed of coke is used by atomising feedstock with steam and the high temperature is maintained by partially burning the coke particles in a fluidised bed burner.

QUESTIONS

1. Why is desalting of crude necessary?
2. What is the operating procedure for an electric desalter?
3. Present a flowsheet diagram for a crude distillation unit and its accessories, indicating the equipment involved and explain their functions.
4. Distinguish between sour and sweet crudes.
5. What is a Doctor's test?
6. What are the various parameters that distinguish different crudes?
7. What are the differences between the extractive and sweetening mercaptan oxidation methods?
8. Would you suggest merox treatment while catalytic hydrogenation is also available?
9. Why is pretreatment of naphtha required before plat-forming? Also, mention the predominant reactions involved in catalytic reforming.

10. Present a flowsheet diagram of a naphtha reforming plant, mentioning the necessity of each piece of equipment present in the flow sheet.
11. Why is debutanisation of the reformate required?
12. What are the methods for manufacturing benzene-free gasoline?
13. What are the various components of gasoline?
14. Distinguish between a thermal cracking unit and a catalytic cracking unit.
15. How do the qualities of gasoline from an FCC and reformer differ?
16. Discuss the qualities of middle distillates obtained from an FCC unit.
17. What are the effects of nickel, vanadium, and sodium present in the feedstock on an FCC unit? What are the necessary precautions required in an FCC unit in case feed contains all these metals?
18. Define the terms *spent catalyst*, *regenerated catalyst*, and *equilibrium coke*.
19. What are the functions of a waste heat boiler in an FCC unit?
20. How is petroleum coke manufactured?
21. What are the advantages and disadvantages of a hydrocracking unit?
22. What is mild hydrocracking?

4 Lubricating Oil and Grease

4.1 COMPONENTS OF FINISHED LUBRICATING OILS

The use of animal fats to reduce friction and wear and tear of mechanical parts has been the practice from time immemorial. However, since the availability of petroleum sources, lubricants are now manufactured using petroleum stocks. Today's lubricating oil is mainly composed of base hydrocarbon oil, lubricating base oil stock (LOBS), obtained from vacuum distillates after treatment in the refinery, as discussed in Chapter 3, with some additives to meet the requirements for its end use. Synthetic base oils, such as polyalphaolefins, alkylated aromatics, polybutenes, and aliphatic diesters, are also used as base oils.

The following additives are blended with the base oil.

Detergents: These are surfactants to cleanse the harmful carbon and sludge deposits on the surface of the metals in contact. Sodium or calcium sulfonates or organic sulfonates are excellent detergent agents in lube oils.

Dispersants: These are used to disperse the oil-insoluble products of oxidation and other formations in the oil phase and does not allow these to deposit on the metallic surfaces of bearing or rolling or sliding metals. Examples of dispersants are succinimides, esters of polysuccinic acid or succinate ester, and hydroxyethyl imide.

Antioxidants and stabilisers: These agents prevent auto-oxidation of hydrocarbon base oils present in the lubricant. This chemical reaction is in three stages: initiation, propagation, and termination, similar to a polymerisation reaction forming resinous layers. Copper soaps are an excellent retardant of such auto-oxidation. Aromatic amines and phenols are examples of antioxidants.

Viscosity index improvers: Polymethacrylates and polyisobutylene are excellent viscosity index (VI) improvers. These agents keep the viscosity of oil nearly unchanged over a wide range of temperature fluctuations.

Friction modifiers: These agents modify the coefficient of friction by adhering to the metallic surface. Examples are amines, amides, their derivatives, carboxylic acids, phosphoric acids and their salts.

Pour point depressants: Polymethacrylates, polyacrylates, and di-tetra paraffin-phenol-phthalate act as the pour point suppressants.

Demulsifiers: Acidic gases, moisture, carbon particles and other products may form at high temperatures, especially in engines, and form an emulsion with the lubricant oils. Sulfonates, alkylated phenolic resins, polyethylene oxide, etc., are good demulsifiers.

Anti-foaming agents: Gases and moisture are responsible for foam formation with the lubricating oil. The most widely used anti-foaming agent is polydimethylsiloxane.

Corrosion inhibitors: Ingress of oxygen and the presence of moisture cause oxide corrosion, and acidic chemicals and mercaptans may cause chemical corrosion aided by high temperatures. Esters or amides of dodycyl-succinic acid, thiophosphates, etc., act as corrosion inhibitors. Anti-oxidants also prevent oxide corrosion.

Thickeners: Sodium or calcium soaps act as thickeners, which are required to retain the film of lubricant over the metallic surfaces in contact and do not allow the metallic surfaces to come in direct contact without the lubricant film within them.

A variety of lubricants are used depending on the type of application, such as automobiles, aircraft, ships, and engines. These are broadly classified as automotive lubricants, aviation lubricants, industrial lubricants, marine lubricants, etc. Under this broad classification, they are further classified as engine oil, gear oil, bearing oil, hydraulic oil or transmission oil, cylinder oil, etc., depending on the field of application. Lubricants used in marine diesel engines are presented in Table 4.4.

4.2 AUTOMOTIVE OILS

The majority of automobiles include vehicles run on motor spirit (petrol) or diesel. Different lubricants are used for petrol engines (which are spark ignition type) and diesel engines (which are compression ignition type). These are known as engine oils, which are suitable for use in high temperatures and the oxidising environment of engines. Load (weight to carry) and speed of the vehicles are also to be taken care of before selecting a lube oil to apply. Usually, the temperature of an engine rises rapidly during the start up and continues at that temperature during motion. Such a wide and sudden change in temperature demands that the lubricant should have a high VI. In addition to temperature fluctuations, lubricants are prone to oxidation and cracking, leading to the formation of cokes, carbons, and gummy substances, which may ultimately deposit on the engine, causing irreparable damage. In addition to engine oils, different lubricants are applicable for other parts of the vehicle, such as the gears, brake, clutch, and bearings. Gear boxes contain the gears immersed in lubricating oil having low viscosity to reduce friction at high speed. The brake and clutch require lubricating oils of low viscosity. Bearings are used in various parts of the automobile from engine to wheels, and require low to high viscous lubricants. At low temperatures and high load, bearings at wheels are lubricated by grease. Since materials of construction and type of engines vary with the make, appropriate lubricants are selected and prescribed by the manufacturers. No single lubricant is therefore applicable for all makes. Finished lubes are classified according to the Society

TABLE 4.1
SAE J 300 Classification of Automotive Lubricants

SAE No.	ASTM D445
Winter grades	Viscosity mm²/sec
0 W	3.8 (min) at 100°C and 3250 (max) at –35°C
5 W	3.8 (min) at 100°C and 3500 (max) at –30°C
10 W	4.1 (min) at 100°C and 3500 (max) at –25°C
15W	5.6 (min) at 100°C and 3500 (max) at –20°C
20 W	5.6 (min) at 100°C and 4500 (max) at –10°C
25 W	9.3 (min) at 100°C and 6000 (max) at –5°C
Summer grades	Viscosity in mm²/sec at 100°C
20	5.6 (min) and 9.3 (max)
30	9.3 (min) and 12.5 (max)
40	12.5 (min) and 16.3 (max)
50	16.3 (min) and 21.9 (max)
60	21.9 (min) and 26.1 (max)

of Automotive Engineers' (SAE) numbers as given in Table 4.1. The viscosity of the lubricants and its variation with temperature (VI) and the pour point are the important parameters to satisfy the compatibility of application of lubes. Winter grades are classified as SAE numbers from 0 to 25W as typical examples of cold temperatures and from 20 to 60 SAE numbers for warming up the cranks of engines. A multigrade lubricant is a blend of more than one type of lubricant. For example, SAE15W 50 is an example of a blend of two grade oils. Usually, polymeric materials, such as ethylene–propylene copolymer, polymethyl acrylate, and butadiene, are added to these multigrade oils. However, rigorous testing of appropriate lubes must be carried out on cars of different makes in the testing laboratory or workshop for their suitability before prescribing them for engines and other parts. Since the performance of these lubes may not be satisfactory after a certain period of time due to degradation because of contamination, reaction, physical and chemical changes in the property of the ingredients or the base oils, it is inevitable that the lube must be drained out and replaced with fresh stock. This drain out period must be specified for the prescribed lubricants. The longer the drain out period, the more attractive the lubricant is in the market.

4.3 INDUSTRIAL LUBRICANTS

Industries use a large amount of lubricants, known as industrial oils, i.e., transmission oils, turbine oils, compressor oils, seal oils, cooling oils, gear oils, bearing oils, hydraulic oils, and cutting oils.

4.3.1 Bearing Lubricants

Bearings used in machineries face either sliding or rolling frictions. Usually, greases or solid lubricants are used to lubricate the small bearing surfaces. Lithium greases or graphite are the common lubricants. However, a solid polymeric lubricant, such as polytetraflouroethylene (PTFE), coating of the sliding bearing surfaces may be used for low load friction. These are popularly known as non-lubricated or self-lubricated bearings. Circulating lubricating oils are commonly employed for bearings of high load and speed where lubricating oil not only reduces friction but also cools the bearing surfaces. For this lubrication system, a thin oil layer between the bearing surfaces must be maintained by high pressure circulating pumps. Gaseous lubricants, such as air, nitrogen, oxygen, and helium, are used in the bearings of aeroplanes and aircrafts. A gaseous lubricant has the advantage that the chemical properties of gas do not change with temperature. At high temperatures, the viscosity of gas increases and, as a result, a gaseous lubricant is preferred over liquid lubricants in high load bearings in aerodynamic applications. Liquid or gaseous lubricants or grease are not applicable for bearings used under vacuum and at very high temperatures when radiation is appreciable. In this case, solid lubricants are used in the bearings. Molybdenum disulfide, graphite, boron nitride, and cadmium iodide powders are used as the lubricants. PTFE rings or lining over the bearing surfaces are also used to lubricate the bearing surfaces. Initially, these solids are not lubricating in nature, but with the heat of friction the lubricating property is manifested.

4.3.2 Hydraulic Lubricants

Hydraulic fluids are viscous liquids used in power transmission for control, braking in automobiles and machineries, raising or lowering loads by multiplying the transmitted force, and so on. In addition to these activities, hydraulic fluids lubricate the mating parts of machines and are used in a wide variety of environments, such as air, water, gaseous, and high and low temperatures. These fluids are practically non-volatile in the temperature of use. These fluids are mostly high viscous, high VI, petroleum-based oil with or without additives. Synthetic oils, such as esters of polyglycols, phosphoric acid, and silicones, are also used as hydraulic fluids. It is desirable that all hydraulic lubricants should also have a high flash point and be flameproof.

4.3.3 Compressor Lubricants

Gases are compressed either in reciprocating or centrifugal compressors. In the reciprocating compressor, the piston and cylinder is lubricated by lubricating oil, which must have a low vapour pressure and a low carbon-forming tendency. Lube vapour, especially petroleum base oil and carbon particles, may contaminate the compressed gas and lead to explosion. Modern reciprocating compressors are lubricated by PTFE polymer rings and lining. For natural gas or refinery gas compressors, polyalkylene glycols are used. For centrifugal compressors, solid lubricants or polymeric synthetic oils are used.

4.3.4 Pump Lubricants

Small water pumps are lubricated by grease in the bearing, but large pumps are lubricated by liquid lubricants either using the gravity falling cup method or by separate circulation using gear or screw pumps. Modern pumps use solid lubricants, like graphite, or polymeric material, like PTFE.

Miscellaneous industrial lubricants, such as transformer oils, cutting oils, and sealing oils, are used, and the specification of each is defined by the user industries. The varieties of industrial grade lubricants are classified as viscosity grades according to ISO 3448 standard, and are presented in Table 4.2.

4.4 AVIATION LUBRICANTS

Some of the major factors determining the selection of lubricants for the engines and bearings of aerodynamic vehicles are air temperature and pressure at various altitudes, type of engines and turbines, load and speed, etc. Air temperature may be as low as −80°C at the highest altitude or as high as +40°C on the ground. Pressure may vary from 1 atm on the ground, falling with an increase in altitude. The engines may be reciprocating or rotary types. Previously, very pure petroleum base oils or castor oil were used without any additives. Now, in the modern turbine-type engines, multigrade oils are being used. The typical specifications of reciprocating and turbine engines oils are presented in Tables 4.3A and 4.3B.

TABLE 4.2
ISO 3448 Standards for Viscosities of Industrial Lubricants

	Viscosity, mm^2/sec at 40°C	
ISO Grades	Minimum	Maximum
2	1.98	2.42
3	2.88	3.52
5	4.14	5.06
7	6.12	7.48
10	9.00	11.00
15	13.50	16.50
22	19.80	24.20
32	28.80	35.20
46	41.40	50.60
68	61.20	74.80
100	90.00	110.0
150	135.00	165.00
220	198.00	242.00
320	288.00	352.00
460	414.00	506.00
680	612.00	748.00
1000	900.00	1100.00
1500	1350.00	1650.0

TABLE 4.3A
Properties of Typical Aviation Lubricants for Reciprocating Engines

	Viscosity Grades				
Properties	**65**	**80**	**100**	**120**	**20W/50**
Viscosity at 100°C, mm²/s	11	15	19	23	20
Viscosity at 40°C, mm²/s	95	130	200	270	140
Viscosity index	110	105	100	100	150
Pour point, °C	−25	−23	−21	−20	−30
Flash point, °C	230	240	250	260	260
Sulfur, %	0.3	0.3	0.4	0.4	0.1
Density at 15°C	0.882	0.889	0.891	0.894	0.878
Ash contents, wt%	0.001	0.002	0.002	0.002	0.001
Total acidity, mg/g	0.02	0.02	0.02	0.02	0.12

4.5 MARINE LUBRICANTS

Marine craft, such as barges, speed boats, small ships, tankers, big ships, and submarines, are run on a variety of fuels, e.g., high speed diesel, light diesel oil (LDO), and residual fuel oil (FO). Small vehicles run directly on fuels using reciprocating engines and bigger vehicles use turbines. Modern cargo ships and crude tankers are run on electricity empowered by their own power plant fuelled by furnace oil or LDO. Hence, the requirements of lubricants vary with the types of vehicles. Large vehicles use lubricants for motors, for reciprocating or turbine engines, bearings, and industrial lubricants wherever applicable. Lubricants used in marine diesel engines are presented in Table 4.4.

4.6 GREASES

Grease is thickened oil containing the base oil and thickening additives. The mineral base oil must meet many different requirements, e.g., viscosity, VI, and oxidation stability, such as that required for making lubricating oil. But the other important properties for base oil required for grease making are the viscosity-gravity-constant (VGC), the aniline point, carbon type, and solubilising property. The thickeners are

TABLE 4.3B
Properties of Typical Aviation Lubricants for Turbine Engines

Properties	**Oil Viscosity Grades**		
Viscosity at 100°C, mm²/sec	3	5	7.5
Viscosity at 40°C, mm²/sec	14	29	34
Viscosity index	>100	>100	>100
Pour point, °C	−65	−60	− 60
Flash point, °C	225	255	235
Sulfur, %	<0.1	<0.1	<0.1
Total acidity, mg/g	0.15	0.2	0.152
Autoignition temperature, °C	400	420	390

TABLE 4.4
Lubricants for Diesel Engines for Marine Vessels

Properties	System Oil	Cylinder Oil	Trunk Piston Engine Oil	Test Method
Grades	SAE 30	SAE 50	SAE 40	ASTM
Viscosity at 100°C, mm^2/s	11.5	19.0	14.0	D445
Viscosity at 40°C, mm^2/s	103	218	138	D445
Flash point, closed cup, °C	225	210	220	D93
Pour point, °C	−18	−12	−18	D97

the sodium, calcium, or lithium soaps of fatty acids. Greases are popularly classified according to the type of thickeners used. The melting point and the water content of the soaps are important parameters.

Sodium soaps (melt at 150°C) are preferred over calcium soaps (melt at 100°C) for their higher melting point. The evaporation of water present in these soaps limits their use much below their melting points. Lithium-12 hydroxystearate soap is an excellent thickener for grease making. This grease is able to maintain its properties at high temperatures without losing its lubricating property up to 140°C along with excellent resistance to mechanical deformation and immunity to the presence of water in the field of application. Modern grease employs a mixture of calcium–lithium and aluminium–barium soaps. A major thickener in the grease industry is lithium 12-hydroxystearate, a soap of 12-hydroxystearic acid, obtained from vegetable oils and animal fats. In place of metallic soaps, thickening agents, such as clays, PTFE powder, and graphite, are also used in grease. But these require frequent replacement of grease to maintain uniform lubricity. Most modern greases also include some additives to modify the property of the grease to suit different environments of applications. These are antioxidants, corrosion inhibitors, anti-wear, extreme pressure additives, adhesives, etc. Grease behaves as non-Newtonian fluids and the majority usually manifest as Bingham plastics. The consistency index or apparent viscosity is measured for grease. Apparent viscosity is measured as per the ASTM D1092 method and consistency is measured by cone penetration index as per the ASTM D1403 method. A worked grease sample, obtained through a grease worker, which is a cylindrical vessel equipped with a piston with holes, is prepared before testing for cone penetration. Standard working is done by forcing the piston back and forth 60 times in 60 sec. The penetration index is measured at 25°C. In addition to these, the drop point is also determined according to the ASTM D556 method. Grease specification is classified according to the National Lubricating Grease Institute (NLGI). The properties and uses of typical greases are presented in Table 4.5.

4.7 LUBE BLENDING AND GREASE MANUFACTURE

Lube blending is the intelligent art of mixing various ingredients with the lube base stock, satisfying the desired physical, chemical, transport and mechanical properties,

TABLE 4.5
Specifications of Typical Greases

NLGI Grades	Thickeners	Penetration Index	Drop point, °C	Uses
1, 2, 3	Sodium soap	250–340	180–185	Bearings and gears
1, 2	Aluminium soap	265–340	90	Gears
0, 1, 2, 3	Calcium soap	220–385	80–90	Bearings, hydraulic presses, water pumps, etc.
2	Lithium–calcium soap	265–295	165–180	Gears, bearings, and water pumps, etc.
0, 1, 2	Lithium soap	265–385	170–190	Gears, bearings, and joints, etc.
2	Lithium compound with synthetic base	265–295	>250	Aviation and automobile gears and bearings

usually known as tribological properties. The blending units consist of kettles with steam or electric heating facilities where the additives are mixed while continuously monitoring the desired properties. Additives, such as anti-wear, antioxidant, anti-rust, anti-foam, and corrosion inhibitors, are given in small quantities and are procured from various sources. Usually lube blending and grease making are carried out in the same plant where large amounts of soaps of fatty oils are blended for grease and hence a saponification unit is part and parcel of a blending plant. A schematic general lube blending plant is shown in Figure 4.1a and a typical calcium grease making plant is presented in Figure 4.1b, where calcium soap is prepared from metered amounts of lime slurry and vegetable oil at the desirable operating temperature and oil/lime ratio in a continuous stirred steam-heated tank or a kettle heater. Modern plants use electric heating in the kettles. A number of kettles may be used to increase the production rate. Fats or vegetable oil are saponified and a soap mixture is pumped to a special type of kettle, known as a Lancaster mixer, where the base oil (usually petroleum base) is mixed at a high speed of about 3000–4000 rpm and at a temperature of 170°C–180°C. In fact, the operating temperature and time of mixing are monitored for the desired consistency of the grease. The effluent from the kettle is then filtered to remove sludge and particulates from the liquid grease before storage or continuous packaging.

4.8 ENVIRONMENTAL IMPACT OF LUBRICANTS

Since it is necessary to replace the lubricating oil and grease after some days or weeks or even after a few months, these must be disposed off. In many applications, such as in two-stroke engines, railways, chains, and in rubber tyres, lubricants are continuously lost to the atmosphere. Most of these losses are due to inappropriate maintenance and leakages, which cause enormous pollution and lead to accidents and fire. Direct draining and disposal in a landfill are also common, leading to pollution. Recycling

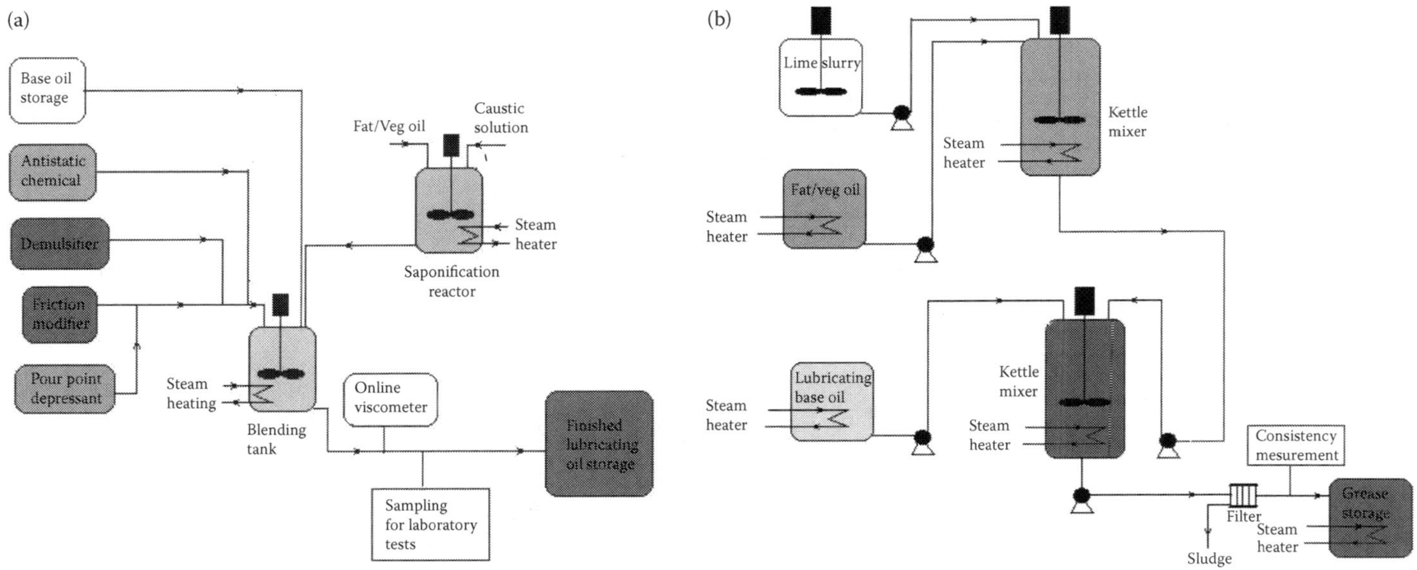

FIGURE 4.1 (a) A schematic lube blending plant. (b) A calcium grease making plant.

is a good practice but costly. It was reported that about half of the lubricants sold in the market are consumed and the remaining half is either recycled or burnt as fuels. The presence of petroleum base stocks and some metal contents are not biodegradable and contamination in water and land may lead to poisonous effects on the aquatic culture and plantation. Polycyclic hydrocarbons (PCH) present in petroleum base oils are potential carcinogens, too. Hence, modern day lubricants are being developed from vegetable or natural base oils rather than from mineral base oils. Vegetable or natural oil-based lubricants are bio-degradable and can be disposed off after biological treatment or as land fill or fertiliser.

4.9 RECLAMATION OF USED LUBRICANTS

As far as the volume of lubricants and grease used and replaced is concerned, recycling of used lubricants after treatment is a standard and economic practice in many countries. There are a variety of reclamation technologies for used lubricants, depending on the quality and quantity of availability. Used lubricants are usually available as a mixture of base oils, additives, and contaminants, such as water, metallic scales, rusts, and inorganic salts. These can be ascertained by bottom sediment and water (BSW) analysis, like crude oil. Hence, the primary treatment involves removal of inorganic matters. The acid-clay process is the oldest and dominant process of re-refining of used lubricants. In this process, after dewatering and desalting of the feedstock, acid is mixed in a stirred vessel. Sludges and oily layers are separated in a settler. Excess acid in the oil mass is then neutralised with a caustic solution and the oily mass is then further clay treated, which is then subject to vacuum distillation. A variety of base oils may be obtained from the distillation step, depending on the quality of the used lubricant feedstock. Acid is recovered from sludges and neutralised with a caustic solution before disposal. The propane deasphalting method is also applied for highly viscous industrial lubricants to recover the base oil and organic matters. Catalytic hydrogenation for desulfurisation, stabilisation of unsaturates, and discoloration may follow for the recovered base oils. The recovery of inorganic additives may not be economically feasible in any of the processes. Recovered base oils are then fed back to the blending plants or sold as FO.

Example 4.1

Calcium stearate soap as a thickener is prepared from fat (mutton tallow) and calcium hydroxide. Determine the amount of lime and fat required to produce 100 tons of soap lye per day in a continuous saponification reactor. Assume that stearic acid content in the tallow is about 30% by weight, purity of lime is 80% by weight, and saponification reaction is 45% complete by weight.

Solution

Assuming saponification is 100% complete,

$$\underset{74\,g}{Ca(OH)_2} + \underset{568\,g}{2C_{17}H_{35}COOH} = \underset{606\,g}{Ca(C_{17}H_{35}COO)_2} + \underset{36\,g}{2H_2O}.$$

Lime required for the production of 100 tons of soap per day is

$$(74/606 \times 100)/0.80 = 15.26 \text{ tons/day.}$$

The fat required is (568/606 × 100)/0.30 = 312.40 tons/day.

For 45% completion of reaction, the amount of lime and fat will be larger and is calculated as

$$\text{Lime: } 15.26/0.45 = 34 \text{ tons/day,}$$

$$\text{Fat: } 312.40/0.45 = 694 \text{ tons/day.}$$

Example 4.2

Determine the VI of the following sample oils. Viscosities at 37.8°C and 98.9°C of the oils are given as

Oil	Viscosity cSt at 98.9°C	Viscosity cSt at 37.8°C
A	9	74
B	14	148
C	17	230
D	40	845
E	45	1017
F	49	1169

Solution

At the viscosity of oil at 98.9°C, cSt, obtain the values of viscosities of reference oils of (100 VI) oil H and (0 VI oil) L from Table A1 of the appendix, as listed below.

Oil	Viscosity of sample and ref oils at 98.9°C, cSt	Viscosity of sample oil at 37.8°C, cSt (U)	Viscosity at 37.8°C, cSt, 100 VI ref oil from table (H)	Viscosity at 37.8°C, cSt, 0 VI ref oil from table (L)	VI = (L – U)/(L – H) × 100 (VI)
A	9	74	71.3	124.5	95
B	14	148	141	281	95
C	17	230	200	424	87
D	40	845	714	1967	89.6
E	45	1017	857	2452	90
F	49	1169	979	2879	90

Example 4.3

Determine the viscosity and the VI of the mixture of oils A, B, C, D, E, and F, from Example 4.1 produced by blending in equal proportions of each of the oils.

Solution

From the blending index (BI) data in Table A2 in the appendix, the BI values are listed below and the viscosity and VI are determined.

Oil	Viscosity of sample oils at 98.9°C, cst	BI from Table A2	Viscosity of sample oils at 37.8°C, cst	BI from Table A2
A	9	25.3	74	32.1
B	14	26.81	148	34.2
C	17	27.4	230	35.55
D	40	30.19	714	38.4
E	45	30.50	857	38.7
F	49	30.75	979	38.9

From the above BI values of the oils, a BI for the blended mixture each of one-sixth of the total volume is, therefore,

$$BI_{mix} \text{ at } 98.9°C = (25.3 + 26.81 + 27.4 + 30.19 + 30.50 + 30.75)/6 = 23,$$

and

$$BI_{mix} \text{ at } 98.9°C = (32.1 + 34.2 + 35.55 + 38.4 + 38.7 + 38.9)/6 = 36.38.$$

From Table-A2 the corresponding viscosities read as,

$$\text{Viscosity of the mixture at } 98.9°C = 23 \text{ cst},$$

$$\text{Viscosity of the mixture at } 37.8°C = 310 \text{ cst}.$$

From Table A1, the values for H and L for the reference oils are read as H = 308.34 cst and L = 711.24 cst at 37.8°C, where U = 310 cst (note that the viscosity of the reference oils and mixture at 98.9°C each having the same viscosity of the mixture at 98.9°C, i.e., 23 cst).

Hence

$$VI_{mix} = (L - U)/(L - H) \times 100 = (711.24 - 310)/(711.24 - 308.34) \times 100 = 99.58.$$

4.10 POWER CONSUMPTION IN A BLENDING TANK

The blending process consists of an agitated tank with a paddle or an impeller-type stirrer. The power required for blending is dependent on various parameters, such as the dimensions of the tank, the dimensions of the paddle or impeller, the viscosity and density of the contents to be blended, and the angular speed of the stirrer. Empirical correlations based on practical data are widely used because theoretical development is not available. Dimensional correlation involving Reynolds number (N_{Re}), Froude number (N_{Fr}), and power number (N_{p0}) is successfully employed for Newtonian fluids. This is presented as

$$N_{p0} / N_{Fr}^{m} = \theta, \tag{4.1}$$

$$N_{p0} = P \cdot g / (n^3 D_a^5 \cdot \rho), \tag{4.2}$$

$$N_{Re} = n \cdot D_a^2 \rho / \mu, \tag{4.3}$$

$$N_{Fr} = n^2 D_a / g, \tag{4.4}$$

where:

P: power required for agitation
g: acceleration due to gravity
n: angular speed of the agitator
D_a: diameter of the agitator paddle or impeller
μ: viscosity of the fluids to be mixed
p: density of the fluids to be mixed
$m = (a - \log(N_{Re})/b$
a and b: parameters depend on tank dimension, paddle position, baffle dimension, etc.

Values of θ are available from the experimental values as a function of N_{Re}, as shown in Figure 4.2. For non-Newtonian fluids, similar correlations are applicable with modified Reynolds number replacing viscosity by the consistency index. Evaluation of power consumption can be explained by the following examples.

Example 4.4

A blending tank is used to blend a soap thickener and a lube base oil with a six-bladed turbine agitator rotating at a speed of 100 rpm. The tank is 2 m in diameter, the turbine is 0.5 m in diameter and positioned 0.5 m above the tank bottom. The tank is filled to a depth of 2 m in an unagitated condition. Determine the power required for the blending operation. Given that the viscosity and density of the fluids to be blended are 12,000 cst and 1,300 kg/m^3, respectively, at the prevailing temperature of the blending operation. Assume that the tank is unbaffled. Given that $a = 1$ and $b = 40$. The values of θ as a function of N_{Re} are given below.

N_{Re}	θ, for unbaffled tank	θ, for baffled tank
1	60	60
10	7	7
100	4	4
1,000	2.7	4.5
10,000	1.5	6.0
100,000	0.0	6.0

*Applicable for a six-bladed tank.

Solution

Evaluation of N_{Re}:

$n = 100/60 = 1.67$ rps,

$D_a = 0.5$ m,

$\mu/p = 12{,}000$ cst $= 12{,}000$ mm²/sec $= 0.012$ m²/sec,

$p = 1300$ kg/m³,

$N_{Re} = (1.67) \times (0.5)^2 \times (1300)/(0.012 \times 1300) = 34.8,$

$\text{Log}(34.8) = 1.54.$

θ from the semi-log plot of Figure 4.2 from the N_{Re} is about 6.00,

$$N_{Fr} = n^2 D_a / g = (1.67)^2 \times (0.5) / 9.81 = 0.142,$$

$$m = \{a - \log(N_{Re})\} / b = (1 - 1.54) / 40 = -0.0135.$$

Hence,

$$P = 6.00 \times (0.142)^{-0.0135} \times (1.67)^3 \times (0.5)^5 \times (1300) / 9.81$$

$$= 118.77 \text{ kg m per sec} = 118.77 \times 9.81 \text{ j/sec} = 1165 \text{ j/sec} = 1.165 \text{ kW}.$$

Example 4.5

Determine the power requirement for Example 4.4 in both the baffled and unbaffled tanks, taking the viscosity and density of the fluids to be blended as 10 cst and 980 kg/m³.

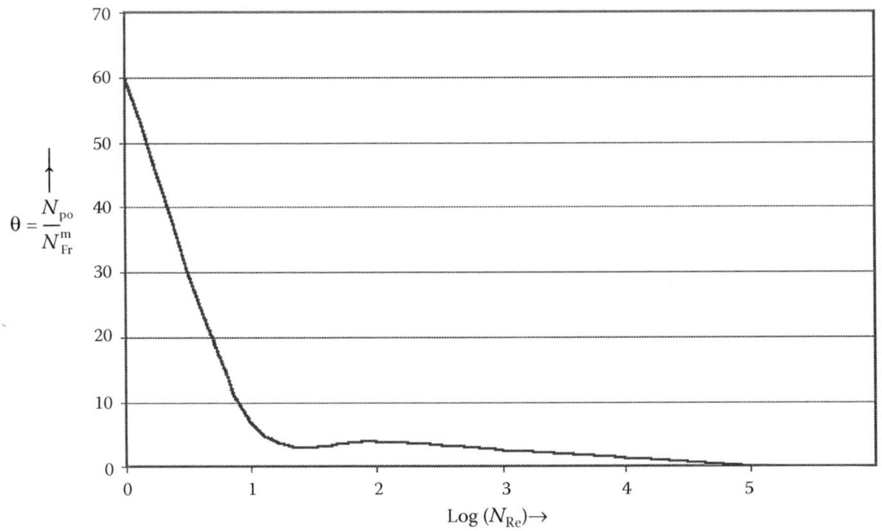

FIGURE 4.2 θ as a function of N_{Re} for a six-bladed agitated tank without baffles.

Solution

For an unbaffled tank:

$n = 1.67$ rps,

$D_a = 0.5$ m,

$\mu/p = 1000 \text{ cst} = 1000 \text{ mm}^2/\text{sec} = 1 \times 10^{-3} \text{ m}^2/\text{sec}$,

$p = 980 \text{ kg/m}^3$,

$N_{Re} = (1.67) \times (0.5)^2 \times (980)/(1 \times 10^{-3} \times 980) = 417.5$,

$\text{Log}(417.5) = 2.62$.

θ from the semi-log plot of Figure 4.2 from the N_{Re} is about 3.6,

$N_{Fr} = n^2 D_a/g = (1.67)^2 \times (0.5)/9.81 = 0.142$,

$m = \{a - \log(N_{Re})\}/b = (1 - 2.62)/40 = -0.0405$.

Hence,

$$P = 3.6 \times (0.142)^{-0.0405} \times (1.67)^3 \times (0.5)^5 \times (980)/9.81$$
$$= 15.7 \text{ kg m per sec} = 15.73 \times 9.81 \text{ j/sec} = 154.36 \text{ j/sec} = 0.154 \text{ kW},$$

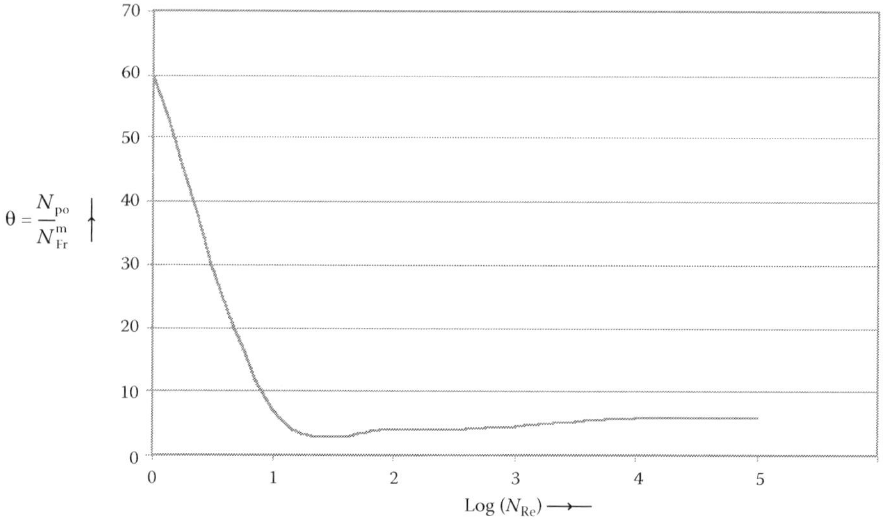

FIGURE 4.3 θ as a function of N_{Re} for a six-bladed agitated tank with baffles.

for baffled tank, $a = 1.7$ and $b = 18$ for the six-bladed agitator,

$$N_{Re} = 417.5,$$

$$\text{Log}(417.5) = 2.62.$$

θ from the semi-log plot of Figure 4.3 from the N_{Re} is about 5.0,

$$N_{Fr} = 0.142,$$

$$m = \{a - \log(N_{Re})\}/b = (1.7 - 2.62)/18 = -0.85.$$

Hence,

$$P = 5 \times (0.142)^{-0.85} \times (1.67)^3 \times (0.5)^5 \times (980)/9.81$$

$$= 382.0 \text{ kg m per sec} = 382 \times 9.81 \text{ j/sec} = 3747.42 \text{ j/sec} = 3.75 \text{ kW}.$$

QUESTIONS

1. What are the functions of lubricants?
2. Mention the various components required to manufacture lubricating oils and greases.
3. What are the basic properties essential for lubricating oils for automotive engines?
4. Discuss the properties of greases required for load bearings.

5. Present a schematic flowsheet diagram of a lube blending plant indicating the necessary equipment involved.
6. How do you design a blending tank?
7. A blending tank is used to blend a soap thickener and a lube base oil with a six-bladed turbine agitator rotating at a speed of 300 rpm. The tank is 3 m in diameter, the turbine is 0.5 m in diameter and positioned 1.0 m above the tank bottom. The tank is filled to a depth of 1.5 m in an unagitated condition. Determine the power required for the blending operation. Given that the viscosity and density of the fluids to be blended are 15,000 cst and 1,200 kg/m^3, respectively, at the prevailing temperature of blending operation. Assume that the tank is unbaffled. Given that $a = 1$ and $b = 40$. The values of θ as a function of N_{Re} are given below.

N_{Re}	θ, for unbaffled tank	θ, for baffled tank
1	60	60
10	7	7
100	4	4
1,000	2.7	4.5
10,000	1.5	6.0
100,000	0.0	6.0

*Applicable for a six-bladed tank.

5 Petrochemicals

5.1 DEFINTIONS OF PETROCHEMICALS

Petrochemicals are chemicals derived from petroleum products. Examples of petrochemicals are plastics, rubbers, fibres, paints, solvents, and detergents. In fact, petroleum products are mixtures of hydrocarbons, whereas the raw materials for petrochemicals are pure hydrocarbons separated and converted to desirable products, such as polymers, solvents, and surfactants, usually in several stages and may be grouped as (1) feedstocks (first-generation petrochemicals), (2) intermediates (second-generation petrochemicals), and (3) finished products (third-generation petrochemicals). Products similar to petrochemicals derived from non-petroleum sources are not strictly petrochemicals. For example, cellulose, natural rubber, natural resins, nylon 11, and ethanol of plant origin are strictly non-petrochemicals. Coal distillation is also a source of varieties of coal chemicals, e.g., benzene, toluene, xylene, and naphthalene. In fact, before petroleum sources were known, coal chemicals were used to produce a variety of products. Many of the chemicals from non-petroleum sources are co-processed with petrochemicals to the finished product. Non-hydrocarbons obtained from petroleum, e.g., hydrogen, carbon monoxide, carbon dioxide, sulfur, and carbon, are also loosely called petrochemicals. Hydrogen, nitrogen and oxides of carbon manufactured from steam reforming and partial oxidation of naphtha are also petrochemicals. These are used for production of ammonia, urea, melamine, fertilizer, etc.

5.1.1 Feedstocks

Feedstocks are the raw hydrocarbons obtained from crude oil refining by distillation and thermal and catalytic processes. For instance, hydrocarbon gases and naphtha are available from atmospheric distillation of crude oil; similarly, benzene, toluene, and xylene, obtained by catalytic reforming and catalytic cracking processes, are the major raw materials for the manufacture of second-generation petrochemicals. Benzene, toluene, xylene, and heavier aromatics are also generated as by-products from petrochemical plants. Thus, the feedstocks for petrochemical plants are either directly obtained from refineries or are further processed to generate them in the petrochemical plant itself.

Natural gas and refinery products are the major source of feedstocks for petrochemicals. A list of the major petrochemicals is given in Table 5.1.

5.1.2 Intermediates

Thermal cracking of ethane, propane, butane, and naphtha produces cracked gases or olefins (ethylene, propylene, butylenes, acetylene, etc.) and liquids (benzene, toluene,

TABLE 5.1
Major Petrochemicals from Petroleum Products

Products from Refinery/Oil Fields	Value-added Petrochemicals
Natural gas (from oil/gas fields) or methane	Carbon black, methanol, acetic acid, formaldehyde, synthesis gas, chlorinated methanes, ethylene glycol, MTBE, acetylene, chloroprene, vinyl chloride, vinyl acetate, etc.
Ethane	Ethylene, polyethylene, ethyl alcohol, ethylene oxide, ethylene glycol, styrene, vinyl chloride, acrylonitrile, etc.
Propane	Propylene, polypropylene, propylene oxide, isopropanol, acrylonitrile, acetone, etc.
Butanes	Isobutanol, butadiene, acrylonitrile, polybutadiene, etc.
Benzene	Phenol, styrene and polystyrene, dodecyl benzene, cyclohexane, caprolactum fibres, nylons, etc.
Toluene	Nitrotoluene, phenol
m-Xylene	p-Xylene
p-Xylene	Terephthalic acid, PET
o-Xylene	Phthalic anhydride
Naphtha	Olefins for PE, PP, LPG, CBFS, pyrolysis gasoline, butadiene, linear alkyl benzene, etc.
Sulfur	Sulfuric acid
Wax	Micro crystalline wax, food grade coating wax, cosmetics, medicinal wax, etc.
Coke	Metallurgical coke, electrode grade coke, etc.
CBFS	Carbon black

xylene, etc.). Olefins are the starting material (monomers) for polyolefin plants. Olefins are also reacted with other hydrocarbons or non-hydrocarbon chemicals to generate vinyl chloride, ethylene glycol, neoprene, ethylene oxide, etc., and these are used as the starting materials (monomers) for the manufacture of a variety of polymers.

5.1.3 Finished Products

Using the above intermediates, a variety of plastics, rubber, fibre, solvent, paint, etc., are manufactured. Polymerisation reactions are carried out for these monomers or intermediates to various polymers, resinous and liquid products. Plastics are available in the form of extrudates, granules, powders, beads, etc., from the manufacturing units as the finished products. These are converted into plastic commodities, such as bags, films, furniture, and products of various shapes and sizes by casting, moulding, or blowing machines, as the marketable products. Plastics are classified into two types, namely, thermoplastic (or thermoplast) and thermosetting plastics (or thermoset). Thermoplasts, usually linear in molecular structure, can be melted (or softened) by heating and solidified (or hardened) by cooling. This heating and cooling cycle can be repeated indefinitely without loss of the original properties. But thermosets will be permanently transformed to a chemically cross linked or non-linear structure and cannot be returned to their original property during a heating and cooling cycle. Plastics are also known as plastomers with a high modulus of elasticity. Synthetic fibres are made from polymers that have a high modulus of elasticity as compared to plastics and rubbers. These polymers are also available in the form of extrudates, powders, and beads, which are converted to fibres in a drawing mechanism and are collected in bales. Rubbers or elastomers are polymers with a low modulus of elasticity. Raw rubber is available from the polymerisation unit in the form of sheets, which are cut and blended with various chemical ingredients along with sulfur (known as vulcanization) to achieve the desired quality for making tyres and other products.

A variety of chemical reactions are involved in the petrochemical manufacturing processes. Most of these reactions are catalytic with heat effects. Examples are dissociation, dehydrogenation, hydrogenation, addition, polymerisation, and condensation. Dissociation reactions occur during the thermal and catalytic cracking process. Dehydrogenation also occurs catalytically or thermally during cracking. Additional reactions, such as oxidation, chlorination, fluorination, and sulfonation of the parent olefin or aromatic hydrocarbons, are required to make intermediates or monomers. Polymerisation occurs mostly in the presence of initiators or catalysts with heat evolution. Two types of polymerisation reactions occur, i.e., addition polymerisation and condensation polymerisation.

Addition polymerisation involves a chain reaction in which monomer molecules join in a chain. For example, olefins or diolefins are polymerised in the presence of initiators, such as free radicals, ionic compounds, or complexes.

Condensation polymerisation involves two monomers, same or different, which combine to form a polymer with the elimination of small molecular weight by-products like water.

A large number of unit operations and processes are involved in a petrochemical plant. Since catalysts play a major role in the synthesis of petrochemicals, research and development of new catalysts is a continuous endeavour by the manufacturers. It is also to be noted that in a polymerisation plant, catalysts may not be recovered rather this is entrained in and become a part of the polymer. The presence of catalyst ingredients in the polymer per unit mass may pose a problem as far as quality is concerned. Hence, a catalyst is selected that requires minimum consumption. The reactors used are tubular, stirred tank or kettle type. These may be packed bed or fluidized bed types, Both single and multiple numbers of reactors are used. The selection of a suitable reactor is a challenging decision in a petrochemical manufacturing unit.

5.2 NAPHTHA CRACKING

Major plastics are manufactured from olefins, which are available either by cracking of gases (gas-based petrochemicals), such as methane, ethane, propane and butane, or from naphtha (naphtha-based petrochemicals). Due to the larger availability of naphtha and the low energy requirement for cracking as compared to gases, naphtha cracking is widely accepted for the manufacture of olefins. Naphtha is a mixture of hydrocarbons boiling in the range of the lowest boiling component (C5) to 150°C, which contains paraffins, naphthenes, and aromatics in various proportions. It is found that paraffins, i.e., straight chain hydrocarbons, yield olefins when heated at a high temperature. The branched chain and aromatic hydrocarbons, on the other hand, become heavier or decomposed to carbon. It is also a fact that the yield of olefins decreases with the increasing molecular weight (hence with boiling point) of the hydrocarbons. For example, if ethane is thermally cracked, it will yield 80% ethylene, while from propane, butane, naphtha, and gas oil as feedstocks, if cracked separately the yields will be 45%, 37%, 30%, and 25%, respectively. Hence, the desirable feedstock naphtha should have a very high content of paraffins and be in the lower boiling range. A suitable boiling range for feedstock naphtha for olefin production is below 100°C and should have a paraffin content of more than 75%. Usually, naphtha in the boiling range of 90°C–150°C is catalytically reformed (as discussed in Chapter 3) in a refinery either to produce gasoline or aromatics. Hence, in the refinery, C5–90°C cut is separated in the naphtha redistillation unit and is sold to the petrochemical industry. Naphtha cracking is carried out in a tube-still furnace at a temperature above 800°C. Due to thermal cracking, the following cracking endothermic reaction takes place,

$$\begin{aligned} C_nH_{2n+2} = {} & CH_4 + C_2H_4 + C_3H_6 + C_4H_8 + C_5H_{10} + C_2H_6 + C_3H_8 \\ & + C_4H_{10} + C_5H_{12} + \cdots + H_2. \end{aligned} \tag{5.1}$$

Light hydrocarbons obtained from the initial cracking reactions further crack into lighter olefins and propagate until the reaction temperature is brought down (quenched). If the reaction is continued indefinitely, branched and cyclised heavy hydrocarbons will be produced and coke will be generated as the ultimate product. Therefore, the cracking reaction is carried out in a very short residence time, i.e., the naphtha feed passes the heater tubes at very high speed to avoid undesirable heavy end products and coke. Usually, residence time is maintained at <1 sec in the traditional cracker furnaces and it is of the order of a few milliseconds in the modern millisecond furnaces. Since a coke layer develops inside the tube surface, the heat transfer rate is rapidly reduced, causing reduced cracking and poor olefins yield. Steam is introduced with the feed to remove the coke layer on the tube surface by converting coke into carbon monoxide and hydrogen by water gas reaction,

$$C + H_2O(\text{steam}) = CO + H_2. \tag{5.2}$$

However, coke cannot be removed completely by steam and the thickness grows during the operating period of the furnace. When the coke layer reaches the point at which cracking operation shows poor yield, the furnace is taken out of service and decoking is carried out with air and steam to remove coke to the maximum extent. Thus, a cracker furnace operates cyclically between the cracking and decoking operations. Excess steam may also partially convert some of the hydrocarbons or naphtha components to carbon monoxide and hydrogen and reduce the yield of olefins. Table 5.2 shows the operating parameters for a typical naphtha cracker.

TABLE 5.2
Typical Operating Conditions in a Naphtha Cracking Furnace

Process Variable	Values
Temperature	800°C
Pressure	Atmospheric
Steam to naphtha ratio	0.6 kg/kg
Reaction time	<1 sec
Cycle time for cracker	20 days
Yield of ethylene	30% of the feed

By-products of cracking are propylene, butylene, butadiene, and aromatics, such as benzene, toluene, xylene, isomeric paraffins, naphthenic components, and polynuclear aromatics. Products from the cracking furnace are quenched by hot oil and cooled before they are sent to the separation units. The product mixture then flows through a series of separators, e.g., demethaniser, de-ethaniser, depropaniser, and debutaniser units, which are multiplated distillation columns.

5.2.1 Primary Fractionator or Stabiliser

Cracked liquids and gases are separated in a fractionating column the bottom product of which is the heavy cracked oil rich in high boiling aromatics. This heavy oil is also partly used as the quenching medium for the products from the furnace and partly sold as the carbon black feedstock (CBFS) due to its heavy aromatic contents. Cracked gases containing hydrocarbons, both saturated and unsaturated, from methane to C_7 hydrocarbons emerge from the top of the column, which is then compressed and amine (or caustic) washed to remove hydrogen sulfide and carbon dioxide gases. This is illustrated in Figure 5.1.

5.2.2 Hydrogen Separator

Amine or caustic washed gases are then passed through a flash separator vessel, where liquid hydrocarbon gases are separated from hydrogen at high pressure and low temperature. Hydrogen from this vessel is used in the hydrogenating units, such as pyrolysis gasoline and butadiene hydrogenation, or as a fuel.

Demethaniser: Liquified gases from a hydrogen separator are then separated from methane in a distillation column where methane (C_1) emerges from the top and is used as a fuel for the cracking furnace. The bottom of the column is then passed to a de-ethaniser.

De-ethaniser: It is also a distillation column that separates ethane and ethylene mixture (C_2 mixture) as the top product from the rest of the liquified gases containing propane, propylene, butane, butylenes, etc.

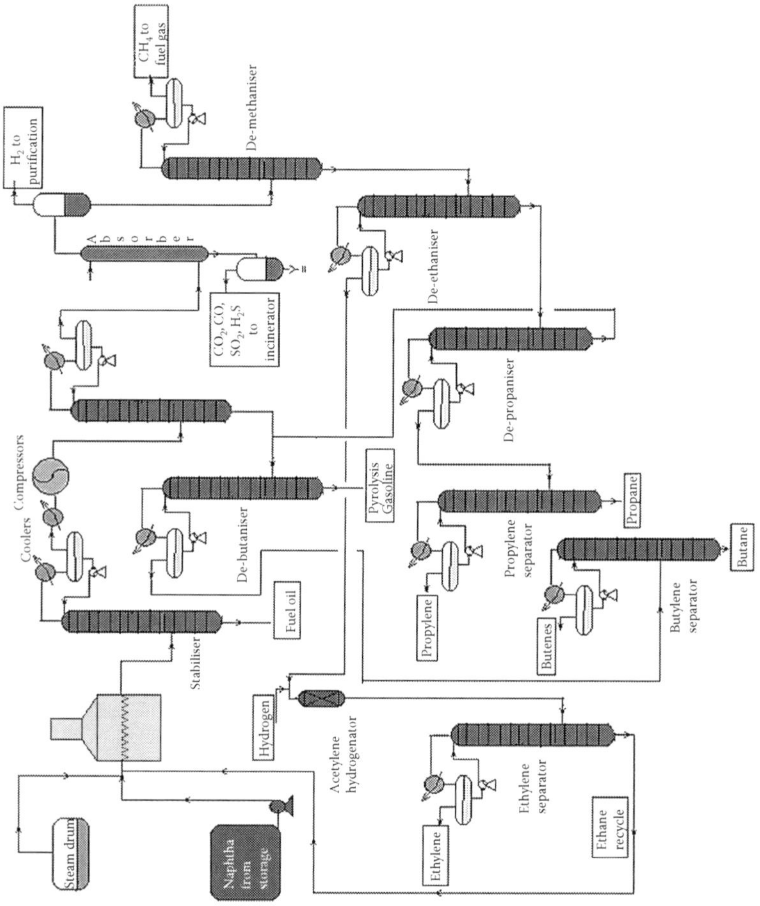

FIGURE 5.1 A simplified naphtha cracking unit.

Ethane-ethylene separator: A C_2 mixture from the top of the de-ethaniser column is then passed through another distillation column that separates ethylene as the top product and ethane as the bottom product. Ethylene is sent to storage and is used up in the polyethylene (PE) synthesis plant. Ethane from this column is recycled to a small cracking furnace to yield additional ethylene.

Depropaniser: The liquified gas mixture from the bottom of the de-ethaniser is separated from propane and propylene (C_3 mixture), which leaves from the top of the column and enters the propane–propylene fractionator. The bottom product contains the butanes, butenes, butadiene, and heavier components, which are then separated from the butane–butene mixture (C_4 mixture).

Propane–propylene separator: In this column, propylene is recovered as the top product and propane as the bottom product. Propylene is stored and used for manufacturing polypropylene, and propane is sold as a domestic fuel—liquified petroleum gas (LPG).

Debutaniser: Butane, butenes, and butadiene (C_4 mixture) are recovered as the top product and components heavier than the C_4 mixture, i.e., C_5 and heavier, are recovered as the pyrolysis gasoline (bottom product). Pyrolysis gasoline is catalytically hydrodesulfurised before it is sold or blended as a gasoline component. A gasoline hydrodesulfurisation unit is similar to a naphtha pretreatment unit described in Chapter 3. A simplified flow sheet of a naphtha cracking unit (NCU) is shown in Figure 5.1. Typical yields of products from a naphtha cracking unit is presented in Table 5.3.

5.3 CONVERSION PROCESSES FOR SELECTED PETROCHEMICALS

5.3.1 Polyethylene

Polyethylene (PE) has evolved as a major plastic and is obtained by polymerising ethylene. Three main types of PE plastics are obtained—low density polyethylene (LDPE), linear low density polyethylene (LLDPE), and high density polyethylene (HDPE)—depending on the type of polymerising process, which are described in the following sections.

PE is the polymer (poly + monomer) of ethylene molecules. This product is used to make a variety of plastics. Polymerisation of ethylene molecules into heavy molecular weight PE is a reaction in which a chain of macromolecule, is produced by the combination of ethylene molecules. Ethylene is a highly reactive monomer that starts combining with other molecules of ethylene in the presence of a catalyst (Ziegler–Nutta catalyst) under a certain pressure and temperature. The reaction steps are in three stages, namely, initiation, propagation, and termination. A radical molecule is formed in the presence of the catalyst in the initiation step. The radical then starts combining with the monomers repeatedly in the propagation stage, which continues indefinitely as long as the monomer molecules are available during reaction until quenched at the termination stage. The properties of a polymer vary with the operating pressure, temperature, and time of reaction. The reaction is exothermic and hence it is essential to control temperature by a proper heat removal system.

TABLE 5.3
Typical Yields from a Naphtha Cracking Unit

Products	Yield % wt
Methane	10–15
Ethylene	25–31
Propylene	12–16
Butadiene	4–4.5
Butenes	3–8
BTX	10–13
C_5+	9–17
Fuel oil	3–6
Hydrogen and other gases	5–6

Three classes of PE, namely, LDPE, HDPE, and LLDPE, are produced in different processes. LDPE is produced in a very high-pressure process, and HDPE and LLDPE are obtained in moderately low-pressure processes. A list of PE manufacturing processes is given in Table 5.4.

5.3.1.1 Low Density Polyethylene

In a tubular reactor, pure liquid ethylene (99.99%) is mixed with hydrogen peroxide and forced through the tubular reactor at very high pressure (3500 atm), surrounded by a cooling medium to extract heat of polymerisation reaction. Oxygen is used with hydrogen peroxide as the initiator.

The reaction takes place in solution. The heat of the reaction is given as 3650 kj/kg,

$$nC_2H_4 = (-\,C_2H_4\,-)_n - 3650 \text{ kj/kg}.$$

The temperature in the reactor is controlled above 200°C to avoid crystallisation of LDPE, which will otherwise damage the reactor tube. Conversion per pass is

TABLE 5.4
Polyethylene Manufacturing Processes

Reactant	Polymer	Type of Polymerisation	Catalyst	Temperature, °C	Pressure, atm
Ethylene	LDPE	Solution or bulk, tubular reactor, liquid phase	H_2O_2	200–240	3500
Ethylene	HDPE	Suspension, stirred tank reactor, liquid phase	$TiCl_4$	60–80	20–35
Ethylene and 1-butene	LLDPE	suspension, fluidised bed reactor, vapour phase	$TiCl_4$	100	7–20

about 20%. Effluent from the reactor, consisting of the product and the unconverted monomer, is separated by high and low pressure separator vessels. Ethylene is recycled to the reactor. The overall conversion achieved by recycling is about 95%–97%. Finally, the molten LDPE is withdrawn from the low pressure separator and extruded, followed by cooling, drying, and pilling. As tubular reactors are prone to plugging and poor heat transfer problems, a thick-walled, stirred tank vessel reactor is used. High-speed agitation helps in good heat transfer. If the agitator stops, the temperature will rise to such an extent that explosion may takes place because of the presence of oxygen in the reaction mixture. Figure 5.2 is a schematic flow diagram of an LDPE manufacturing process. LDPE is used for making films, sheets, tubes, blocks, insulation, hoses etc. But, it cannot be used at temperatures above 80°C because deformation will occur. It has excellent dielectric properties, good elasticity up to −60°C, and it is anti-corrosive. It is soft and waxy and is used to make films.

5.3.1.2 High Density Polyethylene

HDPE is manufactured by the suspension polymerisation method. In this method, high purity ethylene is introduced into the reactor vessel in which a catalyst (Ziegler-Nutta catalyst, $TiCl_4$ in alkyl aluminium) is suspended in benzene at a pressure of 20–35 atm and at a temperature of 60°C–80°C. The ratio of alkyl

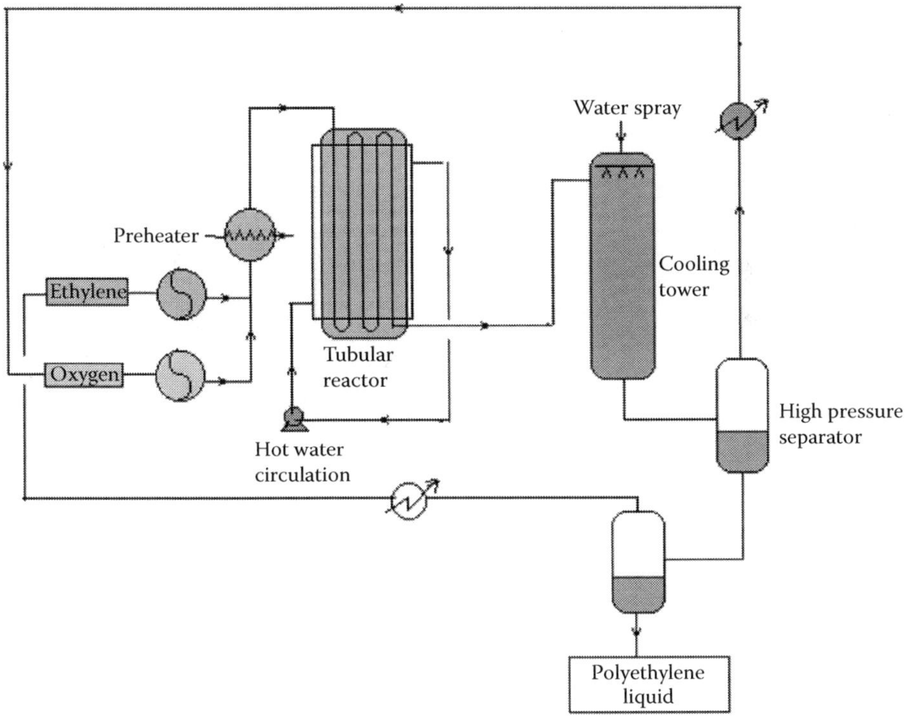

FIGURE 5.2 A high pressure low density polyethylene (LDPE) plant.

aluminium and titanium chloride determines the size of the polymer. The greater the ratio, the greater the molecular weight of the polymer. After the reaction, the polymer mixture is separated from ethylene and inerts in a flash drum. The polymer is water washed and filtered to recover the catalyst (water soluble) and reused. However, recovery of the catalyst from the polymer is not complete. Catalyst consumption is about 1 g of titanium (Ti) per 1500 kg of polymer. This is due to the high residual presence of the catalyst within the polymer. The presence of this impurity limits the HDPE's applications. A modern catalyst has been developed that can yield 6000 kg polymer per gram of Ti, where the contamination of Ti metal in the polymer is a few parts per million. A flow diagram of a HDPE plant is shown in Figure 5.3. HDPE has a rigid and translucent property and is suitable for making electrical goods, bottles, ropes, etc. The main economic advantage of HDPE is that it can be manufactured at much lower pressure as compared to LDPE. HDPE has a higher melting temperature than LDPE and has almost the same glass transition temperature.

5.3.1.3 Linear Low Density Polyethylene

LLDPE is a copolymer of ethylene and 1-butene having linear structure. Today, it is produced by a low pressure fluidised bed process where the temperature and pressure are 100°C and 7–20 atm (0.7–2 MPa), respectively. Gaseous monomers are

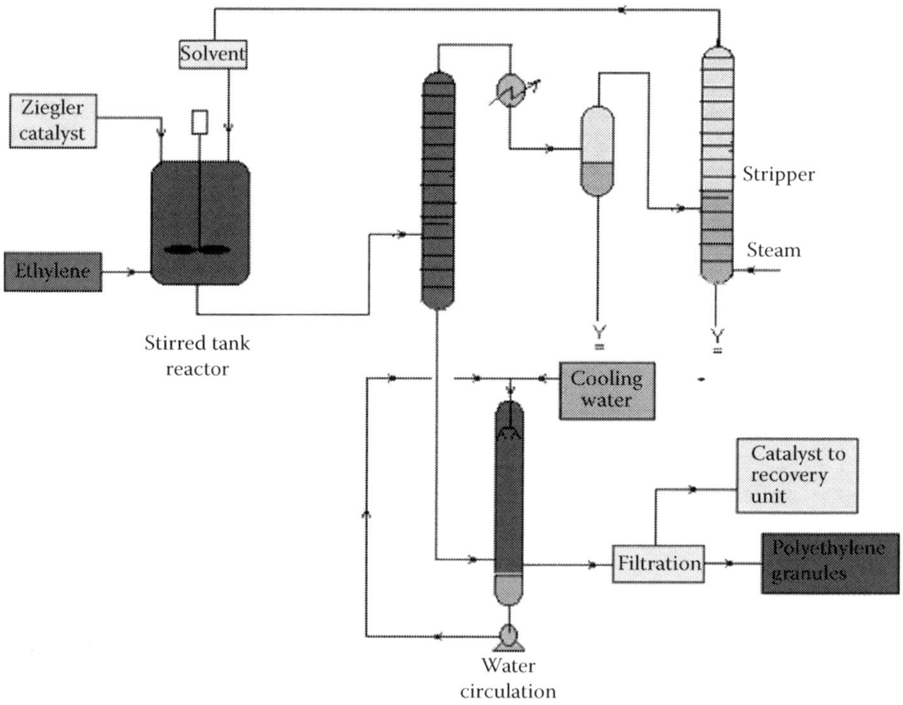

FIGURE 5.3 Low pressure Ziegler process for high density polyethylene manufacture.

used to fluidise the polymer granules previously prepared. During polymerisation, additional polymers are produced of specific sizes depending on the seed polymers present. Unreacted monomers are separated from the effluent and recycled to the reactor. A long residence time in the range of 3–5 h is required for a reaction. Such a process is shown in Figure 5.4. HDPE is also produced by a similar fluidised method in modern plants.

5.3.2 Polypropylene

After PE, polypropylene is a valuable polymer and is used as plastic for making pipes, ropes, fibres, etc. It is manufactured by catalytic reaction in a stirred tank reactor, where Ti and aluminium halides are used as catalysts at a temperature of 60°C–70°C and a pressure of 1–2 MPa. An unreacted monomer is recycled after it is separated from the catalyst and polymer mixture in a flash chamber under vigorous stirring conditions. The mixture of polymer and catalyst is then passed to a centrifugal separator where a catalyst and polypropylene polymer is recovered. Further processing of the spent catalyst in the presence of alcohol is carried out to recover the active components of the catalyst for its reuse. A polypropylene plant is shown in Figure 5.5.

5.3.3 Polyethylene Terephthalate

Polyethylene terephthalate (PET) is also known as polyester. This is produced by an esterification reaction between ethylene glycol and terephthalic acid. This is a two-stage polymerisation process. In the first stage, monomer-ester is produced and in the second stage, polymerisation takes place. Such a plant is shown in Figure 5.6.

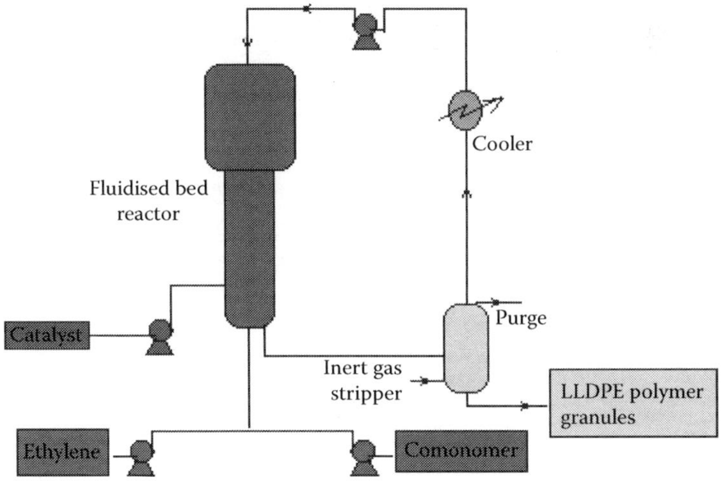

FIGURE 5.4 A fluidised bed LLDPE manufacturing unit.

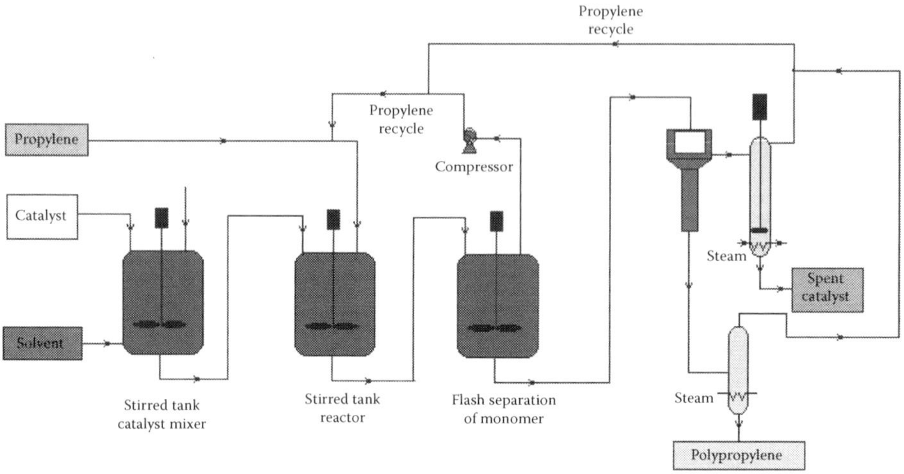

FIGURE 5.5 A polypropylene manufacturing unit.

5.3.3.1 Terephthalic Acid

This is the one of the raw materials used in the production of PET, as described in the previous section. As shown in Figure 5.7, a terephthalic acid manufacturing plant is described. In this process, p-xylene is used as the feed, which is oxidised in the presence of a cobalt-sulfate catalyst supplied as a solution of acetic acid to a reactor

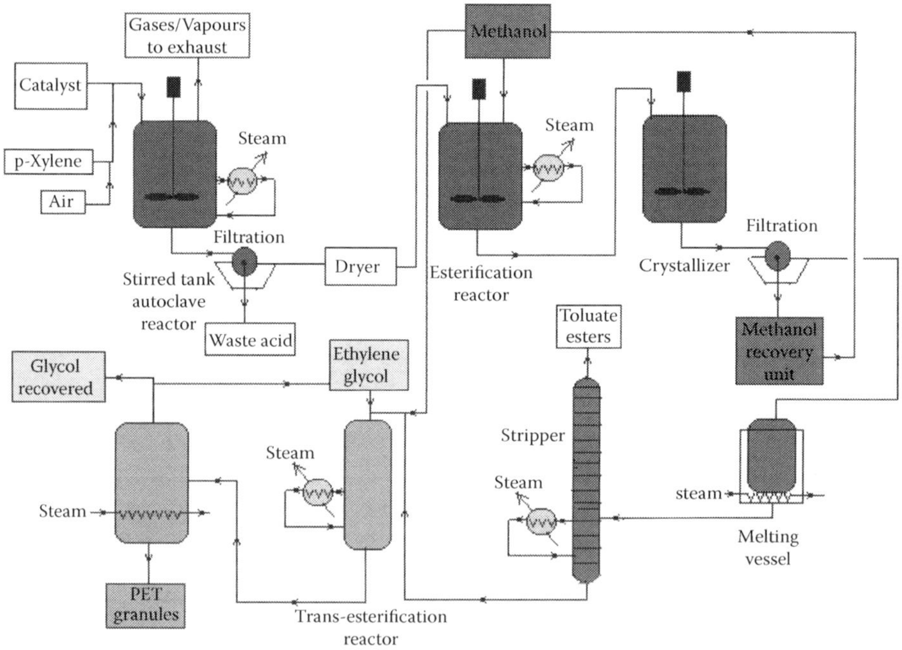

FIGURE 5.6 A polyethylene terephthalate manufacturing unit.

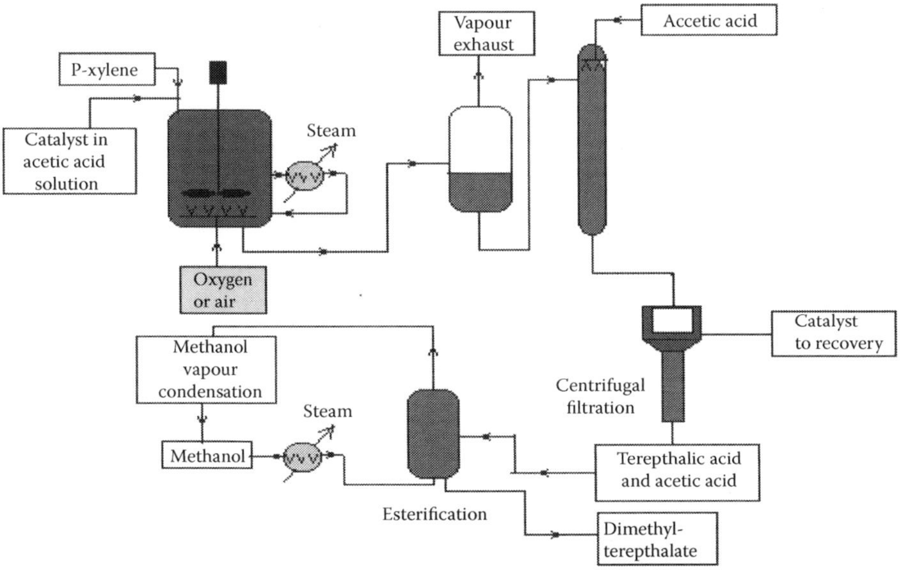

FIGURE 5.7 Terephthalic acid from p-xylene.

where air or pure oxygen is blown through a sparger. The product mixture is then passed to a vapour-liquid separator. Liquid containing terephthalic acid and catalyst slurry is then separated by a centrifugal separator.

After drying, dimethyl terephthalate (DMT) ester is drawn as the final product, which can be converted to acid form in the presence of water in the PET plant. Terephthalic acid either in the form of DMT or terephthalic acid in the absence of water and acetic acid is also known as purified terephthalic acid (PTA). The reaction temperature varies with the source of oxygen, i.e., whether from air or pure oxygen. Usually, air is used as the cheapest source of oxygen and a temperature of 200°C at a pressure of 2–3 MPa are employed in the reactor. Filtrate containing terepthalic acid esterified by methanol is used to remove acetic acid from DMT.

The following reaction takes place.

Oxidation step:

$$H_3CC_6H_4CH_3 + 2O_2 = HOOCC_6H_4COOH \text{ (terephthalic acid).}$$

Esterification reaction:

$$\text{TPA + Methanol = DMT + Water}$$

5.3.3.2 Ethylene Glycol

Ethylene glycol is manufactured by catalytic oxidation of ethylene followed by hydration to glycol. High purity ethylene is converted to ethylene oxide in the presence of silver oxide as the catalyst in a tubular reactor at 250°C–300°C and 1 atm pressure. Air or pure oxygen may be used for the reaction. Ethylene dichloride is dosed in the reactor in very small amounts to avoid ethylene combustion. About 1 mol ethylene to 10 mol air ratio is maintained in the reactor. About 60%–70% of conversion occurs in a reaction time of only 1 sec. Ethylene oxide and unconverted ethylene are scrubbed with water and unconverted ethylene is recycled. The reactions are

$$C_2H_4 + \frac{1}{2}O_2 = C_2H_4O,$$

$$C_2H_4O + H_2O = C_2H_4(OH)_2.$$

Polyglycols are formed as the by-products. A flow sheet of such a plant is shown in Figure 5.8.

A tubular reactor with circulating water in the shell side is employed to remove heat of reaction generating steam. The product mixture from the reactor is water scrubbed to absorb ethylene oxide, which forms glycols. Unconverted ethylene and oxygen is then passed to another reactor for additional conversion. Aqueous solutions of glycols are then separated by vacuum distillation columns.

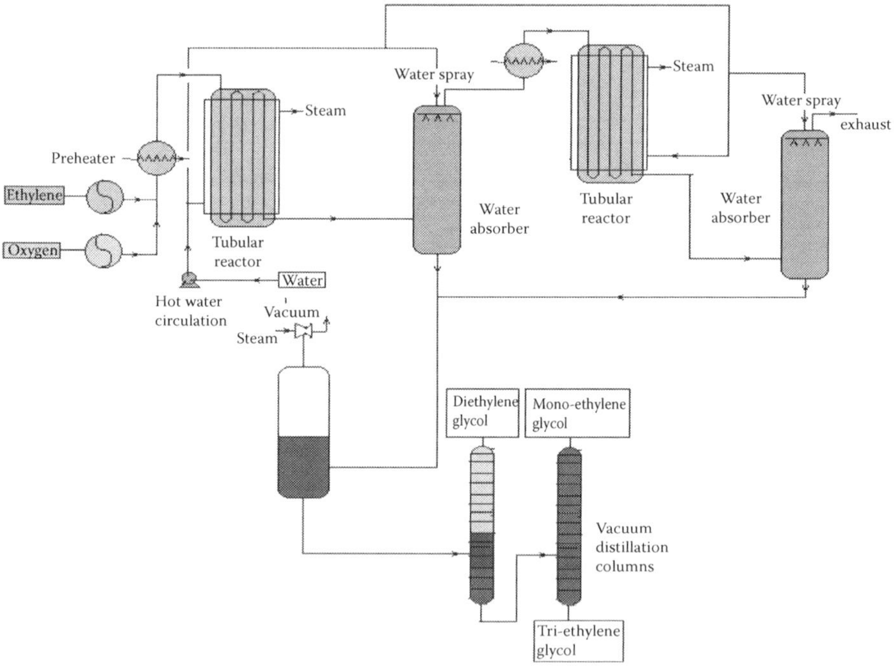

FIGURE 5.8 Ethylene glycol manufacturing.

5.3.4 Polyvinyl Chloride

Polyvinyl chloride (PVC) is a polymer of vinyl chloride,

$$n\mathrm{CH_2 = CHCl = (-CH_2 - CHCl - CH_2 -)_n}.$$

It is a thermoplastic, widely used for many applications, and is commonly produced by the emulsion polymerisation method. An aqueous emulsion is produced in a mixer with ammonium salts of fatty acids and a vinyl chloride monomer and the reaction is allowed to continue in a stirred tank reactor. A very low pH is maintained during the reaction. The product containing the polymer latex and unreacted monomer are separated in a gas–liquid separator. The separated vapour of the monomer is recycled after scrubbing with caustic solution to neutralise the acidic vapours. The polymer from the separator is then dried for making powder of polymer. A PVC manufacturing plant is shown in Figure 5.9.

5.3.5 Polystyrene

Polymerisation of styrene can be done by mass, solution, emulsion, or suspension polymerisation methods. In the mass polymerisation method, styrene is heated at 80°C–85°C in a batch reactor under nitrogen pressure. The conversion is 30%–40% in a long residence time, ranging from 40 to 50 h. This is suitable for small production units. Since no catalyst or initiator is used, this polymer is suitable for electrical insulation. In the solution polymerisation method, styrene and a solvent are mixed in a stirred tank blender followed by pumping through a series of reactors with heating facilities.

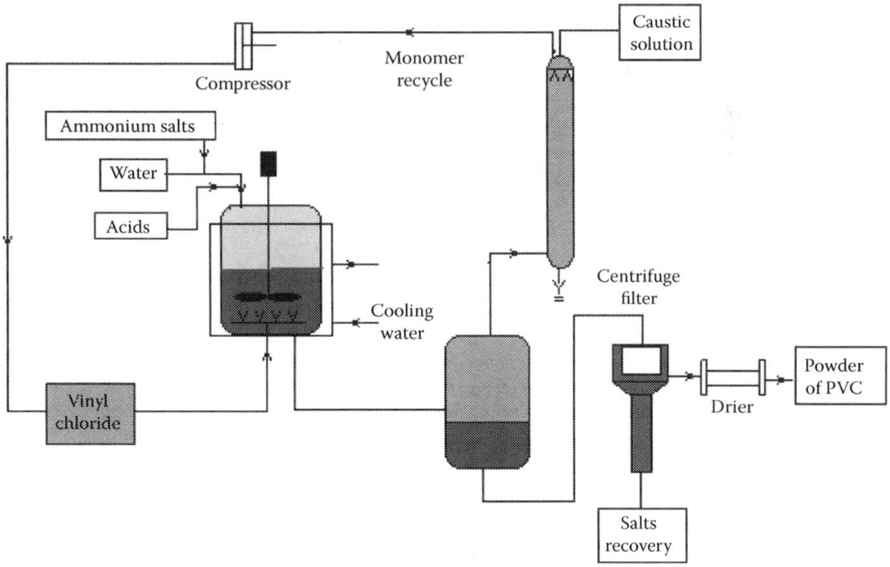

FIGURE 5.9 A PVC manufacturing plant.

In a batch suspension polymerisation process, drops of styrene are dispersed in water in the presence of a benzoyl peroxide initiator. In the emulsion polymerisation method, a stirred tank autoclave is employed, in which styrene and water are emulsified with an initiator. The product is then separated from unreacted monomers, which are then recycled.

5.3.6 Polybutadiene

Butadiene (molecular formula $CH_2 = CH - CH = CH_2$) is obtained as the by-product from an NCU. By catalytic (Ziegler–Nutta) polymerisation of butadiene, a product that resembles natural rubber is obtained. A solvent is used to keep the catalyst and the monomer in the solution during the reaction. The product is separated from its unconverted monomer, which is recycled. The solvent is evaporated, stripped, and recycled. The polymer from the evaporator is then dried to get the final product. Such a plant is shown in Figure 5.10.

5.3.7 Acrylonitrile Butadiene Styrene

Acrylonitrile butadiene styrene (ABS) plastic is obtained by polymerising styrene, acrylonitrile, and polybutadiene in a stirred tank reactor at about 50°C–60°C in the presence of potassium persulfate as a catalyst in an emulsion of sodium stearate.

5.3.8 Styrene–Butadiene Rubber

Styrene can be copolymerised with butadiene to yield styrene–butadiene rubber (SBR). Polymerisation is carried out in a stirred tank reactor at around 50°C in the presence of free radical initiators, e.g., potassium sulfate, in an ammoniacal aqueous solution.

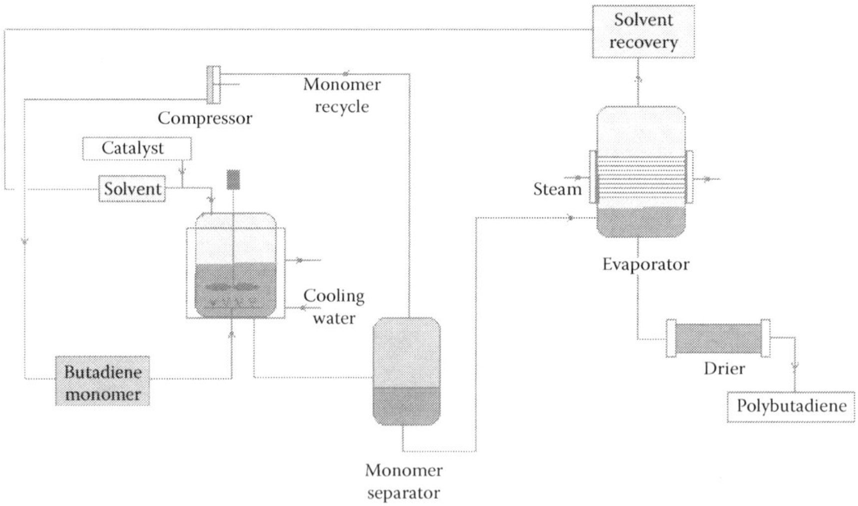

FIGURE 5.10 Manufacturing polybutadiene from butadiene.

5.3.9 Poly Methyl Metha Acrylate

This is a polymer of the methyl metha acrylate monomer, which is produced from acrylonitrile (CH_2CHCN) and sulfuric acid at a temperature below 90°C to yield methyl acrylate (CH_2CONH_2) followed by reacting with methanol to convert methyl acrylate to methyl metha acrylate ($CH_2CHCOOCH_3$). This methyl metha acrylate monomer is then polymerised in a stirred tank autoclave under nitrogen pressure. Any of the solution, emulsion, mass or suspension polymerisation methods are used.

5.3.10 Polytetrafluoroethylene

Polytetrafluoroethylene (PTFE) or *teflon* is a polymer of the tetrafluoroethylene (C_2F_4) monomer. This monomer is prepared from methanol chlorination followed by fluorination. The reactions are carried out in three reactors. Trichloromethane produced in the chlorinator is reacted with hydrofluoric acid in the first reactor at about 65°C in the presence of $SbCl_5$ as the catalyst and is converted to produce mono-chloro-difluoromethane (CHF_2Cl) in the presence of $AlCl_3$ as the catalyst in the second reactor. In the third reactor, CHF_2Cl is then converted to C_2F_4 by catalytic pyrolysis at a temperature of 650°C–800°C in the presence of platinum (Pt) as the catalyst. Polymerisation of C_2F_4 is then carried out in a batch reactor preferably by the suspension polymerisation method at a temperature of 200°C and at a pressure of 1000 psi. The reaction time is about 1 h.

5.3.11 Nylons

Polymers of amine or polyamides are excellent for making fibres. These are well known as nylons. A list of a few nylons and their monomers are presented in Table 5.5. Condensation polymerisation is carried out with the monomer/monomers in autoclaves or stirred tank batch reactors.

Nylon 6,6: This is produced by polymerising adipic acid $[(CH_2)_4(COOH)_2]$ and hexamethylene diamine $[(CH_2)_6(NH_2)_2]$ in an autoclave in an aqueous solution at about 200°C and a pressure of 250 psi for a reaction time of 1–2 h. Such nylon-making steps are shown in Figure 5.11.

Nylon 6: This is obtained from caprolactam $[(CH_2)_5(CO)(NH)]$, which is polymerised in a water solution at about 250°C under nitrogen pressure in a batch autoclave. The reaction time is about 8–12 h.

TABLE 5.5
Nylons and Their Monomers

Nylon	Nylon 6	Nylon 6,6	Nylon 6,10	Nylon 11	Nylon 12
Monomers	Caprolactam	Adipic acid and hexamethylene diamine	Sebacic acid and hexamethylene diamine	11-Amino-undecanoic acid	Laurolactam

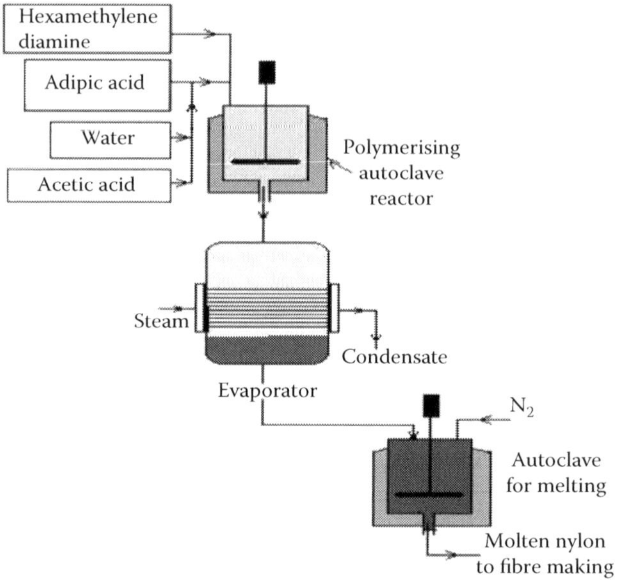

FIGURE 5.11 Manufacturing steps for nylon 6,6.

Nylon 6,10: This is produced from sebacic acid [$(CH_2)_8(COOH)_2$] and hexamethylene diamine [$(CH_2)_6(NH_2)_2$], which are polymerised in a batch reactor where condensation takes place during the reaction.

Nylon 11: This is obtained from 11-amino-undecanoic acid [$(NH_2)(CH_2)_{10}(COOH)$], which is obtained from castor oil, by polymerising in an aqueous medium at a temperature of 220°C in an autoclave.

Nylon 12: This nylon is a product of the polymerisation of laurolactam [$(CH_2)_{11}(CO)(NH)$] at a temperature of about 260°C in an autoclave.

Monomers required for the manufacture of different varieties of nylons are either purchased or manufactured in their respective plants.

Manufacturing processes for the monomers of nylons:

Adipic acid: Commercial adipic acid is made from benzene. Benzene is catalytically hydrogenated to cyclohexane (C_6H_{12}) by Raney nickel as the catalyst at a temperature of 200°C–300°C under a hydrogen pressure of 3–4 MPa. The reaction takes place in a stirred tank reactor. Cyclohexane is then oxidised to a mixture of cyclohexanol ($C_6H_{11}OH$) and cyclohexanone ($C_6H_{10}O$) by cobalt naphthenate as the catalyst at a temperature of 145°C–175°C and at a pressure of 0.8–1 MPa in an oxygen atmosphere. Both the cyclohexanol and cyclohexanone are then further oxidised to adipic acid in the presence of copper and ammonium metavandate as the catalyst in the presence of nitric acid. The oxidation is carried out in a stirred tank reactor at a temperature of 60°C–80°C and atmospheric pressure. The reaction time is about

5–7 min. The product adipic acid $[(CH_2)_4(COOH)_2]$ is then further purified by steam distillation.

Hexamethylene diamine: This is obtained from adipic acid in two stages. In the first stage, adipic acid is converted to adiponitrile $[(CH_2)_4(CN)_2]$ by reacting with gaseous ammonia in the presence of the catalyst, boron phosphate, and at a temperature of about 340°C in a fixed bed reactor. The product adiponitrile is then catalytically hydrogenated to hexamethylene diamine $[(CH_2)_6(NH_2)_2]$ by a cobalt catalyst at a temperature of 100°C–135°C and at a hydrogen pressure of 60–65 MPa in a fixed bed reactor.

Sebacic acid: This acid is commercially produced from castor oil saponification with a caustic solution at about 250°C. In the reactor, the castor oil (glyceride) is split into ricinoleic acid $[CH_3(CH_2)_{13}CHOH(CH)_2COOH]$ and glycerine $[(CH_2OH)_2CHOH]$. Ricinoleic acid is separated from glycerine and further heated in a separate reactor to yield sebacic acid $[(CH_2)_8(COOH)_2]$.

11-Amino-undecanoic acid: This is also produced from castor oil. The oil is first esterified by methanol to yield methyl ester of recinolic acid, followed by pyrolysis at about 500°C. A mixture of products of heptyl alcohol, ester of undecylenic acid, heptaldehyde, etc., is obtained. Hydrolysis yields the undecylenic acid $[CH(CH_2)_9COOH]$, which is then brominated by hydrobromic acid to bromo-undecanoic acid $[CHBr(CH_2)_{10}COOH]$. Treating bromo-undecanoic acid with ammonia yields 11-amino-undecanoic acid $[NH_2(CH_2)_{10}COOH]$.

Caprolactam: Benzene is converted to cyclohexane followed by oxidation to cyclohexanol and cyclohexanone, as described in the production of adipic acid. Cyclohexanol from the mixture is further catalytically dehydrogenated to cyclohexanone in the presence of zinc at a temperature of 400°C at 1 atm. The cyclohexanone is then converted to cyclohexanone oxime $[C_6H_{10}NOH]$ by reacting with hydroxylamine sulfate $[(NH_2OH)_2H_2SO_4]$ at about 18°C–25°C. Cyclohexanone oxime is then further treated with sulfur trioxide in the presence of sulfuric acid to yield caprolactum $[(CH_2)_5(CO)(NH)]$ by a reaction mechanism known as the Beckmann rearrangement. The temperature of this reaction is maintained near 140°C–150°C.

Laurolactam or dodecyl lactam: Butadiene is first catalytically trimerised in a liquid phase reaction by the Ziegler catalyst ($TiCl_4$) in a stirred tank reactor. The product stream containing the desired trimer, 1,5,9-cyclo-dodecatriene, is separated from the by-products polybutadiene, cyclo-octadiene, and vinyl cyclohexene by distillation. This trimer is then catalytically hydrogenated using a nickel catalyst at 200°C and a hydrogen pressure of 1.4 MPa. The reaction is carried out in the liquid phase in a stirred tank autoclave. The hydrogenated product is cyclododecane $[(CH_2)_{12}]$. Cyclododecane is then oxidised in the liquid phase at about 150°C in jacketed stirred tank reactors in the presence of boric acid as the catalyst. The oxidised product stream, containing cyclododecanol $[(CH_2)_{11}OH]$ and cyclododecanone $[(CH_2)_{10}O]$, is separated by distillation. Cyclododecanol is further dehydrogenated to yield cyclododecanone in a separate reactor at 200°C in the liquid phase and in the presence of a copper catalyst supported on alumina. Finally, cyclododecanone is oximated by hydroxylamine hydrogen sulfate [$(NH_2OH)_2H_2SO_4$] at a temperature of 100°C and converted to cyclododecanone oxime [$(CH_2)_{10}NOH$]. Finally, cyclododecanone oxime is converted to laurolactam [$(CH_2)_{11}(CO)(NH)$] ammonium sulfate and concentrated sulfuric acid. The Beckmann rearrangement reaction takes place during the formation of laurolactam.

5.3.12 Phenol Formaldehyde

Phenol (C_6H_5OH) and formaldehyde (HCHO) in the presence of an aqueous ammoniacal medium condense to a phenol-formaldehyde resin. The reaction is carried out in a kettle-type reactor at a temperature of 160°C and at atmospheric pressure. The reaction time is less than 1 h. The product from the reactor is usually blended with hexamethylene triamine and other fillers to make finished resins ready for moulding processes. Depending on the ratio of phenol to formaldehyde, a variety of grades of this resin may be made. This is a thermosetting-type resin.

5.3.13 Urea Formaldehyde

Urea $(CONH_2)_2$ and formaldehyde (HCHO) also readily react to polymerise into urea-formaldehyde resin. Polymerisation is carried out in a glass-lined (or stainless steel) stirred tank reactor in an ammoniacal or weak alkaline medium. The polymerisation reaction takes place at room temperature and atmospheric pressure. This resin is also a thermosetting resin.

5.3.14 Melamine Formaldehyde

Melamine [$C_3N_3(NH_2)_3$] and formaldehyde (HCHO) combine readily to form melamine-formaldehyde resin, which is also a thermosetting resin. The temperature of the reaction is 100°C in a neutral medium, i.e., at a pH of 7 in the aqueous medium at atmospheric pressure. Melamine formaldehyde resins are thermosets.

Manufacturing steps for the monomers for formaldehyde resins:

Formaldehyde: This is produced from methanol by catalytically oxidising over silver at about 620°C and a pressure of 30–70 kPa. Two reactions, oxidation and dehydrogenation, take place simultaneously, yielding formaldehyde in both reactions. In the catalytic oxidation, heat is generated and formaldehyde is formed. Due to heating, an endothermic dehydrogenation reaction takes place. Thus, the reactions become autothermal, both producing formaldehyde. The reactor is a water jacketed stainless steel vessel containing the catalyst.

Methanol: Methanol is commercially manufactured from synthesis gas, i.e., from carbon monoxide and hydrogen. Catalytic hydrogenation of carbon monoxide is carried out over a copper-based catalyst at about 300°C and 2000–3000 MPa pressure to get a 50%–60% conversion. Methanol is usually manufactured in a fertiliser plant where synthesis gas is also used for manufacturing ammonia and urea.

Urea: This is manufactured by synthesising carbon dioxide and ammonia in a tubular reactor at a temperature of 180°C and 14 MPa without a catalyst (Stamicarbon process).

Melamine: Commercial melamine is produced from urea by high-pressure synthesis. Urea in the presence of ammonia pressure while undergoing dehydration yields melamine [$C_3N_3(NH_2)_3$]. In fact, methanol, urea, and melamine are available from fertiliser plants.

Phenol: Phenol can be manufactured by a variety of methods, such as sulfonating benzene in the presence of caustic soda, chlorination of benzene followed by catalytic conversion to phenol, toluene oxidation, and cumene oxidation. A commercial cumene oxidation method is described in brief below.

Cumene: [$C_6H_5CH(CH_3)_2$] is first manufactured from benzene and propylene in the presence of phosphoric acid in a fixed bed reactor. The phosphoric acid is deposited on kieselguhr at a temperature of 190°C–200°C and a pressure of 3–4 MPa. Liquid hourly space velocity with respect to benzene is about 1.5. Cumene is then oxidised to cumyl hydroperoxide and acetone (by-product). The reaction is carried out in cylindrical steel columns agitated by blowing air through cumene in the alkaline solution as an emulsion containing soluble salts of heavy metals like Mn, Co, or Cu. A reaction temperature of 80°C–120°C and a pressure of air at 0.5–0.8 MPa are maintained in the reactor. The reaction time is about 1–2 h. The product stream containing cumyl hydroperoxide is separated from unconverted cumene, which is recycled. Cumyl hydroperoxide is then split into phenol and acetone in the presence of a catalyst, e.g., strong sulfonic acid or ion exchange resins, in a stirred tank reactor at a temperature of 70°C–80°C and a pressure of 0.1–0.2 MPa. Phenol is then purified from by-products, e.g., acetone and methylstyrene, by distillation.

5.3.15 Polyurethane

These are thermosetting resins obtained by condensation polymerisation of polyol (diols or triols) and a diisocyanate. Examples of monomers are toluene diisocyanate [$(CH_3)(C_6H_4)(NCO)_2$] and dihydric alcohol or glycol [$(CH_2OH)_2$]. The polymerisation is carried out in a stirred tank autoclave reactor at a temperature of around 200°C under nitrogen pressure. Molten polymer is then taken directly to the fibre spinning unit.

5.3.15.1 Toluene Diisocyanate

This is produced from toluene in three stages. In the first stage, toluene is nitrated to dinitrotoluene; in the second stage, dinitrotoluene is reduced by catalytic hydrogenation to toluene diamine; and in the third stage, toluene diamine is converted to toluene diisocyanate. In the first stage, toluene is treated with a mixture of concentrated sulfuric acid and nitric acid in a series of agitated reactors at a temperature from 50°C to 65°C. In the second stage, nickel is used as the catalyst to hydrogenate the dinitrotoluene to diamine in a series of stirred tank reactors at a temperature of 170°C and a hydrogen pressure of 8–9 MPa. In the third stage, phosgene ($COCl_2$) is treated with toluene diamine to produce toluene diisocianate. Phosgene is manufactured by chlorination of activated charcoal at a temperature of 50°C.

5.3.16 Silicone

This is a polymer of siloxanes $(R_3SiO)_3$. Industrially, it is produced by alkylating with an organic halide, CH_3Cl, or C_6H_5Cl with elemental silicon (Si) in the presence of a catalyst (Cu and CuO mixture) at about 200°C–300°C and at a pressure of 100–200 kPa for a long reaction time of about 48 h. The reactor is a cylindrical column fitted with a screw-type agitator and a jacket. The di $[(CH_3)_2SiCl_2]$ and tri silanes $[(CH_3)_3SiCl]$ produced are separated and hydrolysed separately, each at a temperature of 30°C–50°C to yield silicone polymers $[(CH_3)_2SiO]_n$. The lower molecular weight polymers are oil, whereas the higher molecular weight polymers are solids. Silicon (Si) is obtained from silicon dioxide by a reduction reaction with carbon at about 1300°C in an arc furnace.

5.4 PETROCHEMICAL COMPLEX

Considering the multiplicity of raw materials, intermediates, and products, it may be necessary to build different large or small plants producing a limited number of products catering to the needs of other petrochemical manufacturing units or plants. In fact, the varieties of petrochemicals cannot be achieved from a single plant, but rather from a group of plants, each group being owned by the same or a different company. Such a group of plants is called a petrochemical complex.

5.4.1 Downstream Units

Finished products from a petrochemical complex are utilised by a large number of manufacturers of domestic and industrial commodities, such as bottles, films, ropes, garments, automotive parts, and paints.

5.4.2 Petrochemicals' Hub

Since various petrochemical plants are needed to build a petrochemical complex, it is inevitable that feedstocks from the mother unit must be utilised by the intermediate manufacturers in a complimentary mission in a common framework of business procedures agreed by them. In order to maintain a healthy flow of finished and semi-finished products under a common umbrella of marketability involving the downstream manufacturers of finished commodities, it is essential to accommodate all of the manufactures in a common area, a petrochemical hub. Such a holistic concept can reduce the considerable cost of storage, transportation, manufacturing, selling, and distribution, and the price of products. In such a hub, industries will be able to share facilities, such as power, transport, water supply, and civic amenities, which will ultimately help to promote sustainable industrial activities.

5.5 PROCESSING OF PLASTIC, RUBBER, AND FIBRE

Plastics, fibres, and rubber are the major synthetic polymers. Plastics are also known as plastomer, which is characterised by a low moduli of elasticity of about 700–7000 kPa. However, elasticity varies with temperature. Plastics soften with an increase in temperature and because of this property, plastics are used for making furniture, ropes, films, pipes, and a variety of domestic and industrial goods, which are used at low temperatures. Rubbers are polymers of lower elasticity with a moduli ranging from 70 to 700 kPa. They are also called elastomers. Hard rubbers are mainly used in making tyres for automobiles, aeroplanes, tractors, bikes, etc. Soft rubbers are used for lining furniture, sheets, bags, shoes, etc. Fibres are the polymers with the highest moduli of elasticity, ranging from 7,000 to 70,000 kPa. Fibres are used mainly as staple fibres for making clothing.

Thermoplastics are polymers that can be melted by heating and solidify after cooling repeatedly without any change in its plastic property. Plastics soften when heated, but there is a temperature at which the viscous liquid mass becomes like glass and brittle while it is cooled from molten state. This temperature is called the *glass transition temperature*. The higher the glass transition temperature, the more suitable the plastic is for use as it becomes brittle at or below this temperature. Hence, additives, known as placticisers, are mixed during the processing of plastics to increase this transition temperature. For instance, polyethylene has a glass transition temperature of −30°C or lower, whereas teflon has a glass transition temperature at slightly above 100°C. A few examples of the glass transition temperatures of plastics are given below.

Plastic	Glass transition temperature in °C
Polyethylene	–30
Polypropylene	–15
Polyisobutylene	–70
Polyvinyl chloride	79
PTFE (Teflon)	109
Nylon 6,6	50
PET	70

The melting or softening point and the density are other important properties that determine the end use of plastic. Elasticity is determined by the percentage elongation of solid plastic for different shapes and molecular weights.

Thermosetting plastic softens when heated, and when cooled it does not return to its original soft stage, but hardens and cannot be remelted without damaging the original properties. Examples are phenol formaldehyde, urea formaldehyde, and melamine formaldehyde. Bakelite, which is used for making electrical switches and accessories, falls into this category. Thermosetting plastics are harder but more brittle than thermoplastics. These are cross-linked polymers and usually require reinforcement with other materials, such as glass fibres, synthetic fibres, cotton, and paper, to obtain the required toughness as engineering plastics. Phenol formaldehyde and urea formaldehyde are the most common thermosetting plastics. Basic raw polymer materials, e.g., powder or pellets of plastics, are obtained from the moulding processes as described in the next sections, which are then mixed with other additives and processed at high pressure, hot compression, blowing, etc., to yield the desired mechanical properties, size, and shape, depending on the type of end uses.

Rubbers are softer materials as compared to plastics with very low elasticity and can be easily stretched longitudinally and restored to its original shape without distortion. Very low mechanical energy is required for such stretching and hence low elasticity. This property makes rubber suitable for making high-pressure tyres and tubes, pipes, balloons, membranes, balls, etc. A good quality tyre made from a basic polymer, such as polyisoprene, neoprene, and polybutadiene with additives, is capable of withstanding compression and expansion repeatedly. They are also good heat and electrical insulators. Mechanical operations like mixing process known as compounding of rubber necessary for desired properties of finished material. Natural rubbers from some plant origins, lacking in necessary mechanical strength, are also mixed with synthetic rubbers in the compounding process.

5.5.1 Moulding of Plastics

Molten matter can be poured into the cavities of templates or moulds of desired objects and cooled to the desired shapes. This process is called moulding, which is carried out in different ways, such as extrusion, injection, blowing, and compression, for making a variety of objects of different shapes, such as bottles, sheets, billets, fibres, chairs, tables, tanks, doors, windows, and many products for domestic and industrial use.

5.5.2 Extrusion Moulding

An extruder is a machine with a horizontal cylindrical barrel through which a shaft with a helical groove wound over. An electric heating coil is wound around the barrel for heating. Plastic powder is poured into the feed entry hopper at one end of the screw and it is then carried away by the helical groove and the annulus of the shaft and the barrel. The powder material softens due to heating and is forced out at the other end of the extruder. Finally, the softened plastic is forced through a die of desired object.

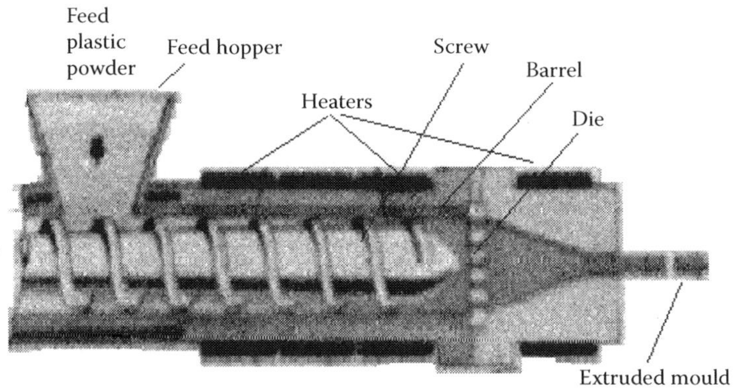

FIGURE 5.12 An extrusion moulding machine.

Construction of a typical extruder is shown in Figure 5.12. Most of the polymers decompose due to uniform or prolonged heating. Hence, uniform and short time heating should be applied during the extrusion process. Extrusion is widely used for blending thermoplastics with additives. Colouring agents and pigments are also used during blending.

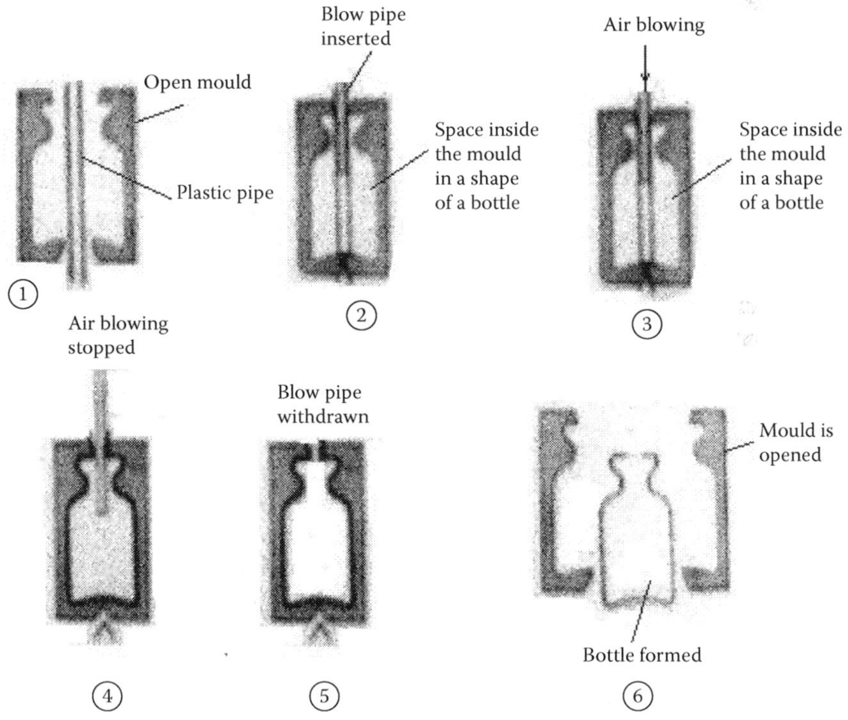

FIGURE 5.13 A blow moulding method for plastic bottle manufacture.

5.5.3 Blow Moulding

A blow moulding scheme is explained in Figure 5.13. In this method, hollow objects, such as bottles, pipes, and drums, are produced. At the start, preformed plastics by extrusion or injection are made and hot preform is then blown into a hollow mould and the desired shape is made. PE, polypropylene, and PET bottles are made by this method.

5.5.4 Compression Moulding

In this method, two pieces of a mould are used to make the desired object. One of the pieces contains the raw plastic powder, while the other is punched at very high pressure to close the mould under heated conditions. Usually, thermosetting plastic materials of complicated shapes are made by this method.

5.5.5 Thermal Moulding

Thermal moulding is a process by which a preform of the plastic sheet of a desired thickness and weight is placed over a specially prepared mould or die with the provision of a vacuum suction and air pressurising system, as shown in Figure 5.14. At the beginning, the sheet is heated to a certain temperature and a vacuum is applied until the sheet takes the form of the contour of the internal space of the die. This is followed by cooling and the product is withdrawn by pushing with the air pressure of the vacuum suction and air pressurising system, as shown in Figure 5.15.

5.5.6 Injection Moulding

In this method, a plunger (which is a thick solid cylindrical rod) drives the plastic in the heated groove of the die and brings back the perform in the return stroke

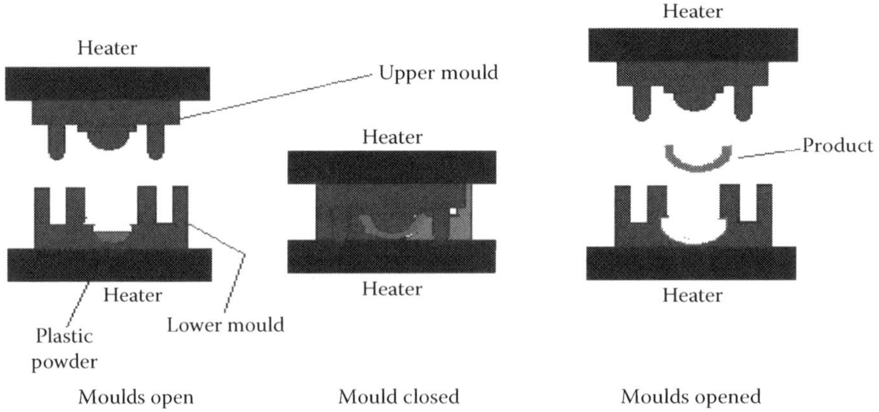

FIGURE 5.14 The compression moulding method.

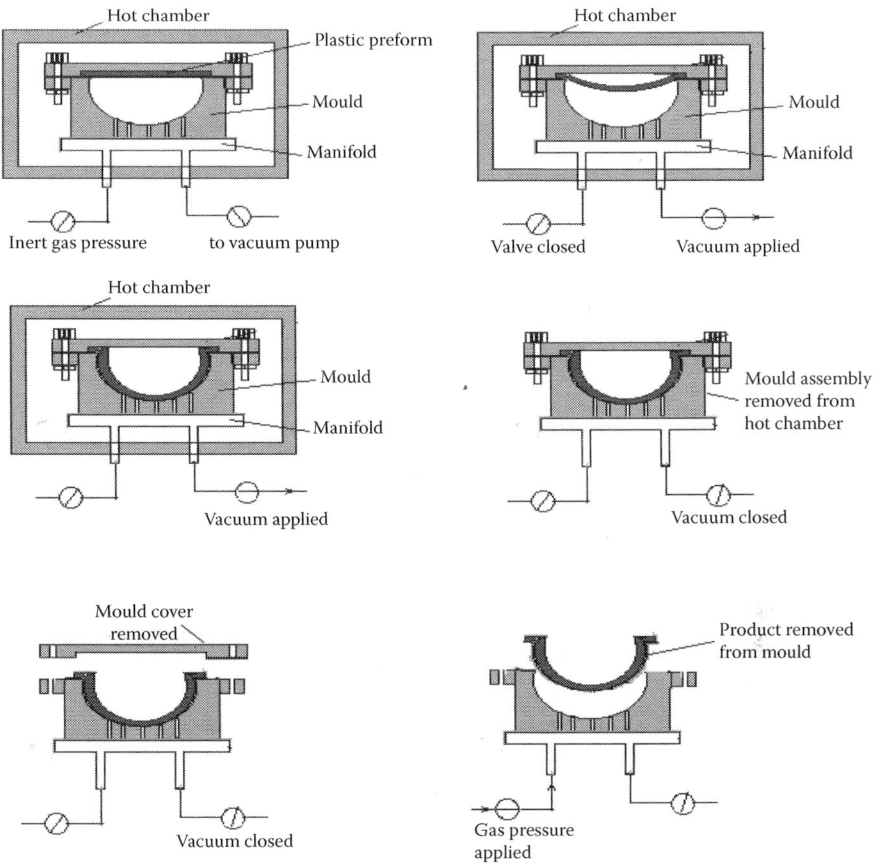

FIGURE 5.15 Stages of the thermal moulding method.

in a reciprocating motion. The entire process takes little time, thus a large number of preforms are produced by a single machine. Such a machine is shown in Figure 5.16. A large plant uses a number of injection moulding machines for the preforms followed by blow moulding for making bottles of various shapes and designs. These plants also supply preforms that are required by small plants, which do not have injection moulding facilities, to produce bottles and other materials.

5.5.7 Rubber Compounding

Rubber is used for making tyres, belts, and many domestic and industrial products. Finished rubber is a mixture of polymers with additives, such as graphitic carbon, sulfur, lubricants, and certain metallic oxides, to impart good elasticity, mechanical strength, abrasive resistance, and other properties that should be retained for long use especially for making tyres. Polymer structure of rubber can be changed by reacting with sulfur during vulcanization and properties of rubber

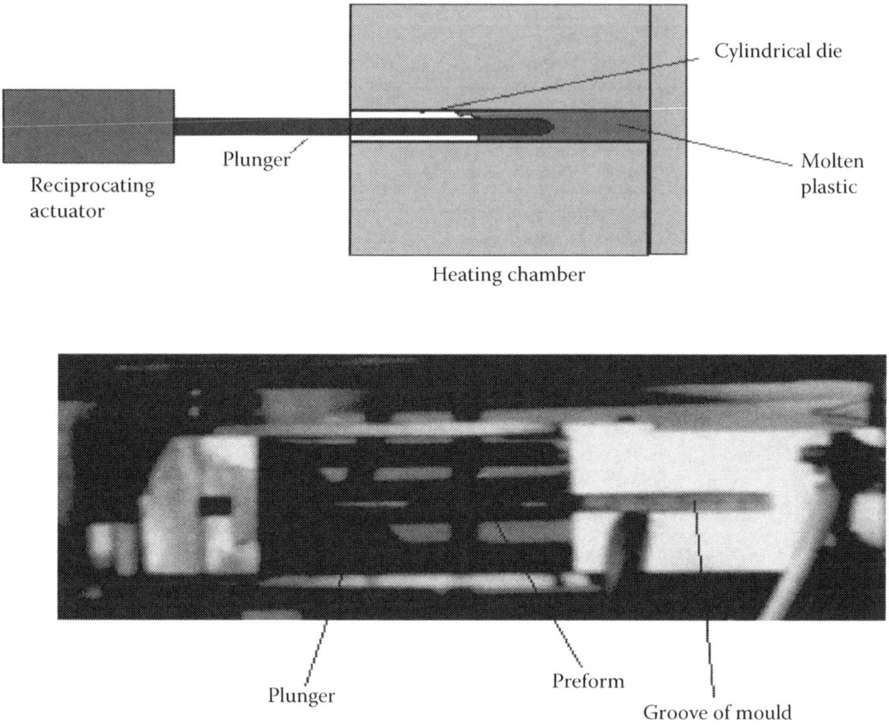

FIGURE 5.16 An injection moulding machine making preform for a bottle.

can be changed as desired by the manufacturer. The process of treating rubber polymer with sulfur is called vulcanisation. Rubber is applied for making coated (calendering) materials, which are moulded or extruded to make a variety of products. Rubber generated or available as scraps are also reused by the reclaiming process and converted back to finished rubber, known as reclaimed rubber. High-pressure and high-speed mixers, like a two-roll mill or Banbury mixer, are commonly employed for mixing rubber components. The clearance between the rolls (or the stator-rotor in a Banbury mixer) is extremely small to help intense mixing at high pressure. A Banbury mixer is shown in Figure 5.17.

QUESTIONS

1. Compare the advantages and disadvantages of ethane cracking and naphtha cracking for ethylene manufacture.
2. Distinguish between synthetic plastic, rubber, and fibre.
3. What are the various methods for polyethylene manufacture?
4. What are the raw materials used for the manufacture of polyethylene terephthalate? Mention the methods of manufacturing these raw materials.

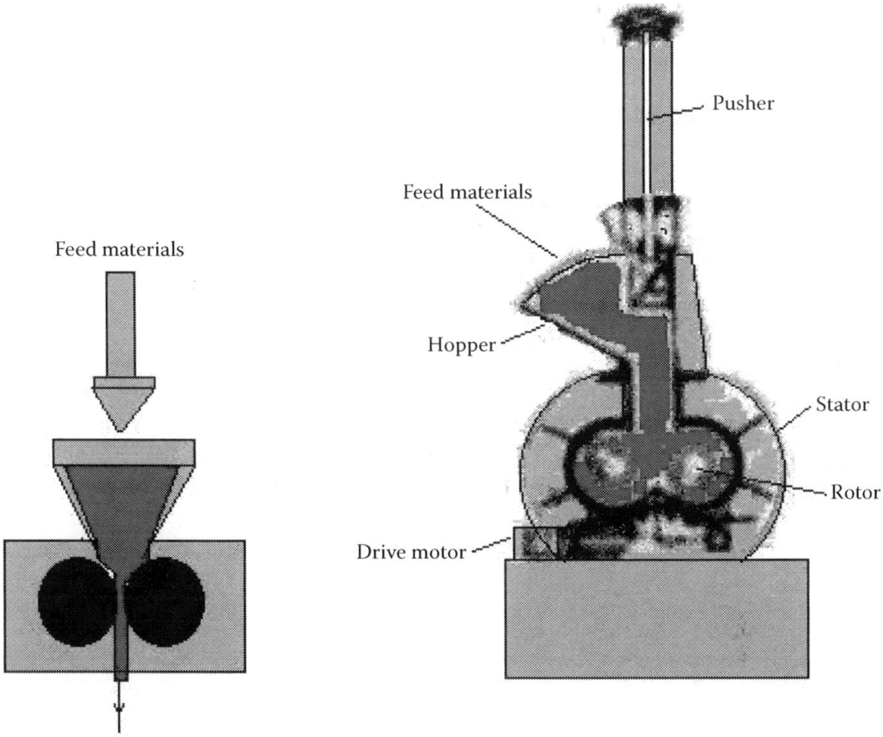

FIGURE 5.17 A Banbury mixer for rubber compounding.

5. What are the various types of nylons available? Mention the method of manufacture of nylon 6,6.
6. What are the various types of moulding methods used for plastic goods? What is a preform?
7. What do you understand by vulcanisation of rubber?

6 Offsite Facilities, Power and Utilities

6.1 LAYOUT OF PETROLEUM AND PETROCHEMICAL PLANTS

A refinery or a petrochemical plant consists of various plants and equipment suitably grouped and located in various sites known as *processing units*. Other facilities include *off sites*, *power plant*, *cooling tower*, *water conditioning plant*, *quality control laboratory*, and *service centres*, such as canteen, first aid, firefighting facilities, attendance registration office of employees, and administrative offices. Typical layouts for a refinery and a petrochemical plant are presented in Figures 6.1a and b.

6.2 PROCESSING UNITS

Various processing units in a refinery and a petrochemical plant have been discussed in Chapters 3 and 5, respectively. The number of processing units is determined by various factors, such as type, availability and price of feedstocks, demand for products, and environmental restrictions. Common units in a refinery are crude distillation units (CDU), both atmospheric and vacuum, followed by stripping and separation processes, such as solvent extraction and crystallisation. Distillates are recovered from residues either by vacuum distillation or by solvent extraction. Thermal and conversion processes follow next. In fact, each conversion unit also consists of separation units, such as flash separation, distillation, and extraction. Processing units, e.g., CDUs, may house some other auxiliary units, such as a gas processing plant, a stabiliser, and a merox unit. A vacuum distillation unit (VDU) is usually separately located or included in a CDU. Other units are a furfural extraction unit (FEU), a propane deasphalting unit (PDA), naphtha hydrodesulfurisation (NHDS), kerosene hydrodesulfurisation (KHDS), a reformer, fluid-catalytic cracking unit (FCC), diesel hydrodesulfurisation (DHDS), a vis-breaking unit (VBU), a bitumen treatment unit (BTU), a solvent dewaxing unit (SDU), and a lube base stock hydrofinishing unit (HFU). In a naphtha-based petrochemical plant, the mother unit is the naphtha cracking unit (NCU), which includes cracking furnaces and product recovery units. Polymerisation units, e.g., linear low density polyethylene (LLDPE), high density polyethylene (HDPE), polypropylene (PP), polybutadiene, and other units, such as a benzene-toluene-xylene extraction (BTX) unit, a pyrolysis gasoline hydrogenation (PGH) unit, and a hydrogenation unit for butenes and butadienes (C4-hydrogenation), follow next. Each processing unit is separated by a physical boundary line from the other units is known as the *battery limit* of the unit. In this boundary, the incoming and outgoing lines of raw materials and products, chemicals, water, steam lines, etc., are suitably located for easy access during operation.

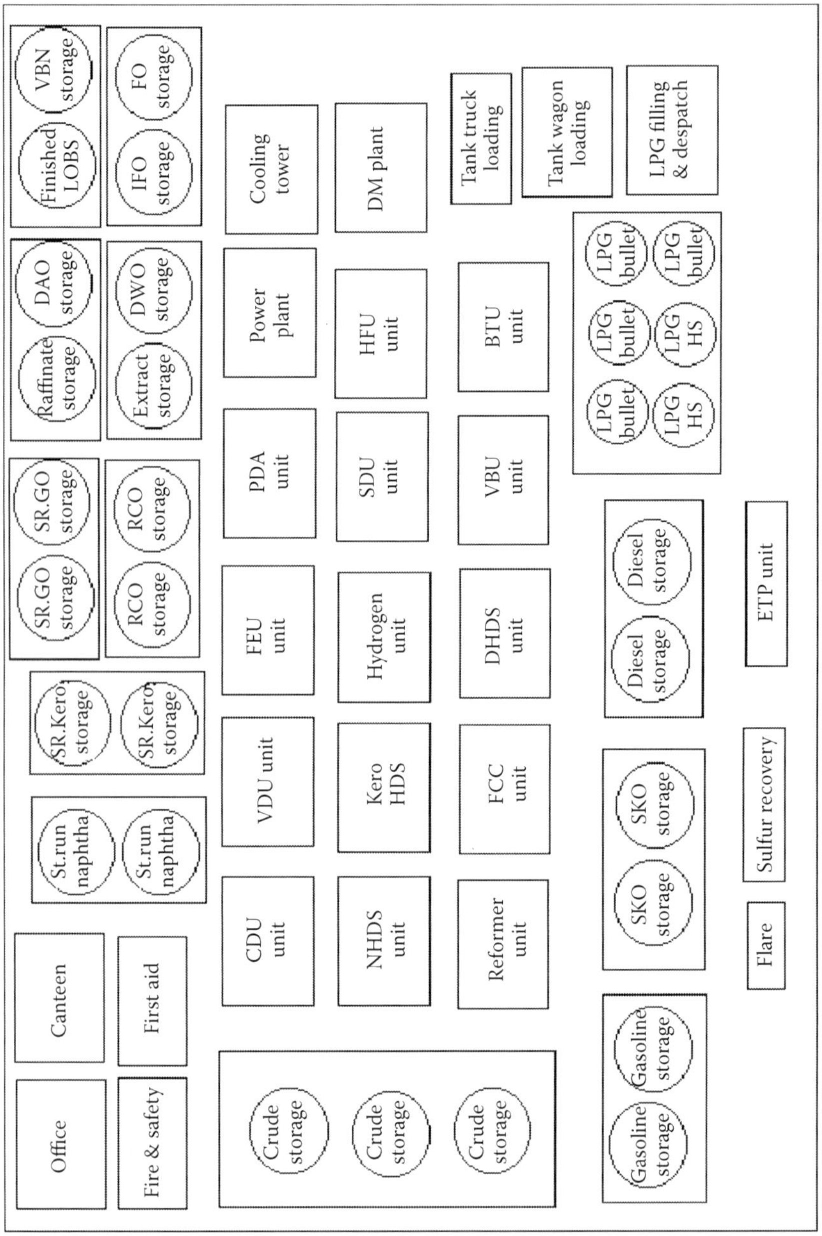

FIGURE 6.1a A typical refinery layout.

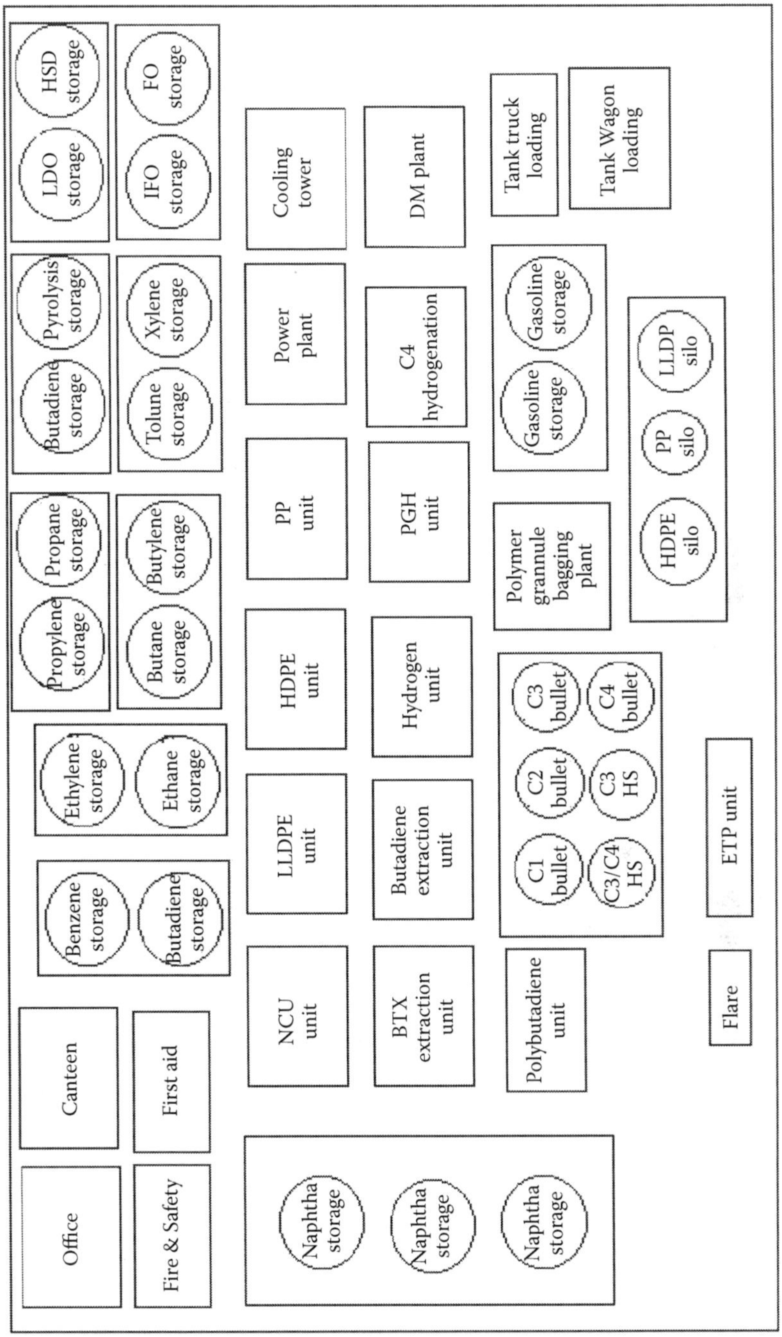

FIGURE 6.1b Layout of a typical petrochemical plant.

The distance between the battery limits of these units is determined by safety and maintenance regulations for easy accessibility during maintenance and in the event of an emergency. Roads and lanes are usually provided between the battery limits. Pipelines carrying gas and liquid are either overlaid or humed (underground) across the space between the battery limits.

6.3 OFFSITE FACILITIES

Off sites are areas where feedstock and its products are stored along with other chemicals necessary for processing. Off sites include facilities in addition to storage, blending, filling, packaging, loading, despatches, and effluent treatment facilities. Usually, it is common practice to store raw materials and intermediate products near the user unit. For example, storage tanks for crude oil and straight run products are located near the CDU; similarly, storage of reduced crude oil (RCO) and vacuum distillate tanks are located near the VDU unit. Storage facilities are strategically laid out so that the pump houses are as near as possible to feed the units. About 70% of the off sites are occupied by storage tanks, which are suitably housed in tank farms. Each tank farm contains a number of tanks storing similar products, finished or semi-finished, surrounded by a cemented bund to contain accidental leakage and drainage of oily water and rain water from the tanks. The advantage of such a tank farm is that it facilitates recovery of valuable products. Each tank farm is provided with a dyke or pit to facilitate the drainage of oil and water to its respective recovery unit through separate channels, e.g., open drains, or to channels leading to the effluent treatment plant (ETP). In a refinery, the filling of liquified petroleum gas (LPG) in cylinders and bitumen in drums are the most common filling operations. Similarly, storage of polymer granules and bagging and despatch are carried out in the off sites of a petrochemical plant. Bulk loading of various products in tank trucks, rail wagons, barges, and tankers are also included in the offsite facilities. Some of the offsite facilities also include fuel pumping and gas distribution network maintenance, effluent water treatment, off gas treatment, and flaring. Since storage constitutes the major function off site, a brief discussion of storage tanks follows. For feedstocks, products, and chemicals that have a high vapour pressure in ambient temperature, storage of these are usually carried out in floating roof or pressurised tanks or cryogenic conditions to avoid loss due to vaporisation. For example, crude oil, naphtha, and gasoline are stored in floating roof tanks where the vapours are contained and liquified by the floating heads and thus do not allow vapour loss to the atmosphere. LPG is stored in pressure vessels (known as bullets) and spherical pressure vessels or Horton spheres. Ethylene is also stored in a pressure vessel at low temperature to keep it in a liquid state. Some of these storage tanks are discussed below.

6.3.1 Floating Roof Tank

As shown in Figure 6.2a, a floating roof tank consists of a floating deck made of multiple chambers, which floats over the surface of the liquid stored. The deck (1) may be made of pontoon or multiple hollow steel chambers. The deck is supported by the buoyant force exerted by the liquid and floats up and down with the changing liquid

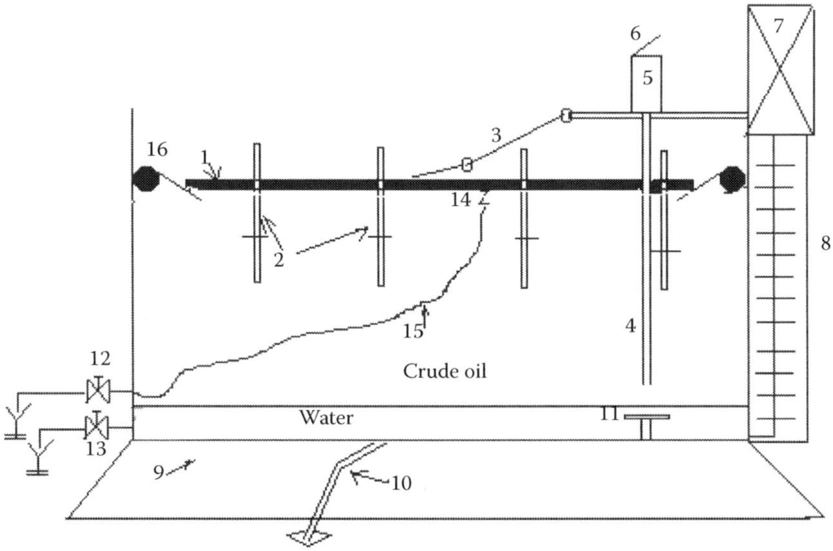

FIGURE 6.2a A floating roof tank for crude oil storage.

level. The circumference of the deck and the shell of the tank are sealed by spring-loaded rolling covers (16), which roll along the tank shell with the up and down movement of the deck. Accessories include a dip pipe (4) running vertically downward from the top of the tank, a flexible steel drain pipe (15) attached to the deck for roof draining, a water drain pipe (13) at the bottom of the tank, and a swing ladder (3) over the deck hanging from the top (7) of the tank shell. The deck is also equipped with hanging supports (2, known as legs or spacers) for when the deck reaches the lower most position of the tank, allowing sufficient space between the tank bottom and the deck for inspection and maintenance. Each leg passes through the floating deck with proper sealing. There are four or more numbers of legs which support the floating roof and these legs are hung on their upper stops while deck is floating. Deck rests on the lower stops of the legs while it is not floating. During this de-floated situation legs hold the entire deck on their lower stops providing work space above the bottom surface of the tank for cleaning and other maintenance jobs.

6.3.2 Fixed Roof Tank

Products such as kerosene, diesel, vacuum distillates, and furnace oil (FO) are liquid at atmospheric conditions because of low vapour pressure. Therefore, these are stored in fixed roof tanks. A typical fixed roof tank is presented in Figure 6.2a, where the roof is supported over the top of the cylindrical shell of the tank, hence these roofs are also called shell or self-supported roofs. The roofs are not exactly flat, but rather conical to avoid the accumulation of rain water. Hence, these are also known as cone roof tanks. For larger tanks, the weight of roof will be enormous, thus a separate bridge and column structure is provided to support these roofs.

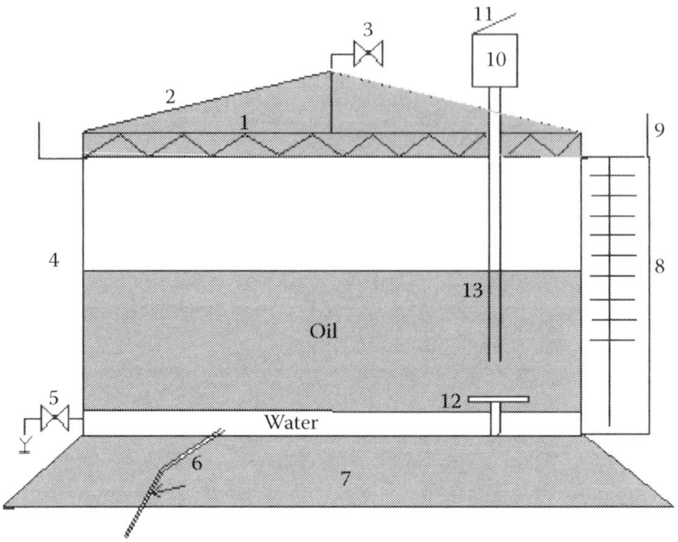

FIGURE 6.2b A typical fixed roof storage tank for diesel oil.

6.3.3 Pressure Vessels

Gases, such as hydrogen, ethylene, propane, butane, and LPG, are stored in thick-walled pressure vessels. Gases that do not liquify can be stored in either horizontal or vertical bullets, but liquified gases are stored in horizontal vessels to allow a large disengagement space for vapour within the tank. Typical pressure vessels are shown in Figure 6.2c.

6.3.4 Horton Sphere

The cylindrical shape has a surface to volume ratio of 4/d and for the spherical shape it is 6/d, where "d" is the diameter. Hence, for spherical tanks larger surface area for cooling is available for the same volume to store. Usually, large spherical vessels are used for storing liquified gases. These vessels are also known as Horton spheres. A typical Horton sphere tank is presented in Figure 6.2d.

6.3.5 Accessories

Each storage tank is provided with a *datum plate* slightly above the bottom of the tank to facilitate the standard measurement of the level in the tank by dip tape or dip rod. For small tanks, straight dip rods graduated with markings (in milimetres or centimetres or metres) are used to measure the level of liquid in the tank, but for large tanks, steel tapes, known as dip tapes, are used. For liquified gases, direct level gauges or differential pressure gauges are employed to measure the liquid level. For gases, pressure and temperature are measured to calculate the quantity of gas stored in a pressure vessel. In fact, temperature is an important parameter to measure for all

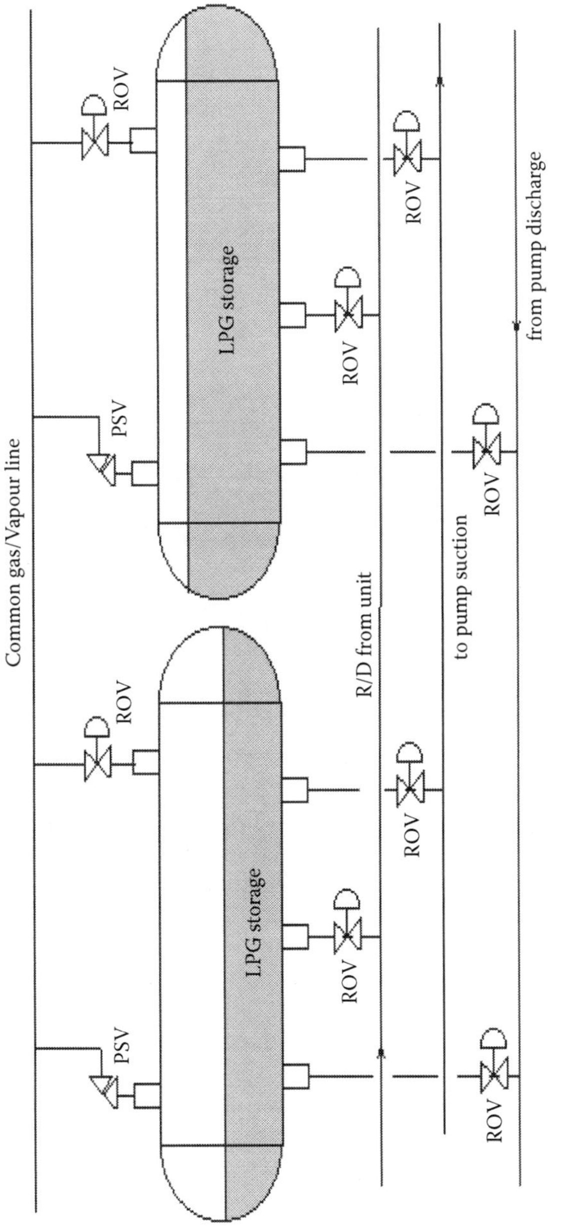

FIGURE 6.2c A typical horizontal pressure vessel for storing propane.

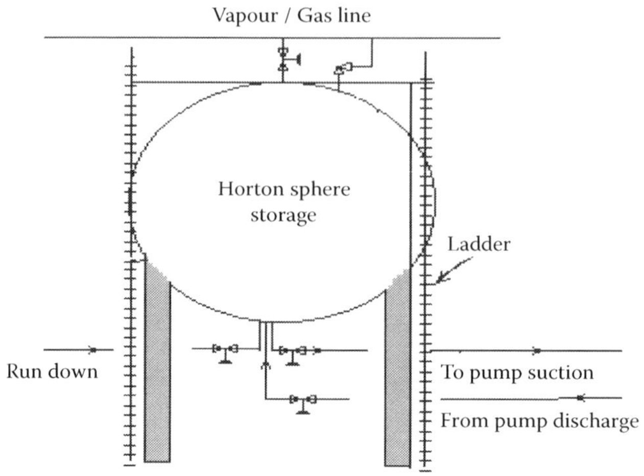

FIGURE 6.2d A Horton sphere storage tank for gases and liquified gases.

types of products, whether they are liquids or gases. A dip tape is lowered through a pipe that terminates slightly above the datum plate. To maintain the gravity of fall, a bob made of brass or aluminium is attached to the tip of the dip tape. The dip pipe and the bob facilitate the straightening of the dip tape and do not allow swings sideways to avoid an error in level measurement. For white oils, such as naphtha, gasoline, kerosene, and diesel, which are not black and sticky, a special type of chemical (known as oil finding paste) is applied to the dip tape to identify the oil level. The presence of water in the tank is also measured and deducted from the total quantity. A separate chemical (water finding paste) is applied to the dip tape for measuring the level of water below the oil. In this dipping method, the dip tape is lowered until it touches the datum plate and the liquid level is read directly from the tape. This method is called the innage method of dipping. For viscous and sticky oils, such as FO, asphalt, and wax, the dip tape is lowered until it touches the surface of the liquid in the tank, i.e., the height of the empty space is measured rather than the height of the liquid, which is determined by deducting this from the reference height from the datum plate marked over the dip pipe or tank. This type of measurement is called the outage method of dipping. Each tank also bears marking of the safe filling height, cautioning the limit of maximum filling without overflowing. Asphalt, RCO, FO, wax, etc., solidify at room temperature. Hence, a steam coil is provided for maintaining the liquid phase. Crude oil tanks are also provided with steam coils and side mixers to maintain uniform composition and avoid solidification of residuous and waxy matters. Anti-static protection is maintained by providing continuous earthing plates connected with the tank shell and grouted (concreted) underground. Each tank is placed over a prepared land and concrete pad, known as a tank pad, which supports the dead and dynamic loads of the tank. Modern tank pads are constructed to withstand vibration and earthquakes. For gases and liquified gases, low temperature is desirable for safe storage. Partial vaporisation and recompression cycles are also employed for cooling highly volatile liquids.

6.3.6 Blending Operations

Various intermediate products, such as straight run and vacuum distillates, reformats, and desulfurised distillates, are received as rundown streams, many of which are later blended with other suitable streams in order to adjust the desired properties. For example, motor spirit is a blend of a variety of streams, which are debutanised reformate naphtha, FCC cracked gasoline, thermally cracked gasoline, vis-broken naphtha, pyrolysis gasoline from petrochemical plants, etc. These are blended in proper proportions for adjusting the octane number, vapour pressure, oxidation stability, etc. During the winter season and for low temperature climates, motor spirit is also blended with butane to adjust the vapour pressure. Similarly, FO is a mixture of vis-broken vacuum distillates, asphalt, short residue, etc., and blending is carried out to correct the viscosity and the flash point. Bitumen blending is also sometimes necessary to correct the penetration index, flash, and softening points. Internal fuel oil (IFO) for consumption within a refinery is a mixture of asphalt, residue, wax, etc. Proper blending may be necessary for these components to be used in the furnaces of process units and the power plant.

6.3.7 Filling, Loading, and Despatch Operations

Bulk filling, cylinder filling, drum filling, bagging, etc., are carried out in any refinery or petrochemical plant. For example, filling of LPG in cylinders is carried out in a separate filling plant where an overhead cage conveyor carries empty cylinders to the rotary filling machine, followed by a roller conveyor for carrying the filled cylinders for sealing, labelling, weighing, and safety checking and, finally, the cylinders are lifted by overhead conveyor to stacking or despatching. LPG is also directly filled in bulk to trucks carrying pressure vessels with a cooling arrangement meant for dispatches to filling plants located away from the refinery or petrochemical plant. Liquid fuels, such as motor spirit, kerosene, aviation turbine fuel (ATF), high speed diesel (HSD), and FO, with low vapour pressure are filled in tank wagons in a facility called a tank wagon gantry. This facility consists of a shed containing overhead lines of products placed above the wagons, which are filled by filling hoses and remote controlled pumps. Similar facilities are also provided separately for tank truck loading. Products that freeze at room temperature and melts above 100°C, e.g., bitumen and grease, are filled in steel drums and dispatched in trucks or box wagons. Solid pellets or granules of polymers are filled in bags in a separate bagging plant. Modern bagging plants have automatic continuous filling, labelling, and weighing facilities and carry the bags by a conveyor to a stack yard where they are forklifted to box wagons or hoisted to ships. Large-scale loading of bags of solid products filled in special types of box wagons called tipplers, which are emptied by turning the tippler (the container) upside down to the containers of a ship.

6.3.8 Pipeline Transport

Though despatches of crude oil, petroleum, and petrochemical products are carried out by rail, road, and sea, a large quantity of these are transported through pipelines. Products like petrol, kerosene, ATF, HSD, and naphtha are transported through the same pipeline in batches (known as parcels) where each parcel of product is separated by the parcels of other products in sequence. For example, gasoline is transferred with naphtha parcels preceding the gasoline parcel at the upstream and following the gasoline parcel at the downstream. Similarly, kerosene parcels precede and follow ATF parcels; kerosene also precedes and follows HSD parcels as well. The sequence is maintained for continuous pumping operation for a single or multiple products in the same pipeline and without any break. In fact, interfacing plugs of two products are formed and can be drawn off following programmed timing and pressure maintenance. The above sequence is also helpful to dispose of mixed products in small quantities, which can be routed to a finished tank with a minimum adjustable specification, e.g., a kerosene–diesel mixture can be upgraded as kerosene or diesel. Separate pipelines are earmarked for liquified natural gas (LNG), crude oil, lubricating oil, etc. However, to maintain very high quality and to reduce intermittent operations, hard rubber ball separators (known as pigs) are introduced into the pipe to separate the products of different qualities. Of course, pig operations requires high cost of operation and maintenance, initial investment cost is much cheaper than the cost of construction of separate pipe line for each product. These are practiced for the same pipeline. In order to maintain uniform pressure and flow, booster pumping stations are provided at various suitable distant locations.

Pipeline transfer is the most economic as compared to other modes of transport because of continuous transport in large quantities and the shortest delivery time. Sub-sea pipelines are the biggest means of transport of liquids and gases from the offshore producing wells to onshore storages. A pipeline transport mechanism is explained schematically in Figure 6.2e.

6.3.9 Effluent Water Treatment

A refinery or a petrochemical plant consumes huge amounts of water for processing, steam raising, cooling, fire fighting, washing, etc. The water consumed is groundwater, river water, or sea water. Since sea water is saline, it can only be used after desalination. Ground and river water are usually sweet and require softening or demineralisation before use. Used water from a refinery or petrochemical plant is contaminated with the hydrocarbon oils, acids, alkali, salts, and chemicals used. This contaminated water (waste water) is hazardous to plants, vegetation, and aquatic animals, and makes natural water unfit for consumption. Hence, waste water must be treated within the refinery or petrochemical plant to recover or remove a substantial amount of the contaminants before discharging to surrounding land or actuaries. A modern ETP involves three types of treatment—physical, chemical, and biological. In the physical treatment step, most of the oils and suspended matter are separated by gravity settling, API chain-grate separation, skimming operation, packed plate separation, dissolved air floatation, etc. Oil collected from these steps is known as slop, which is usually reprocessed along with crude oil. Following physical treatment, chemicals such as ferrous sulfate, lime, fuller's earth, or, recently, poly-electrolytes are mixed in a high-speed mixer and later

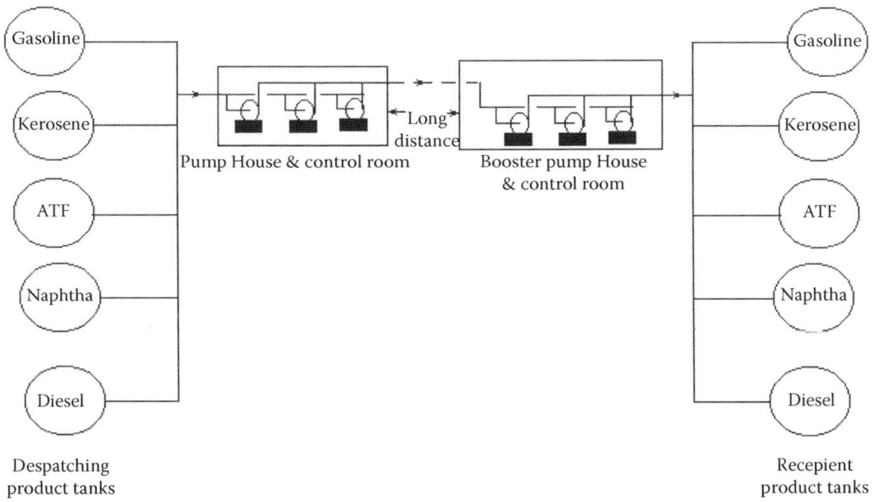

FIGURE 6.2e A schematic pipeline transfer station for products.

flocculate the remaining oil and fine particles to settle by gravity. Physical and chemical treatments are also known as primary treatment. Biological treatment is known as secondary treatment. After primary treatment, dissolved oil and organic matter and salts are treated in a biological treatment unit. A modern biological treatment unit consists of a variety of units, such as the trickle bed filter, aerator, and lagoon. In the biological treatment step, dissolved oil and organic matter are converted to innocuous carbon dioxide by microbial organisms (microbes), which eat these oil and organic matter dissolved in water. The mass of microbes grown are later disposed of as fertiliser, land fill or incinerated in a furnace. Treated water is then stored in a separate lake to monitor the quality as per the specifications laid down by the pollution control board of the country before discharge to land or water resources. In some plants, attempts are being made to reuse the treated water for processing and even steam raising for the boilers, and little or no water is discharged outside the plant. For this *zero discharge policy*, tertiary treatment of effluent using reverse osmosis, membrane separation, electrodialyses, ozonation, etc., are carried out to make the water suitable for reuse. A modern ETP is presented in Figure 6.2f. The permissible limits of pollutants in discharged water are presented in Table 6.1.

6.3.10 Off Gas Treatment

Gaseous effluents or off gases containing hydrocarbon gases, such as methane, ethane, ethylene, propylene, and butane, along with hydrogen sulfide, mercaptans, hydrogen, etc., in various proportions need to be exhausted to atmosphere for processing corrections, sudden safety releases, storage limitations, accidental release, etc. These off gases are burnt in furnaces for heat generation and excess gases are exhausted and burnt through a flaring stack. The presence of sulfur compounds, usually

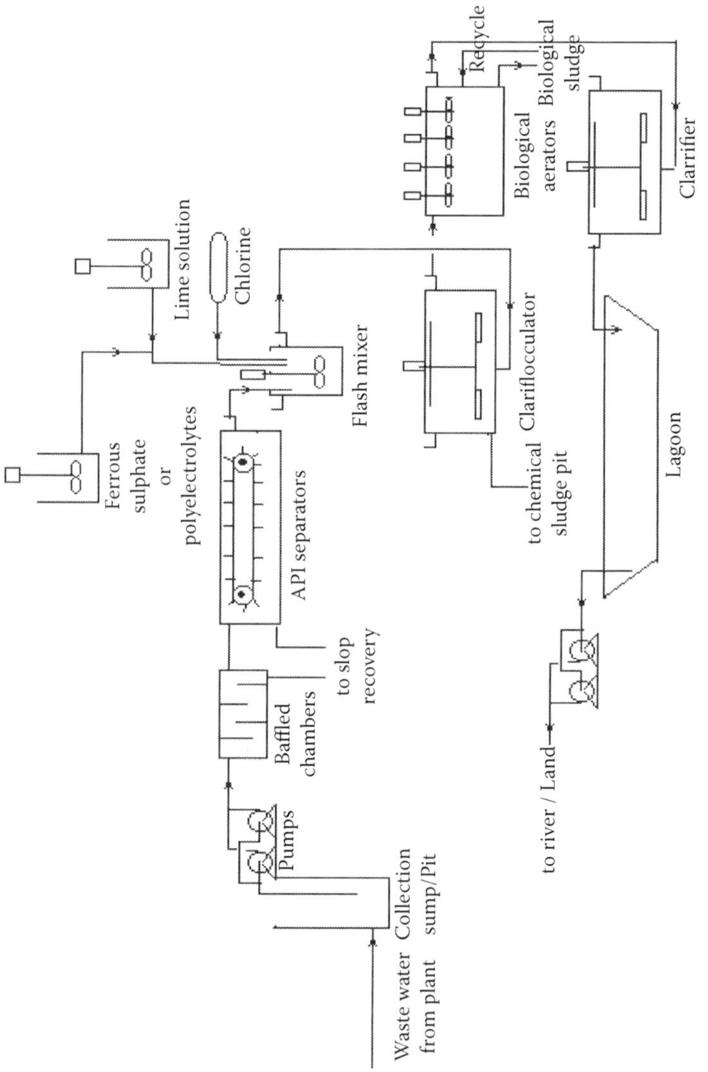

FIGURE 6.2f A simplified scheme of an effluent water treatment plant.

TABLE 6.1
Permissible Limits of Typical Pollutants in Discharged Water

Parameters	Inland Surface Water	Public Sewer	Irrigation Water	Coastal Water
SSP, mg/L, max	100	600	200	100
pH	5.5–9.0	5.5–9.0	5.5–9.0	5.5–9.0
Oil and grease, ppm, max	10	20	10	20
BOD, mg/L, max	30	350	100	100
COD, mg/L, max	250	250	250	250
Phenol, ppm, max	1	5	5	5
Pb, ppm	0.1	1.0	1.0	1.0
Hg, ppm	0.01	0.01	0.01	0.01
As, ppm	0.2	0.2	0.2	0.2

hydrogen sulfide, which is a foul-smelling pollutant, are also burnt through a flaring stack. Since flared off gases produce carbon monoxide (CO, for incomplete combustion), oxides of sulfur (SOx) and nitrogen (NOx), which are major pollutants of natural air, steps are taken by the refiners or petrochemicals manufacturers to abate such pollution. Sulfur recovery from sulfur compounds and oxides has become a modern practice. Catalytic oxidation of hydrocarbons is also practised for complete combustion of hydrocarbons to reduce CO generation. A common flare stack is shown in Figure 6.2g. Off gases are separated from the condensates and liquids in a drum before the gases enter the flare stack. Gases are then burnt at the top of the stack. Burners at the top are kept lit by a flame produced by electrically sparking a mixture of butane or propane and air at the bottom of the stack. The generated flame fronts

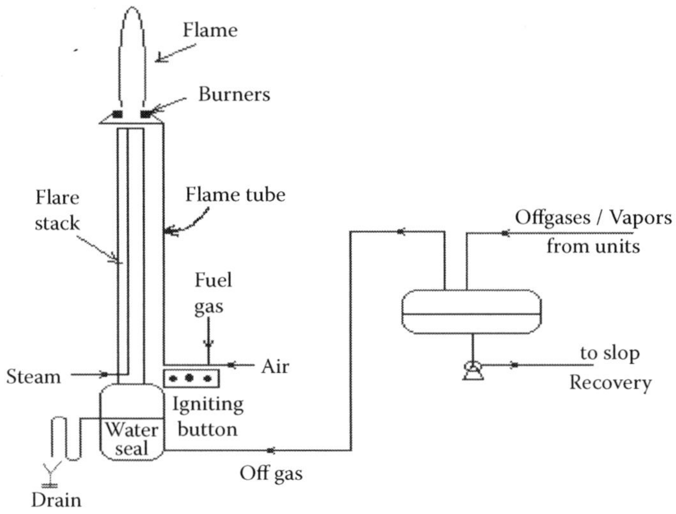

FIGURE 6.2g Off gas flare of a hydrocarbon processing plant.

TABLE 6.2
Permissible Limits of Typical Air Pollutants

			Central Pollution Control Board, India	
Pollutants	**EPA, USA**	**WHO**	**Industrial**	**Residential**
Carbon dioxide, ppm, max	50	500	500	500
Sulfur dioxide, ppm, max	5	60	120	80
Hydrocarbons, ppm, max	500	500	500	500
Ozone, ppm, max	0.05	0.05	0.05	0.05
NO_x, ppm, max	30	80	120	80
SPM, ppm, max	60	90	500	200
Pb, mg/m^3, max	0.2	0.2	0.2	0.2

travel up the burners through separate flame tubes leading to the burner tips on the top of the stack. A list of criteria for gaseous pollutants and their permissible limits are presented in Table 6.2.

6.3.11 Internal Fuel Oil Circulation

Another important operation for any plant is the continuous supply of fuel to all the furnaces. Usually, in refineries and petrochemical plants, off gases such as methane, ethane, propane, butane, olefinic gases, and hydrogen are continuously produced. The excess of off gases during production is directly routed to the furnaces and the rest is flared for want of storage space. Liquid hydrocarbons, such as FO, asphalt, and wax, are also used as IFO, when off gases are not available in sufficient quantities. Most modern furnaces are provided with combination firing, i.e., either or both gas and liquid fuels can be burnt in these furnaces. IFO storage and circulation is a part of the offsite facilities. Since IFO is a mixture of low-valued hydrocarbons, such as asphalt, short residue, vis-broken heavy distillates, and wax, the viscosity and melting point are high. Hence, steam-heating facilities are provided for storage tanks, and pipelines are also wound with steam coils (steam traced) to avoid congealing of the fuel oil. Positive displacement pumps, e.g., gear, screw, or lobe pumps, are employed to handle these viscous oils. To maintain constant pressure, it is essential to have a circulation circuit with a back pressure control arrangement. This is explained in Figure 6.2h.

6.4 POWER AND STEAM GENERATING PLANT

Since power consumption is enormous in refineries and petrochemical plants, it is essential that these plants should have captive power plants. Usually, unwanted or excess hydrocarbons, both gas and liquid (known as internal fuel), are burnt in the power plant furnaces to raise steam for both electricity generation and for process steam requirements. A water tube boiler is used to raise high-pressure superheated steam for turning turbines to generate electricity. Demineralised water from

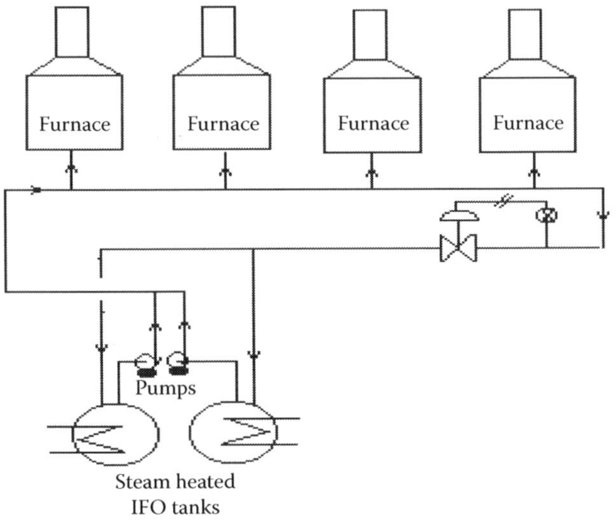

FIGURE 6.2h IFO circulation in a refinery.

a demineralisation (DM) plant is first deaerated in a steam purging vessel where water is cascaded down over a series of plates countercurrently with the rising steam. Deaeration is essential to remove dissolved oxygen, which causes corrosion due to galvenic cell reactions, as given below, with the iron of the pipe and boiler walls.

$$Fe(s) \rightarrow Fe^{+2}(aq) + 2e,$$

$$O_2(g) + 2H_2O + 4e \rightarrow 4OH^{-1}(aq),$$

$$Fe^{+2}(aq) + 2OH^{-1}(aq) \rightarrow Fe(OH)_2,$$

$$Fe(OH)_2 \rightarrow FeO + H_2O,$$

$$FeO + O_2 \rightarrow Fe_2O_3 \text{ (rust)}.$$

Chemicals like sodium sulfite and hydrazine hydrate are also injected in very small amounts to absorb the remaining dissolved oxygen in water.

The oxidation reactions are

$$Na_2SO_3 + O_2 \rightarrow Na_2SO_4,$$

$$N_2H_4 + O_2 \rightarrow N_2 + 2H_2O.$$

Deaerated water then passes through the economiser chamber where hot flue gas from the furnace heats up the incoming deaerated water, which is then flashed into a steam drum to produce dry steam followed by superheating with hot gases from the fire box immediately after combustion. Superheated steam then enters the turbine and turns it at high speed. This is summarised below.

1. Hot water from the economiser enters the furnace tubes and then to the boiler drum where steam is separated from the condensates.
2. Steam from the boiler drum then enters the cyclone separators to evolve dry saturated steam and the condensates enter the mud drum where thermosifon takes place by natural convection currents due to the density difference between the condensates and the mud drum temperature. Dry saturated steam from the steam drum then enters the superheater section of the furnace to yield high pressure and high temperature superheated steam (usually above 65 bar and at 400°C).
3. Superheated steam then enters the turbine blades to generate mechanical rotation by converting the pressure energy into kinetic energy. As a result, moderate pressure steam is wasted from the turbine and is finally condensed. The turbine then turns the alternator for conversion of kinetic energy into electricity. Figure 6.3 depicts a typical power plant where hot gas is generated at the fire box by burning fuel and the dry saturated steam is superheated. Hot flue gas then flows countercurrently with steam from the boiler drum, where water in the drum is reheated in the bottom drum and circulates back to the drum. Water concentrated with salts and other inorganic matter accumulates in the mud drum and water concentrated with salt is drained (blown) out time to time from the mud drum. Flue gas is then further cooled by the incoming feed water in the

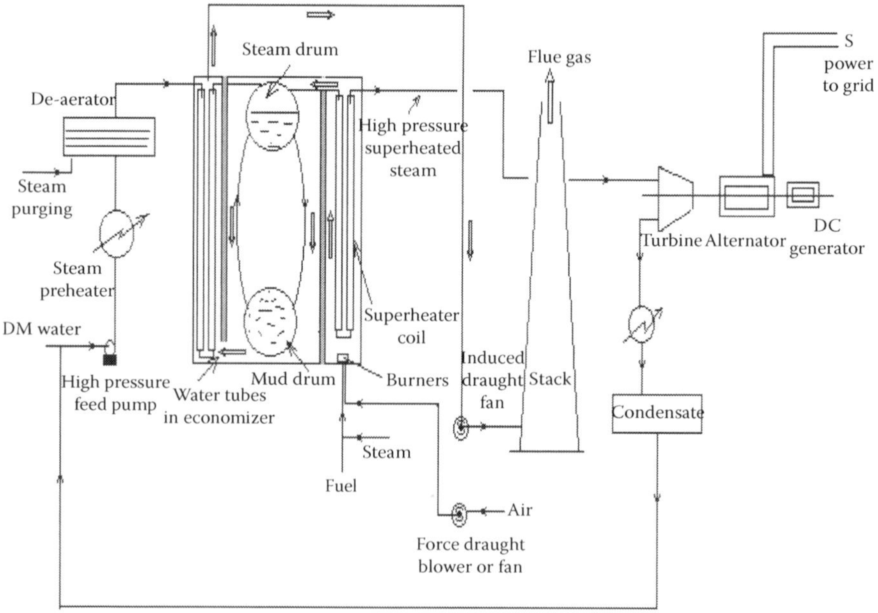

FIGURE 6.3 A schematic arrangement of a thermal power plant.

economiser section and finally leaves through the stack. The draught for the flue gas flow in the furnace is maintained by a combined force draught (FD) and induced draught (ID) fans.

Water is converted to superheated steam, which then moves the turbine at high speed. Steam exhausted from the turbine is cooled and the condensate collected is reused in the boiler. The turbine shaft drives an alternator to produce power of alternating current and voltage, which is then transmitted to the transmission grid for distribution and consumption.

A generator connected to the system supplies direct current to the poles of the alternator. However, steam is generated in a larger quantity than is required for power generation, such that the excess could be used as process steam and for steam-driven pumps/compressors in the plant. Electricity is consumed by pumping, compressing, and driving various rotary machines, electronic instruments and gadgets, lighting, etc. Medium- and low-pressure steam is also generated from high-pressure steam from the boiler through steam-reducing stations by means of bleeding steam to the atmosphere through nozzles. Some of the rotary machines are driven by steam in case of a power shutdown.

6.5 COOLING TOWER

A large circulation of cooling water is required to condense and cool various streams of vapours, steam, gases, and liquids. Usually, the coolant water temperature is around 33°C and the exit hot water temperature from the coolers/condensers is around 50°C. This hot water is cooled by partial vaporisation of the water in a tower, known as a cooling tower, where hot water falls countercurrently against the rising air through wooden chambers. The hot water is distributed over the top chamber before falling down to the chambers. Each chamber has a fan to suck air from the chamber from the top this maintains induced draught (ID) to allow air to vaporise falling water and extract the vapour–air mix to atmosphere. Cold water is collected in a pond below the tower from where it is pumped back to the coolers and condensers. Such an arrangement of a tower chamber is shown in Figure 6.4a. While the demand for quantities of such cooling water is large, a number of chambers are provided. High pressure and high capacity pumps are used to maintain circulation through all the condensers and coolers in the plant. A typical cooling tower containing six chambers is shown in Figure 6.4b. Modern cooling towers use fibre-reinforced plastic (FRP) louvres in place of wood.

6.6 WATER CONDITIONING PLANT

Raw water from river or ground is not suitable as such for boiler or processing purposes as it contains suspended and dissolved matter, hence the treatment of water is essential. Water is classified as hard or soft, containing non-carbonate salts of magnesium, calcium, etc., other than sodium. Water containing carbonate salts can be removed by heating and hardness is termed temporary. Non-carbonates, such as chlorides and sulfates, cannot be removed by heating and that is why they lead to

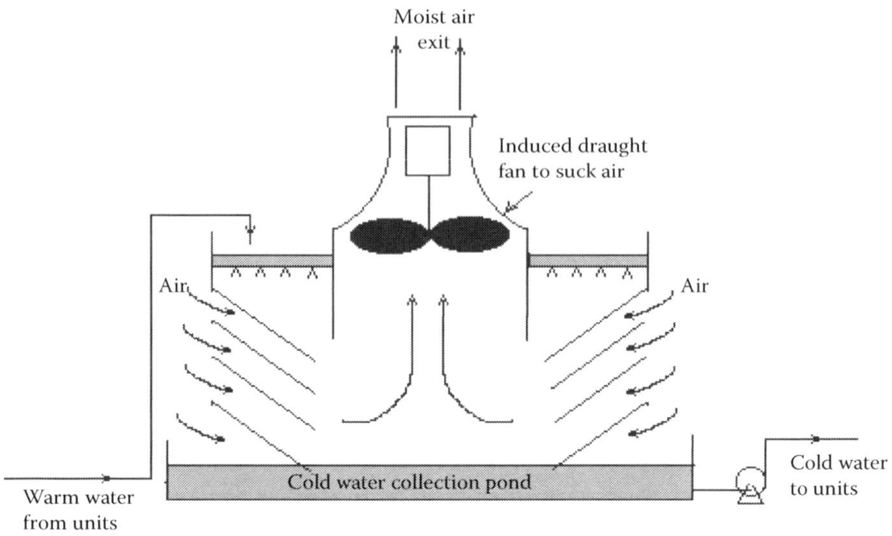

FIGURE 6.4a A single chamber of a cooling tower.

permanent hardness which is softened by using chemicals only like lime and soda. The hardness of water is expressed in terms of calcium carbonate precipitated equivalent to the salts present in the sample water. The presence of these salts gives rise to scale deposition in the equipment, causing damage to the surface due to corrosion and erosion, and the germination of fungi or algae, etc.

Scale formation also causes a reduction in the heat transfer efficiency of heat exchangers. Modern methods employ sand filtration followed by ion-exchange not only for the removal of suspended matter and salts, but also for deionising all the cations and anions producing pure water molecules. A major consumer of deminearalised (DM) water is the boiler, which requires very pure water without ions causing salts. Even dissolved oxygen is not allowed in the boiler feed water. The modern DM plant consists of ion-exchange resin-packed vessels. These resins are capable of replacing the calcium and magnesium ions with hydrogen ions by the following exchange reactions,

$$2H - R + CaCl_2 \rightarrow Ca - R_2 + 2HCl,$$

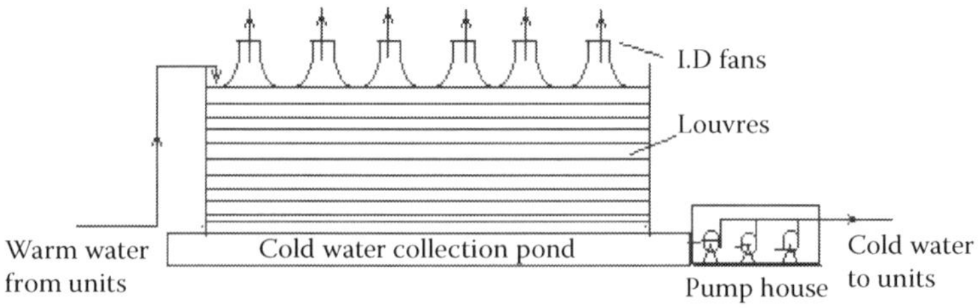

FIGURE 6.4b A typical cooling tower of a refinery.

$$2H - R + CaSO_4 \rightarrow Ca - R_2 + H_2SO_4,$$

$$2H - R + MgCl_2 \rightarrow Mg - R_2 + 2HCl,$$

$$2H - R + MgSO_4 \rightarrow Mg - R_2 + H_2SO_4,$$

$$2H - R + Na_2SO_4 \rightarrow 2Na - R + H_2SO_4,$$

$$H - R + Nacl \rightarrow Na - R + HCl,$$

where H–R is the *hydrogen–cation exchange* resin and R represents the radical of the resin part, usually made from sulfonated-poly-styrene-divinyl-benzene, which replaces the cations (Ca, Mg, Na, etc.) of the salts present in water with hydrogen ion producing acids of the corresponding anions of the original salts. Carbonate salts generate carbon dioxide with the ion-exchange reaction,

$$2H - R + CaCO_3 \rightarrow Ca - R_2 + H_2O + CO_2 \uparrow,$$

$$2H - R + MgCO_3 \rightarrow Mg - R_2 + H_2O + CO_2 \uparrow,$$

$$2H - R + Na_2CO_3 \rightarrow 2Na - R + H_2O + CO_2 \uparrow.$$

Degassification is essential for these reactions. The presence of silica, alumina, chromates, etc., also generate silic acid, chromic acid, etc. Finally, water leaving these cation exchangers become acidic, which is then passed to an anion exchanger for replacing the acidic anions (SO_4, Cl, etc.) with OH ions to generate demineralised or deionised water.

$$R - OH + HCl \rightarrow R - Cl + H_2O,$$

$$R - OH + H_2SO_4 \rightarrow R - Cl + H_2O,$$

where R is the radical of the resin and OH is the anion for exchange. In fact, additional treatment in a mixed bed contains two types of resins, one is a strongly basic anion exchange resin and the other is a weakly basic anion exchange resin. The former removes weak acids, e.g., carbonic, silicic, etc., and the latter removes strong acids (hydrochloric, sulfuric, nitric acids) remaining in the water. Following the above reactions, both types of resins are saturated with the cations and anions exchanged after a certain volume of water is treated. These resins are then regenerated by reversing the reactions usually with sulfuric acid for cation exchanging and with caustic soda for anion exchange resins, respectively. Thus, the treatment process is operated in batches at least in two chains. When one chain of exchangers treats water, the other chain is regenerated to make a continuous production of DM water. A typical DM plant is presented in Figure 6.5. However, tubewell water and filtered sweet waters are only used as potable water. Raw water and treated waste water are used for equipment washing, fire fighting, etc.

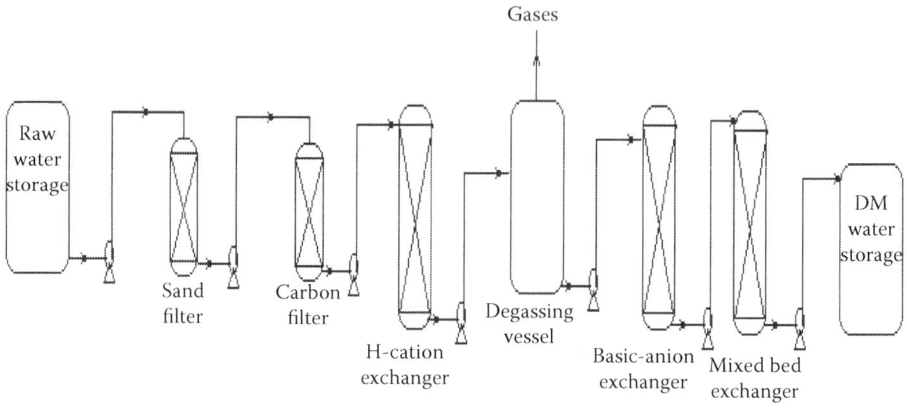

FIGURE 6.5 A single chain of a DM plant.

QUESTIONS

1. Sketch a conceptual layout of a refinery having atmospheric and vacuum distillation units, a gas plant, a naphtha reforming unit with a naphtha pre-treatment unit, off site, and an office building.
2. Draw a layout of a naphtha cracking unit for olefin manufacture showing the necessary off sites and other essential service centres.
3. Define the terms battery limit, tank farm, and utilities.
4. Why does boiler feed water need treatment?
5. Briefly describe the steam raising process in a boiler for power production.
6. Describe the principle of water cooling in a cooling tower.
7. Define hardness of water and its measuring unit. Describe in brief the process of DM water production.
8. How are fuel gas and fuel oil distributions carried out in a refinery?
9. Why does waste water need treatment?
10. How should off gases from a refinery or a petrochemical plant be treated?

7 Material and Energy Balances

7.1 MEASUREMENT OF QUANTITY OF CRUDE OIL AND PRODUCTS

In a refinery, crude oil is received in tanks and the stock is evaluated at a particular temperature in volumetric quantity. The price of crude oil is internationally expressed in US dollars per barrel (1 bbl = 42 gallons = 158.9 L). The stock and processing rate is evaluated in volumetric quantity, usually in kilolitres (m^3). The prices of fuels, e.g., kerosene, diesel, and furnace oil, are rated in volume, while gases and liquified gases, bitumen, lube oil, and wax are rated in weight. Packed products in cylinders or in drums are always rated in weight.

7.1.1 Tank Dipping

The liquid quantity stored in a tank is commonly measured by a rod, tape, or stick and level measuring instruments. Graduated sticks are used to measure depth of liquid in tanks of short heights. For larger tanks, more than 3 m high, graduated dip tapes are used.

These tapes are usually made of steel or brass or aluminium coiled over a wheel and the end of the tape is attached to a metallic ball or bob made of brass or aluminium. For low viscous and non–sticky liquids, a dip tape is lowered into the liquid until it touches the bottom of the tank or the datum plate in the tank. To ensure the correct vertical distance from the top to the bottom, it is essential to check the tape reading against a reference height marked on the tank. The height of the liquid is then read as the difference between the reading of the wetted portion and the bob end. Colour–sensitive pastes are used to mark the beginning of the wetted portion. For example, oil finding and water finding pastes (two different colour–sensitive dyes that change colour as they contact oil and water, respectively) are used. This type of measurement is known as the innage method of dipping. For sticky and viscous liquids or semi–solid or solid materials, outage dipping is done. In this method, the height of the space above the stock of materials is measured by a tape as lowering of the same will be difficult through the liquid. The typical measuring methods for both innage and outage dipping are presented in Figure 7.1. Level measuring instruments, such as float tape, a differential pressure gauge, a displacer level gauge, and electrical level gauges, are also used for measurement and recording in a distant control room. Since measurement of level by these instruments may not be accurate, it is therefore manual dipping, as discussed earlier, that is done for more accuracy over a long period of time, e.g., every day, once in a month or once in a year, where commercial transactions are involved. However, instrument reading is essential for continuous evaluation and recording.

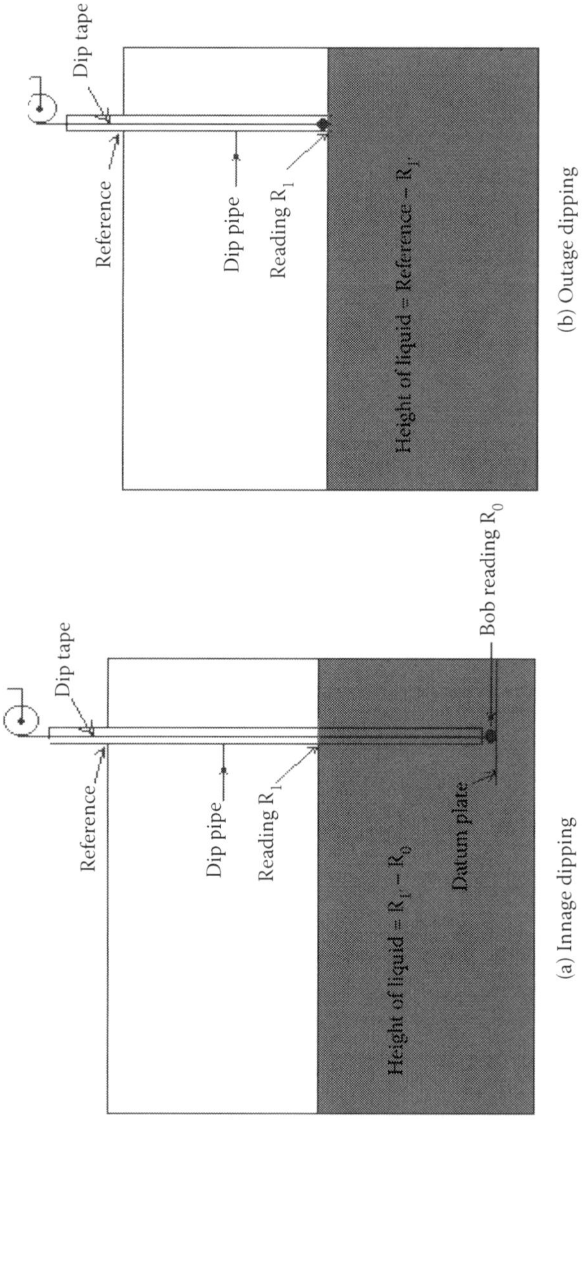

FIGURE 7.1 Dipping of liquids in a storage tank.

7.1.2 Volume Correction

The volume of a liquid increases and decreases with an increase and decrease in temperature. Hence, in the hydrocarbon industry, the volume of liquid is always expressed at a temperature of 15.5°C, which is taken as the standard temperature for measurement. During dipping of liquids in tanks, it is mandatory to measure the temperature of the liquid. For accuracy, a sample that is taken as the equal mixture of top, middle and bottom of the tank must be collected during dipping. This sample is abbreviated as a TMB sample, which is a representative sample of the average of the temperature, density, and composition of the liquid in store. The volume corresponding to the dip of liquid measured is calculated from the tank calibration chart and is reported as the measured average temperature during dipping. Finally, this volume is expressed at 15.5°C with the help of the ASTM volume correction factor, which takes into account the coefficient of thermal expansion of the liquid.

7.1.3 Density Correction

Density is defined as the mass per unit of volume, hence the density of a liquid decreases with an increase in temperature and vice versa. Therefore, in order to measure the mass quantity of a liquid, e.g., kilogram or ton, it is necessary to obtain both the volume and density at the same temperature. Since the temperature of the liquid during dipping and the temperature while density is measured will differ, both the volume and density are reported at 15.5°C with the help of the ASTM correction factors. It is convenient to measure the depth of liquid petroleum in a storage (dipping or gauging) tank and the volume is evaluated from a tank calibration chart. Tables 7.1 and 7.2 present the ASTM table of density and volume corrections for oils at various temperatures. The mass of oil received or transferred is then evaluated by multiplying the volume by the density, both at standard temperature. However, the absolute mass of a stock requires correction of the atmospheric pressure different from the pressure at sea level. Accurate evaluation of the quantity of oil received or transferred is very important for the value of the oil stocks sold or received. The following example will be helpful for understanding the method of evaluation of stocks.

Example 7.1

The following data applies for a receiving tank of crude oil. The tank dip before receiving (the opening dip) was 600 cm with a 1-cm water cut at 35°C and after receiving was complete (the closing dip) it was 800 cm with a 2-cm water cut at 40°C. The densities of the opening and closing oil samples were reported to be 0.8488 and 0.8430 g/cc, respectively. Determine the quantity of crude oil in metric ton received in the tank. The tank calibration chart and the ASTM table for density and volume correction factors are used for obtaining tank volume and correction factors, respectively.

Solution

From the density correction table: At 35°C, the sample density is 0.8450; the corresponding density at 15°C from the table reads as 0.8581 g/cc; and at 40°C,

TABLE 7.1
Density Reduction to 15°C

Observed Temperature, °C	Observed Density 0.840	0.841	0.842	0.843	0.845	0.846	0.849
	Corresponding Density 15°C						
25	0.8466	0.8476	0.8486	0.8496	0.8506	0.8526	0.8556
25.5	0.8470	0.8480	0.8490	0.8499	0.8509	0.8529	0.8559
26	0.8473	0.8483	0.8493	0.8503	0.8513	0.8532	0.8562
26.5	0.8476	0.8486	0.8496	0.8506	0.8516	0.8536	0.8566
27	0.8480	0.8489	0.8499	0.8509	0.8519	0.8539	0.8569
27.5	0.8483	0.8493	0.8503	0.8512	0.8522	0.8542	0.8572
28	0.8486	0.8496	0.8506	0.8516	0.8526	0.8546	0.8575
28.5	0.8489	0.8499	0.8509	0.8519	0.8529	0.8549	0.8579
29	0.8493	0.8502	0.8512	0.8522	0.8532	0.8552	0.8582
29.5	0.8496	0.8506	0.8516	0.8526	0.8535	0.8555	0.8585
30	0.8499	0.8509	0.8519	0.8529	0.8539	0.8559	0.8588
30.5	0.8502	0.8512	0.8522	0.8532	0.8542	0.8562	0.8592
31	0.8506	0.8516	0.8525	0.8535	0.8545	0.8565	0.8595
31.5	0.8509	0.8519	0.8529	0.8539	0.8548	0.8568	0.8598
32	0.8512	0.8522	0.8532	0.8542	0.8552	0.8572	0.8601
32.5	0.8515	0.8525	0.8535	0.8545	0.8555	0.8575	0.8604
33	0.8519	0.8529	0.8538	0.8548	0.8558	0.8578	0.8608
33.5	0.8522	0.8532	0.8542	0.8552	0.8561	0.8581	0.8611
34	0.8525	0.8535	0.8545	0.8555	0.8565	0.8584	0.8614
34.5	0.8528	0.8538	0.8548	0.8558	0.8568	0.8588	0.8617
35	0.8532	0.8542	0.8551	0.8561	0.8571	0.8591	0.8621
35.5	0.8535	0.8545	0.8555	0.8565	0.8574	0.8594	0.8624
36	0.8538	0.8548	0.8558	0.8568	0.8578	0.8597	0.8627
36.5	0.8541	0.8551	0.8561	0.8571	0.8581	0.8601	0.8630
37	0.8545	0.8554	0.8564	0.8574	0.8584	0.8604	0.8633
37.5	0.8548	0.8558	0.8568	0.8577	0.8587	0.8607	0.8637
38	0.8551	0.8561	0.8571	0.8581	0.8591	0.8610	0.8640
38.5	0.8554	0.8564	0.8574	0.8584	0.8594	0.8613	0.8643
39	0.8557	0.8567	0.8577	0.8587	0.8597	0.8617	0.8646

TABLE 7.1 (CONTINUED)
Density Reduction to 15°C

Observed	Observed Density						
	0.840	0.841	0.842	0.843	0.845	0.846	0.849
Temperature, °C	Corresponding Density 15°C						
39.5	0.8561	0.8571	0.8580	0.8590	0.8600	0.8620	0.8649
40	0.8564	0.8574	0.8584	0.8594	0.8603	0.8623	0.8653
40.5	0.8567	0.8577	0.8587	0.8597	0.8607	0.8626	0.8656
41	0.8570	0.8580	0.8590	0.8600	0.8610	0.8629	0.8659
41.5	0.8574	0.8583	0.8593	0.8603	0.8613	0.8633	0.8662
42	0.8577	0.8587	0.8596	0.8606	0.8616	0.8636	0.8665
42.5	0.8580	0.8590	0.8600	0.8610	0.8619	0.8639	0.8669
43	0.8583	0.8593	0.8603	0.8613	0.8623	0.8642	0.8672
43.5	0.8586	0.8596	0.8606	0.8616	0.8626	0.8645	0.8675
44	0.8590	0.8599	0.8609	0.8619	0.8629	0.8649	0.8678
44.5	0.8593	0.8603	0.8612	0.8622	0.8632	0.8652	0.8681
45	0.8596	0.8606	0.8616	0.8625	0.8635	0.8655	0.8684
45.5	0.8599	0.8609	0.8619	0.8629	0.8638	0.8658	0.8688
46	0.8602	0.8612	0.8622	0.8632	0.8642	0.8661	0.8691
46.5	0.8606	0.8615	0.8625	0.8635	0.8645	0.8664	0.8694
47	0.8609	0.8619	0.8628	0.8638	0.8648	0.8668	0.8697
47.5	0.8612	0.8622	0.8632	0.8641	0.8651	0.8671	0.8700
48	0.8615	0.8625	0.8635	0.8545	0.8654	0.8674	0.8703
48.5	0.8618	0.8628	0.8638	0.8648	0.8657	0.8677	0.8707
49	0.8621	0.8631	0.8641	0.8651	0.8661	0.8680	0.8710
49.5	0.8625	0.8634	0.8644	0.8654	0.8664	0.8683	0.8713
50	0.8628	0.8638	0.8647	0.8657	0.8667	0.8687	0.8716

the sample density is 0.84350; the corresponding density at 15°C from the table reads as 0.8504 g/cc.

Densities at 15°C and volume correction factors are obtained from ASTM-IP Tables 7.1 and 7.2 and the corresponding nearest range of densities at 15°C between 0.840 to 0.870 g/cc, at 35°C, are read as

Volume reduction factor	Corresponding to density at 15°C
0.9842	0.855
0.9843	0.860

TABLE 7.2
Volume Reduction Factors

Observed	Density at 15°						
	0.840	0.845	0.850	0.855	0.860	0.865	0.870
Temperature, °C	Factor for Reducing Volume to 15°C						
25	0.9918	0.9919	0.9920	0.9921	0.9922	0.9922	0.9923
25.5	0.9914	0.9915	0.9916	0.0017	0.9918	0.9918	0.9919
26	0.9910	0.9911	0.9912	0.9913	0.9914	0.9914	0.9915
26.5	0.9906	0.9907	0.9908	0.9909	0.9910	0.9911	0.9911
27	0.9902	0.9903	0.9904	0.9905	0.9906	0.9907	0.9908
27.5	0.9898	0.9899	0.9900	0.9901	0.9902	0.9903	0.9904
28	0.9893	0.9895	0.9896	0.9897	0.9898	0.9899	0.9900
28.5	0.9889	0.9891	0.9892	0.9893	0.9894	0.9895	0.9896
29	0.9885	0.9887	0.9888	0.9889	0.9890	0.9891	0.9892
29.5	0.9881	0.9883	0.9884	0.9885	0.9886	0.9887	0.9888
30	0.9877	0.9879	0.9880	0.9881	0.9882	0.9883	0.9884
30.5	0.9873	0.9874	0.9876	0.9877	0.9878	0.9880	0.9881
31	0.9869	0.9870	0.9872	0.9873	0.9874	0.9876	0.9877
31.5	0.9865	0.9866	0.9868	0.9869	0.9871	0.9872	0.9873
32	0.9861	0.9862	0.9864	0.9865	0.9867	0.9868	0.9869
32.5	0.9857	0.9858	0.9860	0.9861	0.9863	0.9864	0.9865
33	0.9853	0.9854	0.9856	0.9857	0.9859	0.9860	0.9861
33.5	0.9848	0.9850	0.9852	0.9853	0.9855	0.9856	0.9858
34	0.9844	0.9846	0.9847	0.9849	0.9851	0.9852	0.9854
34.5	0.9840	0.9842	0.9843	0.9846	0.9847	0.9849	0.9850
35	0.9836	0.9838	0.9840	0.9842	0.9843	0.9845	0.9846
35.5	0.9832	0.9834	0.9836	0.9838	0.9839	0.9811	0.9842
36	0.9828	0.9830	0.9832	0.9834	0.9835	0.9837	0.9838
36.5	0.9824	0.9826	0.9828	0.9830	0.9831	0.9833	0.9835
37	0.9820	0.9822	0.9823	0.9826	0.9828	0.9829	0.9831
37.5	0.9816	0.9818	0.9819	0.9822	0.9824	0.9825	0.9827
38	0.9812	0.9814	0.9816	0.9818	0.9820	0.9821	0.9823
38.5	0.9808	0.9810	0.9812	0.9814	0.9816	0.9818	0.9819
39	0.9804	0.9806	0.9808	0.9810	0.9812	0.9814	0.9815
39.5	0.9799	0.9802	0.9804	0.9806	0.9808	0.9810	0.9812
40	0.9795	0.9798	0.9800	0.9802	0.9804	0.9806	0.9808
40.5	0.9791	0.9794	0.9796	0.9798	0.9800	0.9802	0.9804
41	0.9787	0.9790	0.9792	0.9794	0.9796	0.9798	0.9800
41.5	0.9783	0.9786	0.9788	0.9790	0.9792	0.9794	0.9796
42	0.9779	0.9782	0.9784	0.9786	0.9788	0.9790	0.9792
42.5	0.9775	0.9777	0.9780	0.9782	0.9785	0.9787	0.9789
43	0.9771	0.9773	0.9776	0.9778	0.9781	0.9783	0.9785
43.5	0.9767	0.9769	0.9772	0.9774	0.9777	0.9779	0.9781
44	0.9763	0.9765	0.9768	0.9770	0.9773	0.9775	0.9777
44.5	0.9759	0.9761	0.9764	0.9767	0.9769	0.9771	0.9773

TABLE 7.2 (CONTINUED)
Volume Reduction Factors

Observed Temperature, °C	0.840	0.845	0.850	0.855	0.860	0.865	0.870
	Density at 15°C — Factor for Reducing Volume to 15°C						
45	0.9755	0.9757	0.9760	0.9763	0.9765	0.9767	0.9769
45.5	0.9750	0.9753	0.9756	0.9759	0.9761	0.9763	0.9766
46	0.9746	0.9749	0.9752	0.9755	0.9757	0.9760	0.9762
46.5	0.9742	0.9745	0.9748	0.9751	0.9753	0.9756	0.9758
47	0.9738	0.9741	0.9744	0.9747	0.9750	0.9752	0.9754
47.5	0.9734	0.9737	0.9740	0.9743	0.9746	0.9748	0.9750
48	0.9730	0.9733	0.9736	0.9739	0.9742	0.9744	0.9747
48.5	0.9726	0.9729	0.9732	0.9735	0.9738	0.9740	0.9743
49	0.9722	0.9725	0.9728	0.9731	0.9734	0.9736	0.9739
49.5	0.9718	0.9721	0.9724	0.9727	0.9730	0.9733	0.9735
50	0.9714	0.9717	0.9720	0.9723	0.9726	0.9729	0.9731

Thus, the volume reduction factor corresponding to a density of 0.8581 g/cc at 15°C is calculated by interpolation as

$$0.9842 + \{(0.9843 - 0.9842)/(0.860 - 0.855)\} \times (0.8581 - 0.8550) = 0.984262.$$

Similarly at 40°C, the volume correction factor corresponding to a density of 0.8504 g/cc at 15°C, is by interpolation,

$$0.9800 + \{(0.9802 - 0.9800)/(0.855 - 0.850)\} \times (0.8504 - 0.8500) = 0.980016.$$

The following table shows the calculations for the quantity of crude oil received in the tank:

Product	Gross oil cut, cm	Water cut, cm	Net oil cut, cm	Temperature, °C	Density of samples, g/cc	Volume from calibration chart, m³	Corrected density at 15°C from the ASTM table, g/cc	Volume at 15°C using reduction factor from ASTM table	Metric tons of stocks
Crude	600	1	599	35	0.8450	11,980	0.8581	0.984262	10,118.251
Crude	800	2	798	40	0.8430	15,960	0.8504	0.980016	13,301.153

Hence, the crude oil received is 3,182.9022 t.

Had there been no correction for volume and density, the received quantity of crude could be 15,960x0.8435 – 11,980x0.8450 = 3,339.16 t. Hence, the quantity of loss to the receiver could be could be 3,339.16 – 3,182.9022 = 156.2578 t.

If the price of crude is Rs 15,000 per ton, the loss would be Rs 2,343,867 i.e., around Rs 23.5 lakhs.

7.2 MEASUREMENT OF GASES IN CLOSED VESSELS

Gases that are easily liquified by pressure at room temperature, such as liquified petroleum gas (LPG), propane, and butane, are stored in cylindrical or spherical pressure vessels. The volume that can be stored in a spherical vessel (Horton sphere) is more than the volume stored in a cylindrical vessel of the same dimensions. The volumes of a sphere and a cylinder are $4/3\ \Pi\ R^3$ and $\Pi\ R^3$ (where R = H of the cylinder), respectively, i.e., the volume of a sphere is 4/3 times that of a cylinder. Another advantage of using the Horton sphere is the larger surface area ($4\ \Pi\ R^2$) for cooling compared with the cylindrical vessel of the same dimension ($2\ \Pi\ R^2$); this is essential for keeping a large amount of liquified gas cooled. The quantity of liquified gases is assessed by the level of the liquid and the volume, which is obtained from the calibration chart of the vessel. In fact, the level is either seen through a gauge glass or by a differential pressure gauge. Since gases are stocked and transacted in mass, volume and density must be used subject to temperature corrections. Gases that are not liquified at room temperature, such as hydrogen, nitrogen, and helium, are quantified by an equation of state based on the measured temperature and pressure in the vessel. Of course, the entire internal volume of the vessel must be obtained from the calibration chart.

7.3 MATERIAL BALANCE IN A PLANT

Usually, material and energy balances are carried out at different frequencies, e.g., daily, weekly, monthly, and yearly. In refineries, daily material and energy balances are prepared for the previous 24 h (e.g., from 07:00 a.m. of the previous day to 07:00 a.m. of the next day). Various records, such as tank dips of crude oil, liquid and liquified products, meter readings of fuel gas, and electricity and steam consumption, are taken for making such daily balances. Monthly balances, usually at the end of the month, are prepared by similar methods and reconciliation is carried out with the cumulative daily balances. Annual material and energy balances are similarly prepared reconciled, and audited by authorised auditors for the evaluation of annual profit and loss statement. This is also required for preparing *profit and loss* and the *balance sheet* of the company. There are some statutory obligations for compulsory material and energy balances, whose guidelines are given by the government according to the Companies Act. These are really non-technical topics, but a plant or unit's material and energy balances are the responsibility of the engineers who prepare these balances based on scientific and technical rules. Let us discuss how these balances are prepared.

Example 7.2

Consider a refinery processing crude oil to produce LPG, motor spirit (MS; naphtha inclusive), superior kerosene oil (SKO), high speed diesel (HSD), and atmospheric bottom reduced crude oil (RCO) from the crude distillation unit. Two crude oil tanks, 01 and 02, have been used for feeding the unit during the 24-h operation from 07:00 a.m. the previous day until 07:00 a.m. of the next day. Tank dips are given in the following tables where products were received in the nine tanks (run down tanks) numbers 04–09.

Opening dips on 1.7.05 at 07:00 a.m.

Tank No.	Product	Gross dip, cm	Water cut, cm	Temperature, °C
01	Crude	800	1.0	30
02	Crude	700	1.0	30
03	LPG	64	0.0	20
04	LPG	85	0.0	20
05	MS	701	0.0	25
06	SKO	630	0.0	30
07	HSD	200	0.0	30
08	RCO	300	0.0	100
09	RCO	400	0.0	120

Closing dips on 2.7.05 at 07:00 a.m. Closing dips on 2.7.05 at 07:00 a.m.

Tank No.	Product	Gross dip, cm	Water cut, cm	Temperature, °C
01	Crude	525	1.0	28
02	Crude	677.5	1.0	27
03	LPG	216	0.0	26
04	LPG	200	0.0	22
05	MS	800	0.0	25
06	SKO	900	0.0	30
07	HSD	950	0.0	32
08	RCO	900	0.0	100
09	RCO	664	0.0	120

Present a material balance of the crude distillation unit.

Solution

After consulting the calibration charts of the respective tanks, the ASTM density and volume corrections for the following amounts of stock in the tanks are evaluated.

Opening stocks on 1.7.05 at 07:00 a.m.

Tank No.	Product	Volume at 15°C, kl	Density at 15°C	t
01	Crude	30,737.000	0.8546	26,268.00
02	Crude	26,892.051	0.8532	22,944.299
03	LPG	38.40	0.5500	21.20
04	LPG	51.00	0.5500	28.05
05	MS	5,608.00	0.6800	3,813.44
06	SKO	4,895.10	0.7900	3,867.12
07	HSD	1,554.00	0.8430	1,310.022
08	RCO	1,533.00	0.9000	1,379.70
09	RCO	2,044.00	0.9000	1,839.60
			Total opening stocks	61,471.431

Closing stocks on 2.7.05 at 07:00 a.m.

Tank No.	Product	Volume at 15°C (kl)	Density at 15°C	t
01	Crude	21,000	0.8546	17,946.60
02	Crude	27,100	0.8532	23,121.72
03	LPG	130.00	0.5500	71.50
04	LPG	120.00	0.5500	66.00
05	MS	6,400.00	0.6800	4,352.00
06	SKO	7,200.00	0.7900	5,688.00
07	HSD	3,600.00	0.8430	3,034.80
08	RCO	4,500.00	0.9000	4,050.00
09	RCO	3,320.00	0.9000	2,988.00
			Total closing stocks	61,318.62

Hence, loss in materials = 61,471.431 − 61,318.62 = 152.811 t.

The material balance indicating each product are then obtained for a 24-h operation is presented as

Product	Opening t	Closing t	t/day
Crude throughput	49,212.299	41,068.320	8,143.9789
Products rates:			
LPG	49.25	137.50	88.25
MS	3,813.44	4,352.00	538.56
SKO	3,867.12	5,688.00	1,820.88
HSD	1,310.022	3,034.80	1,724.778
RCO	3,219.30	7,038.00	3,818.70
Off gas/fuel gas/loss		by balance	152.811
Total			8,143.9789

This example shows how stock tables are produced and overall material balance is carried out for losses/off gases. Crude processed and product withdrawal breakup (also known as a pumping sheet) calculated above is also essential for assessing productivity and production targets. Separate material balances as above for each plant must be similarly prepared. Utility material balances, such as water balance, chemical consumption, and steam consumption, are also accounted for proper utilisation of the utilities.

7.3.1 Flow Meter Readings

In a plant like a refinery, the units run continuously for 330 days/year and are shut down once a year for about a month for necessary cleaning and maintenance of the pipes, equipment, instruments, etc. During the operation, the flow rates of the liquids and gases are continuously monitored and recorded for the necessary material and energy balance required for assessing the quality, quantity, and health of the plant and the equipment. The gas flow rates and pressure are measured and controlled through pressure recorder controllers (PRC) and the liquid flow rate is measured in volumetric flow rate (FRC) at a constant temperature of 15.5°C. Since the flow rates may fluctuate widely during the operation, the total flow of the fluids for a particular period cannot be easily calculated and must be integrated over the period. Continuous cumulative flows, usually in mass flow, are recorded by integrators, which indicate the total mass of the fluids that have passed through over a certain period, i.e., a day, a month, or a year.

7.3.2 Fuel Consumption

Hydrocarbon gases or liquids or both are consumed as furnace fuels to supply the necessary heat for unit operations and reactions. Flow recorders and integrators are present to monitor and control the fuel consumption in the process furnaces and power plant boilers.

7.3.3 Steam Consumption

Steam in various forms, such as low pressure (LP), medium pressure (MP), or high pressure (HP), is used as process steam or heating steam or power steam for steam-driven pumps and compressors. Steam flow rates and integrator readings are essential to assess the steam raised, consumed, and returned as condensate. Loss of steam can be calculated by a steam and condensate balance.

7.3.4 Overall Material Balance

It is essential to make a material balance of the entire refinery as well as individual material balances of all the units in the plant, periodically. The overall material balance of the entire refinery is a balance of the total crude processed and the products formed, while the difference indicates fuel and loss for any period. Similarly, fuel and loss for each unit is also evaluated by the overall material balance of that unit. The following example explains how overall material balance is carried out for any unit.

Example 7.3

Present a material balance of a naphtha reforming plant as shown in Figure 7.2 (confirmed ok). From the tank dip data of the naphtha feed and rundown reformate tanks during a test run of a reformer, the flow rates of desulfurised naphtha and debutanised reformate were reported to be 23,528.50 and 20,613.60 kg/h, respectively. Flows of hydrogen rich gas to storage (also for consumption in desulfurising units) and off gases from the debutaniser top were metered (integrator) as 1,684.934 and 166.90 kg/h, respectively.

Solution

With reference to the envelope of input/output streams shown in Figure 7.2, material balance is obtained as follows.

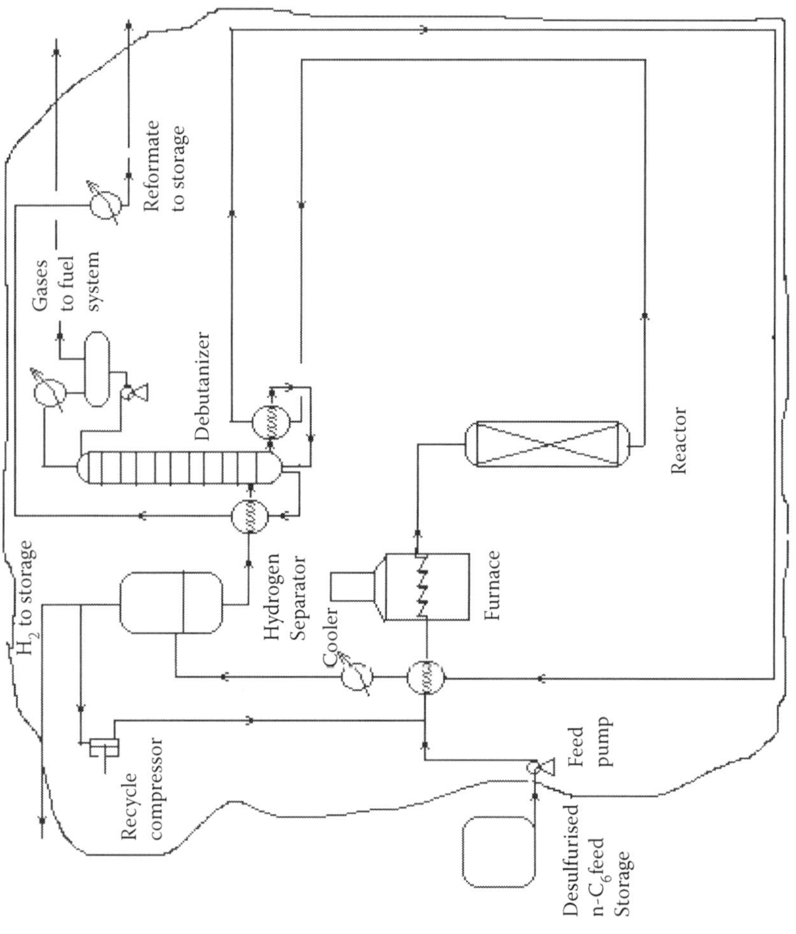

FIGURE 7.2 Input and output streams for material balance in a reforming plant for Example 7.3.

	kg/h
Input streams	
Feed naphtha	23528.50
Total	23528.50
Output streams	
Debutanised reformate	20613.60
Hydrogen to storage	1684.934
Off gas from debutaniser	166.90
Loss (by balance)	1063.066
Total	23528.50

Example 7.4

For the reformer in the previous example, determine the flow rate of hydrogen generated with a flow rate of recycle gas of 11,682 kg/h. A laboratory analysis of gases is given as follows.

Property of gaseous streams:

	Recycle gas	Off gas from debutaniser
Relative density	0.28	0.96
Hydrogen (% vol.)	82.4	12.04

Solution

The relative density is measured with respect to the density of air at the same temperature and pressure as the measurement, i.e., it is the same ratio of molecular weights of the sample to air. Considering the standard average density of air as 28.88, the molecular weights of the gases are determined as

	Recycle gas	Hydrogen rich gas to storage	Off gas from debutaniser
Molecular weight (RD × 28.88)	8.086	8.086	27.72
Kmol/h	1,444.72	208.38	6.021
Hydrogen, kmol/h	1,190.45	171.71	0.725

Total hydrogen generated is 171.71 + 0.725 = 172.44 kmol/h = 344.87 kg/h.

Example 7.5

Referring to the previous problem, determine the ratio of hydrogen per mole of feed that is important to suppress coke formation during reforming. The desired ratio is >5.0 mol/mol. Given that the molecular weight of desulfurised naphtha is about 111.

Solution

Hydrogen in the recycle gas is 1,190.45 kmol/h and the feed rate is 23,528.50/111 or 211.968 kmol/h. Hence, hydrogen to feed ratio is 1,190.45/211.968 or 5.616 mol/mol.

7.4 ENERGY BALANCE IN A PLANT

Energy in the form of heat and electricity is consumed in a plant. Separate balances must be made to monitor the energy consumption in each plant and its equipment. The sources of heat and power in a plant are fuel and electricity (generated or borrowed from outer sources). Because refineries and petrochemical plants require huge amounts of power, captive power plants are run using petroleum fuels. Thus, a large amount of fuel is burnt in both the process furnaces meant for each process plant and in the thermal power plant, which meets the plants' demand for both steam and electricity. The total amount of fuel consumed, utilisation of heat, steam consumption, and electricity consumption for the whole plant per day, per month, and for the year must be assessed.

7.4.1 Heat Balance

Heat is required to raise the temperature of feedstock to the operating temperature of various processes, such as distillation, stripping, extraction, and conversion. A variety of heat transfer equipment, such as heat exchangers, furnaces, and reboilers, are employed in the plant at different temperature ranges. In the heat exchangers, hot and cold fluids are brought over a common metallic surface to mutually exchange heat between them but without any physical contact or mixing. These types of heat exchangers are called recuperators. Although the majority of heat exchangers fall into this category, there are direct contact heat exchangers, too, where hot fluid may come in contact with cold fluid and physically mix up. Most common recuperators are double pipe and shell-and-tube heat exchangers. All these exchangers are discussed in Chapter 8.

7.4.2 Energy Balance in a Heat Exchanger

The amount of heat energy transferred from hot to cold fluid and the heat gained by the cold fluid from the hot fluid must be balanced according to the law of conservation of energy. This law is true when an adiabatic condition exists, that is, no heat is lost to the surroundings from the external surface of heat exchanger. In fact, heat is lost through the insulation to the surroundings and, as a result heat gained by cold fluid is always less than total heat given up the hot fluid. In the case of hot fluid flow in the shell side, heat is lost from the shell to the surroundings. When, cold fluid passes through the shell side, heat is lost from the cold fluid to the surroundings.

Example 7.6

Make a heat balance in a heat exchanger for preheating crude oil at a rate of 312.5 tons/h from 20°C to 51°C by kerosene at a rate of 92.2 tons/h from 193°C to 85°C. Take the countercurrent heat exchange for calculation. Given that the density of crude oil and kerosene is 0.6535 and 0.791 g/cc, respectively, at their corresponding average inlet and outlet temperatures. The specific heat of crude and kerosene at their average temperatures is 0.52 and 0.53 cal/g °C, respectively.

Solution

Heat lost by kerosene is mass flow rate × specific heat × temperature difference between the inlet and outlet, which is $92.2 \times 0.53 \times (193 - 85) = 5.2775 \times 10^9$ cal/h.

Heat gained by crude oil is mass flow rate × specific heat × temperature difference between the outlet and inlet, which is $312.5 \times 0.52 \times (51 - 20) = 5.0375 \times 10^9$ cal/h.

Hence, the heat loss from the exchanger to ambient temperature is $5.2775 \times 10^9 - 5.0375 \times 10^9 = 2.40 \times 10^9$ cal/h.

Efficiency of heat transfer is $5.0375 \times 10^9/5.2775 \times 10^9 = 95.45\%$.

From such a thermodynamic heat balance, the loss of available heat can be measured and action may be taken. For example, in this case, insulation of the outer wall should be rectified to avoid loss of heat to the surroundings. It is to be remembered that the above heat balance is applicable when there is no phase changes in either of the fluids, which are, in fact, liquids in this example.

7.4.3 Energy Balance in a Furnace

Tube- or pipe-still furnaces are commonly employed in petroleum refineries and petrochemical plants. The elements of construction of such furnaces are described in detail in Chapter 8. Gaseous fuel or liquid fuel or both are fired in the furnace through appropriate burners. Heat generated by the combustion of these fuels is transferred to hydrocarbon vapour or liquid feed through the pipes. The rate of heat transfer is governed by radiation and convection. The total amount of heat transferred to the hydrocarbon feed is much lower than the actual amount of heat generated by combustion of the fuel due to heat losses through the stack exhaust, wall loss, and fouling of the inner and outer surfaces of the pipes. These losses can be accounted for by the heat balance of the furnace. This is explained by the following example.

Example 7.7

In a furnace crude oil with a API of 36 is heated from its feed temperature of 260°C–370°C at a rate of 400 tons/h. If the rate of fuel consumption is 3.56 tons/h, present a heat balance of the furnace. Given that the stack gas temperature is 140°C and the ambient air temperature surrounding the furnace is 30°C. The net heating value of fuel is 18,362 kcal/kg and the air to fuel ratio is 20 kg/kg (corresponds to 30% excess air). The specific heat of flue gas is taken as 0.25 kcal/kg

°C. Given that the enthalpies of crude oil at the inlet and exit condition are 126 and 245 kcal/kg, respectively.

Solution

Rate of heat released is fuel firing rate × net heating value = 3.56 × 1,000 × 18,362 = 6.5×10^7 kcal/h.

Heat absorbed by crude oil is throughput × enthalpy difference = 400 × (245 – 126) = 4.76×10^7 kcal/h, which accounts for 73.2% of the heat released by the fuel.

Flue gas generated is (fuel + air) rates = (1 + 20) × fuel rate = 21 × 3,560 = 74,760 kg/h.

Heat lost with the flue gas (stack loss) is the rate of flue gas × cp_{flue} × (exit temperature – 15°C as the base) = 74,760 × 0.25 × (140 – 15) = 2.336×10^6 kcal/h.

Heat lost through outer wall (by difference) is $6.5 \times 10^7 - 4.76 \times 10^7 - 2.336 \times 10^6 = 1.504 \times 10^7$ kcal/h.

Total heat loss through the stack and wall is $2.336 \times 10^6 + 1.504 \times 10^7 = 1.737 \times 10^7$ kcal/h, which accounts for 26% of the heat released by fuel.

Therefore, the efficiency of the furnace accounts for only 74%.

7.4.4 Energy Balance in a Distillation Column

Example 7.8

In a crude distillation column, as shown in Figure 7.3, with the given data, present the heat balance of the unit and find the loss of heat, if any.

For example, data for the crude distillation column

Streams	Temperature, °C	kg/cm²	Tons/h	Density, g/cc at 15°C
Crude	370.0	10.0	315.0	0.6565
Off gases	44		0.820	–
Top draw	44	1.4	50.48	0.56
Top CR	192/95		143.0	0.56
Kero draw	203		72.20	0.79
Kero CR	223/163		271	0.79
HSD draw	288		39.40	0.845
HSD CR	307/247		195	0.845
JBO draw	340		13.60	0.875
RCO draw	344		138.50	0.90
STB, bottom	350	15	2.275	
ST1,kero	203	15	0.20	
ST2,HSD	280	15	0.30	
ST3,JBO	340	15	0.34	

The following table shows the heat balance computations considering the envelope containing the input and output streams as shown in Figure 7.4.

Streams	Tons/h	Temperature, °C	Enthalpy (kcal/kg)	kcal/h
Inputs				
Crude	315.0	370	460	1.449×10^8
STB	2.275	350	786.6	1.789×10^6
ST1,kero	0.20	203	698.4	1.396×10^5
ST2,HSD	0.30	280	744.6	2.2338×10^5
ST3,JBO	0.34	340	780.6	2.654×10^5
Total	318.115			1.473173×10^8
Outputs				
Off gases	0.820	44	33.63	2.757×10^4
Top draw	50.48	44	50	2.524×10^6
Top steam condensate draw	3.115	44	44	1.3706×10^5
Kero draw	72.20	203	133.2	9.617×10^6
HSD draw	39.40	288	183.1	7.216×10^6
JBO draw	13.60	340	216.4	2.943×10^6
RCO draw	138.50	344	206.2	2.8558×10^7
Total	318.115			
Top CR, Q1	143.0	192/95	(220.6 – 109)	1.5958×10^7
Kero CR, Q2	271	223/163	(129.34 – 94.5)	9.441×10^6
HSD CR, Q3	195	307/247	(190.3 – 153.1)	7.254×10^6
Top Qc		127/44		$*1.01 \times 10^7$
Loss (by diff)				5.3540750×10^7
Total				1.473173×10^8

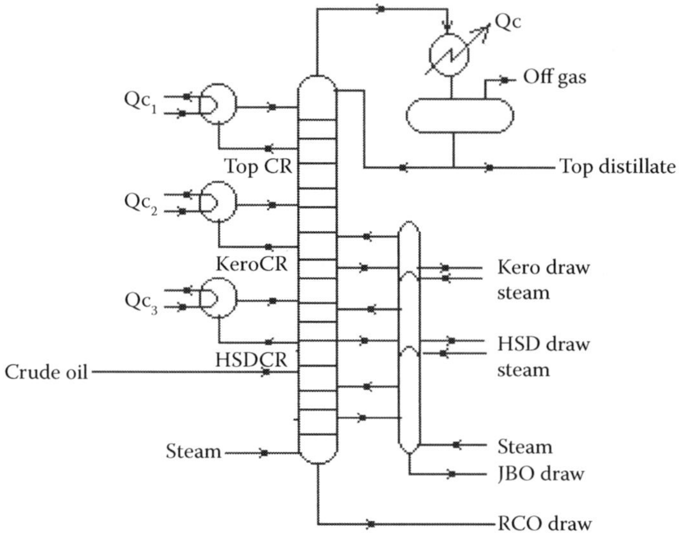

FIGURE 7.3 Streams in a crude distillation unit with a side stripper.

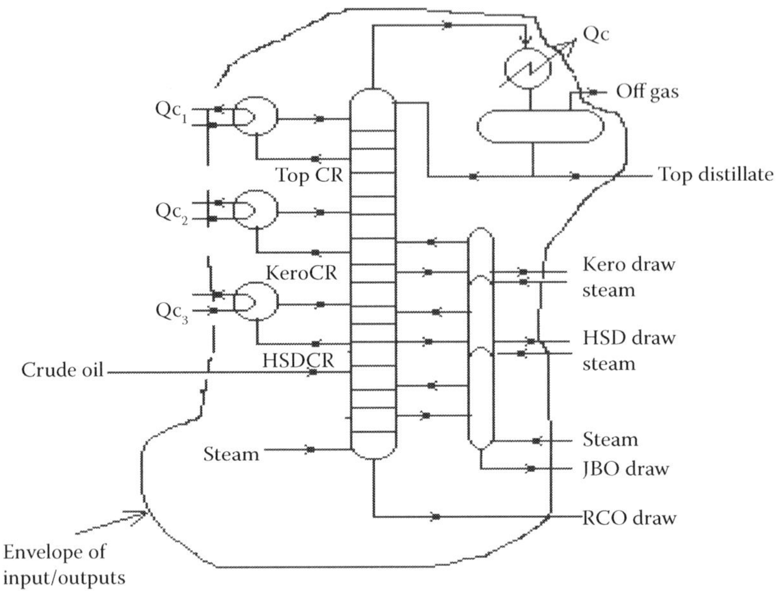

FIGURE 7.4 Input and output streams in the CDU unit.

*(where overhead reflux is 28.5 t/h and Qc is the heat removed by the top condenser as evaluated below:

For condensation of hydrocarbon vapours, the heat removed is $(50.48 + 28.5) \times (154 - 50) = 8.247 \times 10^6$ and for condensing steam, the heat removed is $(2.275 + 0.20 + 0.30 + 0.34) \times (652.72 - 44) = 1.896 \times 10^6$. Total: 1.01×10^7 kcal/h}.

Loss of heat, 5.354×10^7 kcal/h, encountered by the heat balance, as shown in the above table, is thought to be the heat loss from the column walls to the surroundings. In fact, Qc1, Qc2, and Qc3 heat are removed from circulating refluxes in a train of heat exchangers and some in the reboilers of other columns.

7.4.5 Overall Energy Balance

Example 7.9

With reference to the reforming plant, as shown in Figure 7.2, present a heat balance of the plant. The temperatures of the stream are indicated on the flow diagram as shown in Figure 7.5.

Flow rates of streams and temperatures are given as

Streams	Flow rate, kg/h	Temperature, °C
Feed naphtha	23528.50	93
R/d reformate	20613.60	42
H_2 rich gas to:		
Consumption	1684.984	40
Off gas	166.90	43

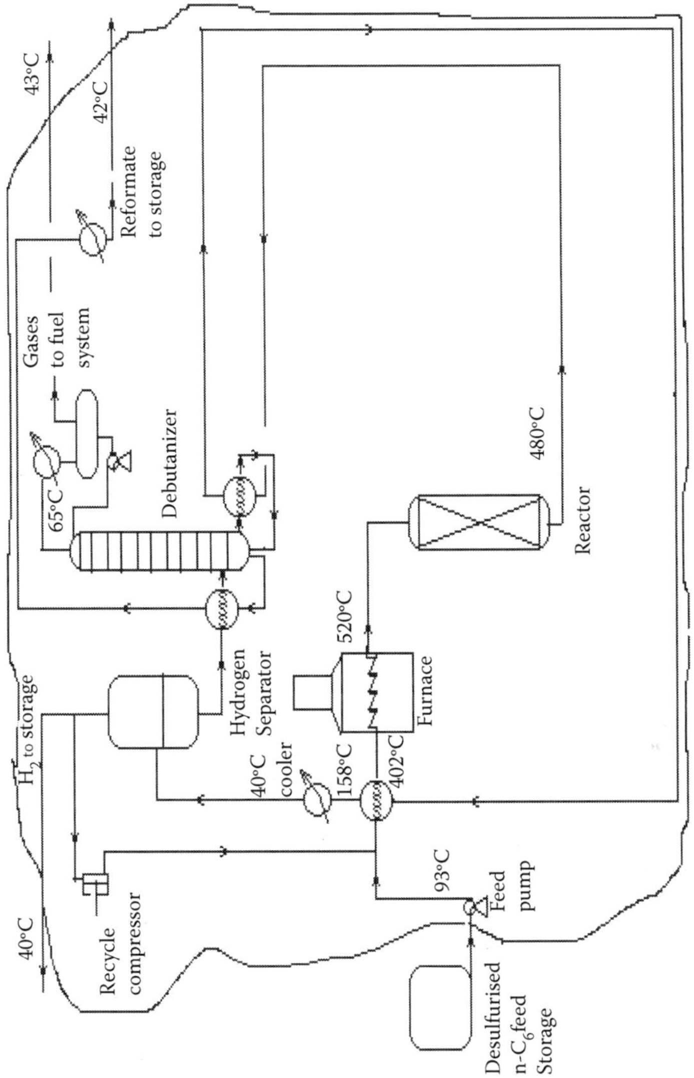

FIGURE 7.5 Input and output streams for heat balance in a reformer.

The temperatures of all the streams are mentioned in Figure 7.5.

Solution

	Flow, kg/h	Temperature, °C in/out	*Enthalpy kcal/kg	kcal/h
Input streams				
Feed naphtha	23528.50	43	44.97	1.058×10^6
Heater duty	23528.50	402/520	265/378.9	2.68058×10^6
Total				3.738×10^6
Output streams				
H_2 rich gas to consumption	1684.984	40	34.68	5.843×10^4
Off gas	166.90	43	33.40	5.574×10^3
Reformate	20613.60	42	21.00	4.328×10^5
Product cooler before separator drum	23528.50	158/40	78.9/20.6	1.3717×10^6
Reformate cooler	20613.60	108/42	52.3/21.54	6.34×10^5
Debutaniser condenser	166.90	65/43	36/23	2.019×10^3
Heat absorbed in the reactor	23528.50	520/480	352/314	8.94×10^5
Loss				3.398×10^5
Total				3.738×10^6

* Enthalpy data are obtained from the Enthalpy Chart.

Hence, heat loss is around 9.09% of the total input, which is mainly due to losses to the surroundings excluding furnace losses.

QUESTIONS

1. Why are material and energy balances essential for a plant?
2. What information must be collected for material balance in a crude distillation unit?
3. How will you carry out a survey in a distillation column?
4. What are the various reasons for material losses?
5. Heat loss cannot be avoided 100%, explain?

8 Heat Exchangers and Pipe-Still Furnaces

8.1 HEAT EXCHANGERS

Heat exchanger equipment exchanges heat between a hot and a cold fluid, i.e., cold fluid is heated by hot fluid without using any fire or combustion. The most widely used heat exchangers transfer heat from hot to cold fluid separated by a metallic surface. These types of heat exchangers are called recuperators, where the fluids do not mix physically, but only the heat flows through the metallic wall surface. Double pipe and shell-and-tube exchangers are common recuperators. In a double pipe exchanger, one of the fluids passes through an inner tube and the other fluid passes through the annular space between the outer and inner tubes. A shell-and-tube heat exchanger consists of a bundle of tubes inside a large diameter outer tube or shell. One of the fluids passes through the inner tubes in the bundle and the other fluid passes through the annular space between the inner tubes and the shell. In both exchangers, fluids may flow either in the opposite (countercurrent) or parallel direction (co-current). In fact, in the shell-and-tube exchanger, flow paths are not exactly counter or co-current, rather they are crosscurrent, while the shell side fluid flows partially at right angles to the tube side fluid. The path equal to the length of the shell travelled by the tube side fluid is called a tube side pass. If the fluid enters at one end of the shell and leaves from the other end, then the number of tube passes is counted as one. If the other end of the tube is returned as a u-tube, the number of tube side passes will be two. Fluid in the shell side travelling a path equal to the length of the shell is similarly counted as a shell side pass.

Usually, circular plates, known as baffles, occupy the annular space between the tubes and the shell, inducing turbulence to enhance heat transfer. Flow in the shell side takes place through an opening of the baffles provided in a sequentially up and down position such that fluid flows at right angles to the tubes. If there is no horizontal baffle in the shell, fluid will pass from one end to the other end, crossing the tube side, thus there will be one shell pass. If there is a single longitudinal baffle (i.e., parallel to the tubes) placed exactly in the middle of the shell, two shell side passes will be generated. Both the double pipe and shell-and-tube exchangers with different tube and shell side passes are explained in Figures 8.1 and 8.2. Industrial heat exchangers may have more than two passes (either or both in the tube and shell sides) to enhance the high rate of heat transfer.

8.2 THEORY OF HEAT EXCHANGE

Heat flows from high temperature to low temperature according to the principle of thermodynamics. According to the law of conservation of energy, heat lost from the

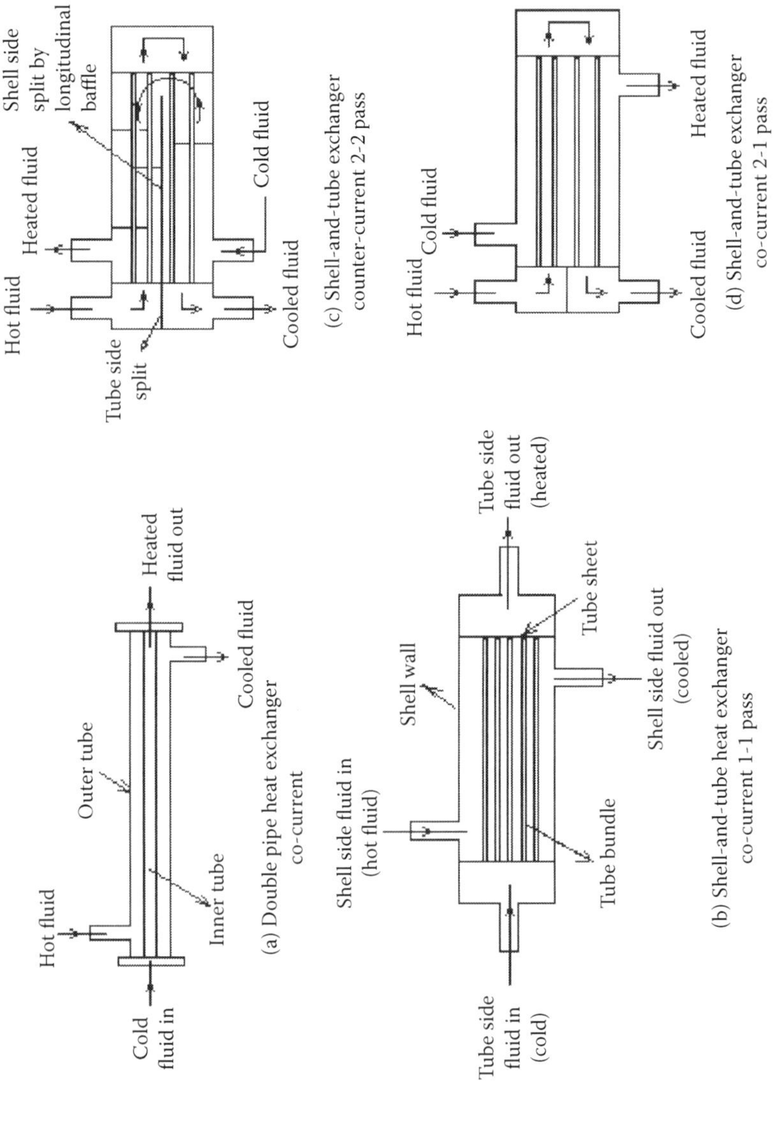

FIGURE 8.1 Typical heat exchangers.

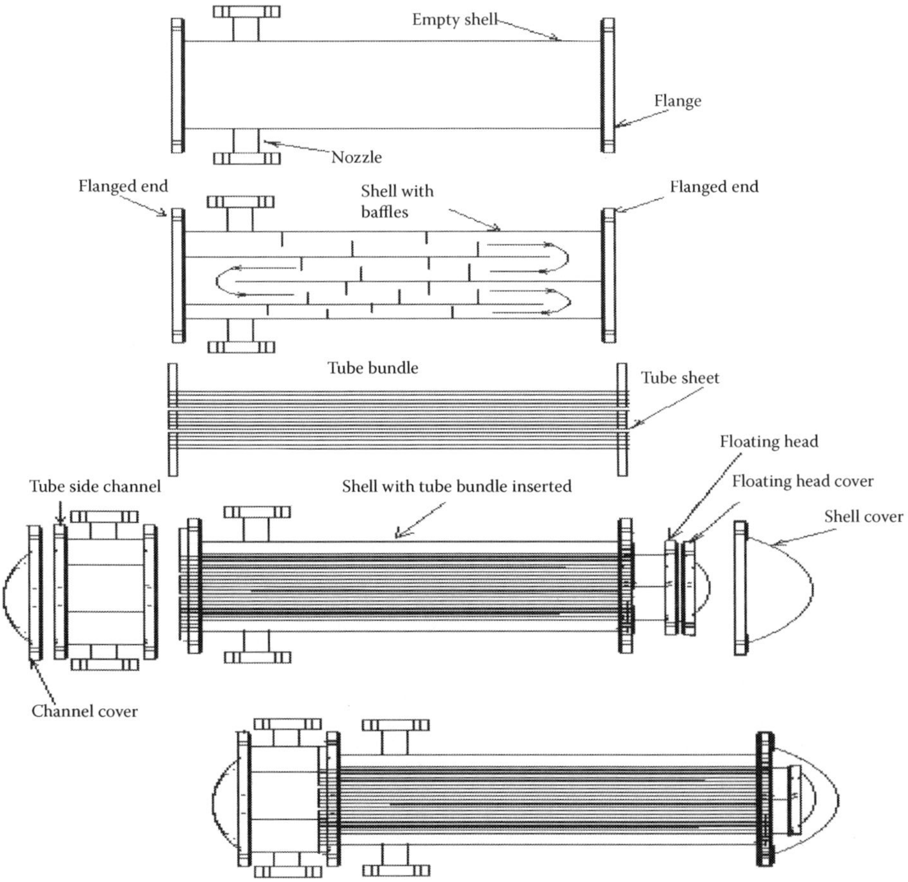

FIGURE 8.2 Sectional views of elements of a shell and a tube heat exchanger.

hot fluid will be gained by the cold fluid, has already demonstrated in Example 7.6 used in the heat balance examples in Chapter 7.

8.2.1 Heat Balance

If the inlet temperatures of the cold and hot fluids are T_{c1} and T_{h1} and the exit temperatures are T_{c2} and T_{h2}, respectively, then the heat balance equation is

$$W_h Cp_h(T_{h1} - T_{h2}) = W_c Cp_c(T_{c2} - T_{c1}). \tag{8.1}$$

Heat flows from the hot fluid through a film of hot fluid, the metallic thickness of the tube wall, and then through a film of cold fluid. Thus, heat flow takes place through a series of three resistances, where heat flows by convection through the fluid films, but by conduction through the metal wall thickness. The overall heat transfer coefficient is designated as U, which takes into account the combined effects of all three resistances and is given by the following relation,

$$1/U_iA_i = 1/U_oA_o = 1/h_iA_i + k_mx_m/A_m + 1/h_oA_o, \tag{8.2}$$

where

U_i: overall heat transfer coefficient based on inside surface area
U_o: overall heat transfer coefficient based on outer surface area
A_i: inside surface area of the tube
A_o: outer surface area of the tube
K_m: thermal conductivity of the metal of the tube
x_m: thickness of the metal of the tube
A_m: mean area of cross section of the metal of the tube
W_h: mass flow rate of heating fluid
W_c: mass flow rate of cooling fluid
h_i: convective heat transfer coefficient of inside fluid film
h_o: convective heat transfer coefficient of outside fluid film

8.2.2 Rate of Heat Transfer

The rate of heat transfer is dependent on the value of U, which is also a function of the film heat transfer coefficients of the fluids (h_i,h_o) and the thermal conductivity of the metal (K_m) of the tube. The greater the value of U, the faster the heat transfer, the greater the fall in hot fluid at the exit, and the greater the exit temperature of the cold fluid. For specific parameters for design, e.g., the tube dimensions and the areas of a heat exchanger, U can be varied by changing the velocity of either or both of the fluids as the film heat transfer coefficient of a fluid is a function of the velocity and properties of the fluid.

For low velocity, while the Reynolds number (NRe) is less than or equal to 2100, the following relation may be used to determine h_i,

$$Nu = 1.86\{(N\text{Re})(N\text{Pr})(D/L)\}^{1/3}(\mu/\mu_w)^{0.14}. \tag{8.3}$$

For higher velocity when the flow is turbulent and NRe is equal to or greater than 10,000, the following relation is applicable for determining h_i,

$$Nu = 0.027(N\text{Re})^{0.8}(N\text{Pr})^{1/3}(\mu/\mu_w)^{0.14}, \tag{8.4}$$

where

Nu: Nusselt number, h_i D/k,
NRe: Reynolds number, DV ρ/μ
NPr: Prandtl number, Cp μ/k
Cp: specific heat of fluid
μ: viscosity of fluid
k: thermal conductivity of fluid
D: inside diameter of the tube
L: length of the tube
μ: viscosity of fluid at the average fluid temperature
μw: viscosity of the fluid at the inside average wall temperature

Though no correlation is accurately applicable for determining h_i for NRe between 2,100 and below 10,000, the range is known as the transition region while the stream line flow becomes turbulent, a generalised chart for all the NRe values is available as shown in Figure 8.3 (J_H vs. NRe).

Where J_H is the Colburn analogy factor

$$J_H = (h/CpG)(Cp\mu/k)^{1/3}. \tag{8.5}$$

For determining h_o, the above equations are applied with the modification of D as the equivalent diameter. For the heat transfer area, the equivalent diameter of the annular space between the shell (or the outer tube) and the inner tube is calculated as four times the hydraulic radius (r_H).

Where

$$r_H = \pi(D_2^2 - D_1^2)/(4\pi D_1), \tag{8.6}$$

and

$$D = 4r_H = (D_2^2 - D_1^2)/D_1, \tag{8.7}$$

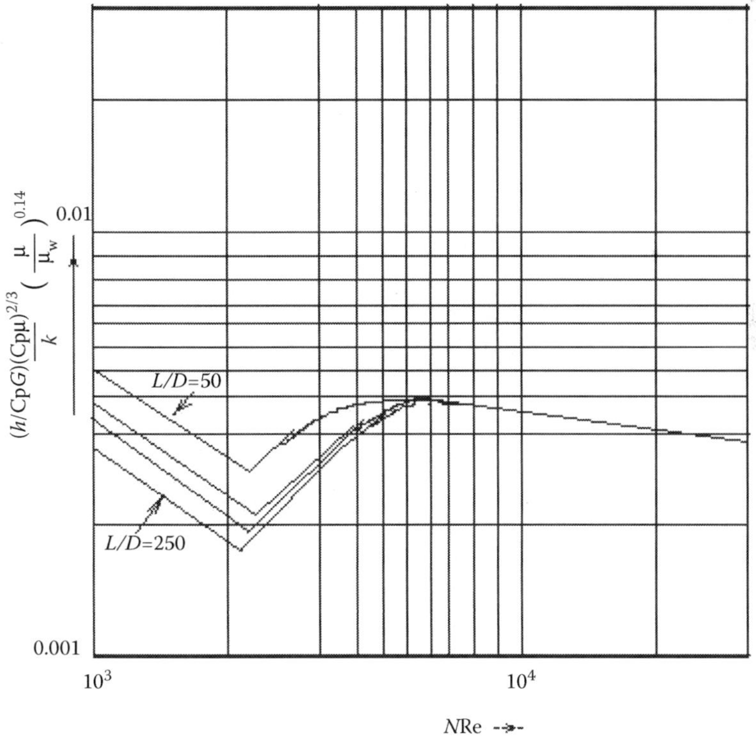

FIGURE 8.3 Colburn factor j_H as a function NRe.

where D_1 and D_2 are the outer diameter of the inner tube and the inner diameter of the shell (or outer tube), respectively. It is to be noted that the pressure drop calculation in the annular space will use a different hydraulic radius as the ratio of the wetted cross section to the wetted perimeter. The wetted cross section will include both the outer surface of the inner tube and the inner surface of the shell, as friction is exerted by both the walls of the inner tube and the shell. This is applicable for evaluating the wetted perimeter as well. Thus, the equivalent diameter for a pressure drop calculation will be

$$(D_2^2 - D_1^2)/(D_1 + D_2) = D_1 - D_2. \tag{8.8}$$

LMTD is the rate of heat exchange (Q) calculated by the following relation,

$$Q = U\,A\,\Delta T. \tag{8.9}$$

Thus,

$$Q = U\,A\,\text{LMTD}, \tag{8.10}$$

where

$$\text{LMTD} = (\Delta T_1 - \Delta T_2)/\text{In}(\Delta T_1 / \Delta T_2), \tag{8.11}$$

and, for countercurrent operation,

$$\Delta T_1 = (T_{h1} - T_{c2}), \tag{8.12}$$

$$\Delta T_2 = (T_{h2} - T_{c1}), \tag{8.13}$$

and for co-current operation,

$$\Delta T_1 = (T_{h1} - T_{c1}), \tag{8.14}$$

$$\Delta T_2 = (T_{h2} - T_{c2}). \tag{8.15}$$

Properties, such as density, specific heat, viscosity, etc., for each fluid are calculated at the average of their respective inlet and outlet temperatures. Exact determination of Q by considering the variation of temperatures along the tube length and the corresponding variations in properties may require a more rigorous method involving computer simulation, which is out of the scope of this book.

Example 8.1

Crude oil at a temperature of 135°C and having an API of 35 is to be preheated at a rate of 20 ton/h with a kerosene circulating reflux available from a crude distillation unit at a temperature of 200°C and a rate of 2904.0 kg/h. If the exit temperatures of kerosene and crude oil are 150°C and 145°C, respectively, determine the overall heat transfer coefficient of the heat exchanger with a total outer tube surface of 13.5 m^2. Assume countercurrent operation. Given that the specific heat of crude oil and kerosene is at average temperatures of 0.53 and 0.79 kcal/kg °C, respectively.

Solution

$$\Delta T1 = 150 - 135 = 15{:}\ \Delta T2 = 200 - 145 = 55 : \text{LMTD}$$
$$- \Delta T1 - \Delta T2/\ln(\Delta T1/\Delta T2) = (55 - 15) / \ln (55 / 15) = 30.78°\text{C}.$$

So, the rate of heat transfer is

$$Q = U\,A\ \text{LMTD} = 20{,}000 \times 0.53 \times (145 - 135) = 106{,}000\ \text{kcal/h m}^2.$$

Hence,

$$U = Q/(A \times \text{LMTD}) = 106{,}000/(13.5 \times 30.78) = 255.1\ \text{kcal/h m}^2\ °\text{C}.$$

It is to be noted that the area given is not specified as inner or outer, hence we assume these to be nearly the same, thus the *U* value for both areas will be the same.

8.3 FOULING

When the heat exchanger is new, the tube surface is very clean. The heat transfer coefficient (*U*) is very high and the exit temperature of the hot fluid attains the minimum possible temperature and, as a result, the cooling fluid is at maximum temperature on leaving the exchanger. However, after a certain period of use, organic or inorganic matter is deposited on the metallic surfaces, called scales or fouling, reducing the heat transfer because of the poor thermal conductivity than that of the metal of the tube. Hence, the exit temperatures of the fluids will not be the same as those obtained in the absence of fouling. Even if the insulation of the exchanger is perfect, i.e., when there is no heat loss to the surrounding atmosphere, the performance of the exchanger will not be as good as the clean exchanger due to fouling. For such a situation, the effect on *U* can be predicted for a countercurrent operation of a heat exchanger from theoretical calculation, as given below.

Combining the heat balance equation (when no loss to surroundings takes place) and the rate equation, *U* is given as

$$U = W_h Cp_h / A(1 - \alpha)\ln\{(Th_1(1 - \alpha) + \alpha Th_2 - T_{c1})/(Th_2 - Tc_1)\}, \qquad (8.16)$$

where

$$\alpha = W_h Cp_h / W_c Cp_c. \qquad (8.17)$$

For given heat exchanger specifications and inlet temperatures, U may be plotted as shown in Figure 8.4.

Example 8.2

Consider the heat exchanger as given in Example 8.1, at the given inlet temperatures and the area of heat transfer, U may be determined as a function of the exit temperature of hot fluid during a countercurrent operation using Equation 8.14 and plotted in Figure 8.4. The maximum U is 400 kcal/m^2 h °C at Th2 = 145°C; where U = 255.1 kcal/h m^2 °C at Th2 = 150°C.

Hence, the perfor`mance of the exchanger at the given temperature of 150°C for the exit hot fluid is (200 − 150)/(200 − 145) × 100 = 91%. The maximum exit temperature of the cold fluid is 146°C.

If the measured exit temperature of cold fluid is 140°C instead of 146°C, then there will be additional loss to the surroundings, and the overall heat absorption by cold fluid is only (140 − 135)/(146 − 135) × 100 = 45.45%.

Hence, in the latter case, we find that the heat absorption reduces to 45% because of loss of heat to the surroundings and fouling.

Example 8.3

Crude oil at a temperature of 135°C and an API of 35 is to be preheated at a rate of 20 ton/h with kerosene circulating reflux available from a crude distillation unit at a temperature of 200°C and a rate of 2904.0 kg/h. If the overall heat transfer coefficient of the heat exchanger with a total outer tube surface of 13.5 m^2 is 255 kcal/m^2 °C, determine the exit temperatures of kerosene and crude oil. Assume countercurrent operation. Given that the specific heat of crude oil and kerosene at average temperatures is 0.53 and 0.79 kcal/kg °C, respectively.

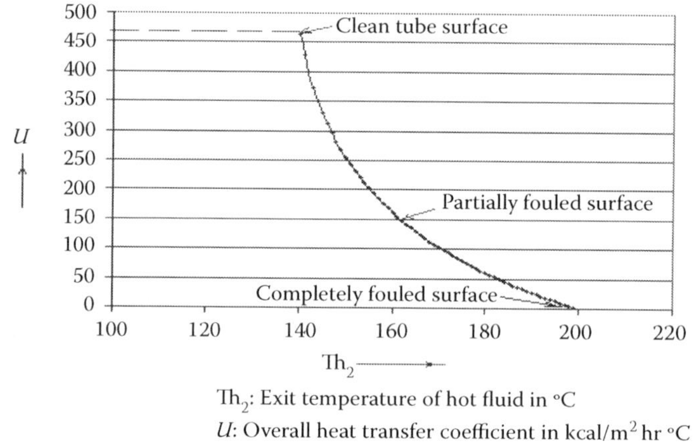

FIGURE 8.4 Theoretical variation of U due to fouling.

Solution

If we assume that there is no heat lost to the surroundings, the thermodynamic heat balance will apply as given in Equation 8.1.

$$Q = W_h Cp_h(Th1 - Th2) = W_c Cp_c(Tc2 - Tc1),$$

where
W_h: kerosene flow, 2904 kg/h
W_c: crude flow, 20,000 kg/h
Cp_h: specific heat of kerosene, 0.79 kcal/kg °C
Cp_c: specific heat of crude, 0.53 kcal/kg °C
Th1: inlet temperature of kerosene, 200°C
Tc1: inlet temperature of crude, 135°C
Hence,

$$Th2 = Th1 - \{W_c Cp_c(Tc2 - Tc1)/W_h Cp_h\}$$

$$= 200 - 20{,}000 \times 0.53(Tc2 - 135)/2{,}079 \times 0.79 \quad \text{(a)}$$

$$= 200 - 6{,}454 \times (Tc2 - 135).$$

Rate equation as
$Q = U\ A$ LMTD
$U = 255$
$A = 13.5$
$\Delta T1$:(200 − Tc2)
$\Delta T2$:(Th2 − 135)
$$\text{LMTD} = (200 - Tc2 - Th2 + 135)/\ln((200 - Tc2)/(Th2 - 135)), \quad \text{(b)}$$
Hence,

$$Q = 20{,}000 \times 0.53(Tc2 - 135) = 13.5 \times 255 \times \text{LMTD},$$

or

$$Tc2 = 135 + 13.5 \times 255 \times \text{LMTD} / 20{,}000 \times 0.53. \quad \text{(c)}$$

Assuming a value of T_{c2}, and with this T_{h2} is evaluated from Equation (a) and LMTD is determined from Equation (b). T_{c2} is recalculated from Equation (c) to check whether the assumed T_{c2} is the same, otherwise a new guess is made for T_{c2} and the calculation is repeated until equality is obtained. A few steps for trial-and-error solutions are given below.

Trial	**Tc2**	**Th2**	**ΔT1**	**ΔT2**	**LMTD**	**Tc2**	**Diff**
1	138	180.6	62	45.6	53.38	152.33	14.3
2	152.33	88.15	47.67	−46.85	–	–	–
3	140	167.77	60.00	32.73	45.00	149.6	9.60
4	141	161.27	38.72	26.27	32.09	145.42	5.42
5	142	154.82	58.00	19.82	35.55	146.5	4.42
6	143	148.36	57.00	13.36	30.08	144.7	1.70
7	144	141.90	56.00	6.90	23.449	142.60	−1.40

Hence, Tc2 (will be between 144 and 143) = 143.5,

$$\text{Th2} = 200 - 6{,}454 \times (143.5 - 135) = 145.14.$$

(A more accurate result is obtained by computer solution, Tc2 = 147.0°C and Th2 = 151.0°C.)

Example 8.4

In a double pipe heat exchanger, crude oil at a temperature of 149°C and an API of 34 is to be preheated to 155°C at a rate of 36,250 kg/h with a vacuum distillate, which is to be cooled from a temperature of 232°C to 172°C at a rate of 3,450 kg/h. Determine the required heat exchanger surface. Assume countercurrent operation.

Given that the properties of the fluids and the metal of the tube are at average temperatures as follows:

Properties	Crude oil	Lube oil	Metal of tube
Density, g/cc	0.855	0.898	Steel tube
Specific heat, kcal/kg °C	0.585	0.615	
Viscosity, Pa·s	8.3×10^{-4}	3×10^{-3}	
Viscosity, Cp	0.83	3.0	
Thermal conductivity, kcal/m² °C h/m	0.1088	0.09983	38.74

Solution

Take the inner tube of *i.d* = 2.067 in (51.675 mm) and *o.d* = 2.38 in (59.5 mm) through which cold fluid will flow and the outer tube of *i.d* = 3.068 in (76.7 mm) and *o.d* = 3.50 in (87.5 mm). Hot fluid will flow countercurrently through the annular space between the outer and inner tubes.

Tube side calculations:
Cold fluid: crude oil

$$\Delta T1 = 232 - 155 = 77\text{: } \Delta T2 = 172 - 149 = 23,$$

$$\text{LMTD} = (77 - 23)/\ln(77/23) = 44.69,$$

$$Q = 36{,}250 \times (155 - 149) \times 0.585 = 127237.50 \text{ kcal/h}.$$

So

$$Q = UA\text{LMTD},$$

or

$$UA = U_0A_0 = U_1A_1 = 127237.50/44.69 = 2847.11. \tag{8.18}$$

Inside area of cross section = $3.14/4 \times (0.051675)^2 = 2.096 \times 10^{-3}$ m^2,

Velocity $(v_i) = 36.250/\{3600 \times 0.855 \times 2.096 \times 10^{-3}\} = 5.621$ m/sec,

$N\text{Re}_i = 0.051675 \times 5.621 \times 855/(8.3 \times 10^{-4}) = 299214.13 > 10{,}000$.

Calculation of h_i:

$$\text{Nu}_i = 0.027(N\text{Re}_i)^{0.8}(N\text{Pr}_i)^{1/3}(\mu/\mu w)^{0.14}, \quad (8.19)$$

where

$N\text{Pr}_i = \text{Cp}_i\mu_i/k_i = 0.585 \times (8.3 \times 10^{-4}/10)/(0.1088/3600) = 16.065.$

Considering $(\mu/\mu w) = 1$

$\text{Nu}_i = 0.027(299214.13)^{0.8}(16.065)^{1/3} = 0.027 \times 24031.766 \times 2.5535 = 165.588,$

or

$$h_iD_i/k_i = 165.588,$$

i.e.,

$$h_i = 165.588k_i/D_i = 165.588 \times 0.1088/0.051675,$$

or

$$h_i = 349.25\text{kcal/h m}^2\ °\text{C}.$$

Shell side calculations:
Calculation of h_o:
Equivalent diameter for flow: D_e,

$D_0 = (D_2^2 - D_1^2)/D_1 = (76.7^2 - 59.5^2)/59.5$ mm = 39.37 mm = 0.031937 m.

Flow area = $3.14/4 \times (D_2^2 - D_1^2) = 3.14 \times (0.0767^2 - 0.0595^2) = 1.838 \times 10^{-3}$.

$V_0 = 3450/(1000 \times 3600 \times 0.898 \times 1.838 \times 10^{-3}) = 0.5803$ m/sec.

$N\text{Re}_0 = D_eV_0\mu_0/\rho_0 = 0.03937 \times 0.5803 \times 898/0.003 = 6838.69 < 10{,}000$.

Hence,

$\text{Nu}_0 = 1.86\{(N\text{Re}_0)(N\text{Pr}_0)(D/L)^{1/3}(\mu/\mu w)^{0.14}$.

Taking $(\mu/\mu w) = 1$ and $L/D = 600$,

$N\text{Pr}_0 = 0.615 \times 0.003/(0.09983/3600) = 66.53,$

$\text{Nu}_0 = (1.86 \times 6838 \times 66.53 \times 600 \times 1)/3 = 260.0,$

$h_0 = k/D_e \times 260 = (0.09983/0.03937) \times 260 = 659.0$ kcal/hm^2 °C.

Metal resistance:

$$x_m = (0.0595 - 0.051675)/2 = 0.0039125 \text{ m},$$

$$D_{lm} = (0.0595 - 0.051675)/\ln(0.0595/0.051675) = 0.0555 \text{ m},$$

$$K_m = 38.74.$$

Overall U_i:

$$U_i = 1/\{1/h_i + x_m D_i/(D_{lm} k_m) + D_i/(D_o h_o)\}$$

$$= 1/\{1/349.25 + 0.0039125 \times 0.051675/(0.0555 \times 38.74) + 0.051675/(0.0595 \times 659)\}$$

$$= 332.8 \text{ kcal/h m}^2 \text{ °C}.$$

So

$$A_i = Q/U_i \text{LMTD} = 127237.50/(332.8 \times 44.69) = 8.6 \text{ m}^2.$$

Example 8.5

A shell and tube heat exchanger of specifications as given below is employed for preheating crude oil by hot kerosene available from a distillation column. Determine the required surface area for heat transfer and also the pressure drops to judge the suitability of the given design specifications of the exchanger.

	Shell Side	**Tube Side**
Fluid	**Kerosene**	**Crude Oil**
Flow rate, kg/sec	5.53	18.10
Inlet temperature, k	472	311
Exit temperature, k	366	350
Inner diameter, m	0.54	0.0205
Outer diameter, m		0.0254
Baffle spacing, m	0.127	
No. of passes	1	4
Square tube, pitch, m		0.03175
No. of tubes		158
Tube spacing, m		0.0063
Length of tube, m		4.877
Properties at average temperatures		
Specific heat, kJ/kg k	2.47	2.05
Specific gravity	0.73	0.83
Viscosity, Pa·sec	0.0004	0.00359
Thermal conductivity, kW/mK	0.0001234	0.3937

Given a combined dirt factor of $R_d = 0.527$ mK/kW and the allowable pressure drop is 69 kPa each. Also assume countercurrent heat exchange. Take thermal conductivity of the metal of the tube as 0.3987 kW/mK.

Solution

Rate of heat exchange, $Q = 5.53 \times 2.47 \times (472 - 366) = 1447.86$ kW, lost by the hot fluid and $Q = 18.10 \times 2.05 \times (350 = 311) = 1447.86$ kW, gained by the cold fluid.

$$\text{LMTD} = (472 - 350) - (366 - 311)/\ln\{(472 - 350)/(366 - 311)\} = 84.1 \text{ k.}$$

(A more accurate calculation correction factor could be used.)

Shell side calculations:

Flow area, $a_o = \text{ID} \times \text{clearance} \times \text{baffle spacing/pitch}$

$$= 0.54 \times 0.0063 \times 0.127/0.03175 = 0.0136 \text{ m}^2,$$

Mass velocity, $G_o = 5.53/0.0136 = 406.6\text{kg/m}^2 \text{ sec}$,

Equivalent diameter, $D_e = 4(\text{pitch}^2 - 3.14 d_o^2/4)/3.14 d_o$

$$= 4(0.03175^2 - 3.14 \times 0.0254^2/4)/3.14 \times 0.0254$$

$$= 0.0251 \text{ m},$$

$$NRe_0 = D_e G_o/\mu_o = 0.0251 \times 406.6/0.0004 = 25514.15 > 10{,}000.$$

So,

$$h_o D_e/k_o = 0.36\,(Nre)^{0.55}(Npr)^{1/3}(\mu/\mu_w)^{0.14}$$

$$= 0.36 \times (25514.15)^{0.55}(2.47 \times 0.0004/0.000132)^{0.33}$$

$$= 186.$$

$$h_0 = k_0/D_e \times 186 = 0.000132/0.0251 \times 186 = 0.9780 \text{ kW/km}^2.$$

Assuming $\mu \approx \mu_w$.

Tube side calculations:

Flow area, a_i = cross sectional area of tube inside × no. of tubes/no. of pass

$$= 3.14\,(0.0205)^2\,158/(4 \times 4)$$

$$= 0.0130 \text{ m}^2.$$

Mass velocity, $G_i = 18.10/0.0130 = 1392.3 \text{ kg/m}^2 \text{ sec}$.

$NRe_i = d_i G_i/\mu_i = 0.0205 \times 1392.3/0.00359 = 7{,}950 < 10{,}000$ (transition zone).

So

$$h_i d_i/k_i = 1.86\,NRe^{0.33}(c_p \mu_o k_o)^{0.33}(d_i/L)^{0.33}(\mu/\mu_w)^{0.14}$$

$$= 1.86 \times 7950^{0.33} \times (2.05 \times 0.00359/0.000137) = 22.61.$$

Also

$$h_i d_i / k_i = 0.36\ NRe^{0.55}(c_p \mu_o / k_o)^{0.33}(\mu / \mu_w)^{0.14}$$
$$= 0.027 \times (7950)^{0.55}(2.05 \times 0.00359 / 0.000137)^{0.33}(0.0205 / 4.877)^{0.33}$$
$$= 134.38.$$

Assuming $\mu \approx \mu_w$.

Taking average

$$h_i d_i / k_i = (22.61 + 134.38)/2 = 78.5,$$

or

$$h_i = k_i / d_i \times 78.5 = 0.000137 / 0.0205 \times 78.5 = 0.5246 \text{ kW/m}^2 \text{ k}.$$

Metallic resistance of tube:

x_m = thickness of tube wall = $(d_o - d_I)/2 = (0.0254 - 0.0205)/2 = 0.00245$ m.

k_m = thermal conductivity of wall material = 0.3987 kW/mK

d_{lm} = log mean diameter = $(d_o - d_I)/\log(d_o/d_I) = (0.0254 - 0.0205)/\ln(0.0254/0/0205)$ = 0.0228 m.

$$U_c = 1/\{1/h_I + x_m d_i / k_m d_{lm} + d_o / d_I h_o) = 1/(1.906 + 0.005525 + 0.825) = 0.3653.$$

$$U_D = 1/\{1/U_c + R_D\} = 1/(1/0.3653 + 0.527) = 0.3063 \text{ kW/m}^2 \text{ k}.$$

Hence, the area required for heat transfer is

$$Q/(U_D \text{LMTD}) = 1447.86/(0.3063 \times 84.1) = 56.20 \text{ m}^2.$$

Whereas, the available area is

$$3.14 \times d_0 \times L \times \text{No. of tubes} = 3.14 \times 0.0254 \times 4.877 \times 158 = 61.457 \text{ m}^2.$$

Hence, the proposed heat exchanger is thermally suitable for the purpose.

Pressure drop calculations:
Shell side pressure drop:

$$\Delta p_0 = f \times G_0^2 \times \text{ID of shell} \times (N+1)/(2 \times g \times \text{density} \times D_e),$$

where

$f = 0.00722$ for $NRe = 25514.15$,

$N + 1 = L/\text{Baffle spacing} = 4.877/0.127 = 38.4$,

$g = 9.81$ m/sec^2.

So

$$\Delta p_0 = 0.00722 \times (406.6)^2 \times 0.54 \times 38.4 / (2 \times 9.81 \times 730 \times 0.0251) = 68.85 \text{ kg/m}^2$$
$$= 0.6885 \text{ kPa} < 68.7 \text{ kPa}.$$

Tube side pressure drop:

$$\Delta p_i = f\, G_i^2\, L \times \text{No. of pass}/(2 \times g \times \text{density} \times d_i) + \text{density} \times (G_i/\text{density})^2/2 \times g$$

$$= 0.0579 \times (1392.3)^2 \times 4.877 \times 4/(2 \times 9.81 \times 830 \times 0.0205) + 830 \times (1392.3/830)^2/(2 \times 981)$$

$$= 774.8\ \text{kg/m}^2 = 7.74\ \text{kPa} < 79\ \text{kPa}.$$

Hence, the proposed heat exchanger satisfies both the heating surface and pressure drop requirements.

Other than double pipe and shell-and-tube heat exchangers, a variety of other heat exchangers are used in the plants, e.g., plate type and extended surface heat exchangers.

8.4 PLATE TYPE HEAT EXCHANGER

Plate type exchangers make use of plates as the recuperator's surface where hot and cold fluids share alternate surfaces of plates connected in such a way that the hot and cold fluids are contacted separately without any physical mixing. Plates and frames are so arranged that box type spaces are created for flow.

Hot and cold fluids share alternate plate-frame chambers. Larger surface area (which can also be varied if required) as compared to double pipe or shell and tube exchangers is available for heat transfer. These are useful for low thermal conductivity fluid, e.g., air or gas, available in large flow rates. This type of exchanger is used for air preheaters with flue gas in furnaces. A plate type heat exchanger is shown in Figure 8.5. A detailed description is out of the scope of this book.

8.5 EXTENDED SURFACE EXCHANGER

When the velocity of one of the fluids is either low or the thermal conductivity is poor, the heat transfer rate can be improved by increasing the area by providing extended fins or studs, as shown in Figure 8.6. The outside surface of the tubes are finned (i.e., thin circular disks) or studded (rectangular studs), as shown in Figure 8.6.

8.6 SCRAPED SURFACE EXCHANGER

This is a special type of double pipe exchanger where the inner pipe has a large diameter to allow a mechanical scraper to clean the inside pipe surface. This type is used in chillers and coolers in a dewaxing plant. Wax deposit inside the surface of the pipe causes a poor cooling rate and may ultimately choke the pipe. The scraper is a rotating screw or blade driven by a shaft. A typical scraped surface chiller is shown in Figure 8.7.

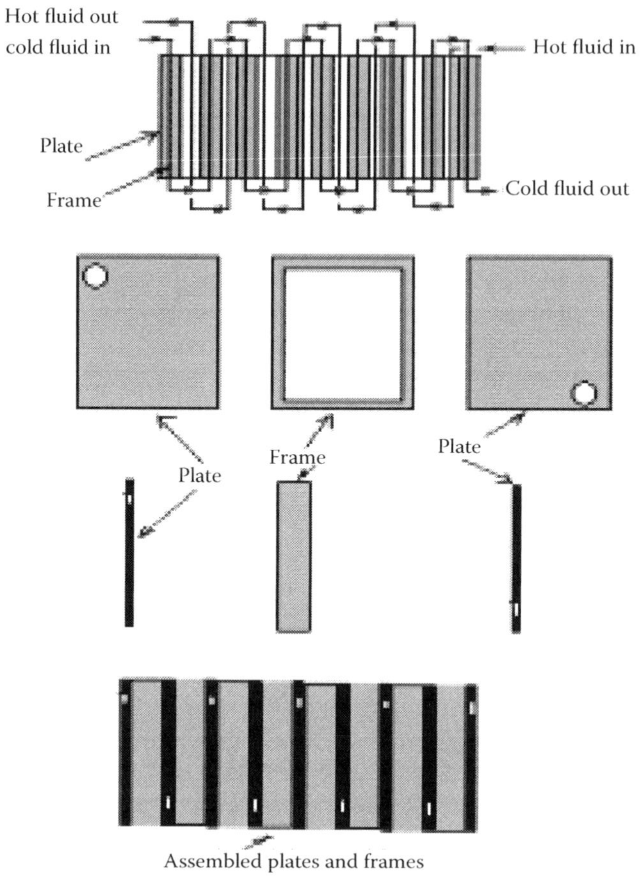

FIGURE 8.5 Components of a plate type heat exchanger.

8.7 HEAT EXCHANGER TRAIN

In order to increase the heat transfer area, more than one heat exchanger may be connected in a series where the tube side fluid passes from one heat exchanger to the other in sequence. Similarly, shell side flows may also be connected in a series. However, series and parallel connections can also be found. Parallel connections in any stream will reduce the pressure drop. A typical heat exchanger train for crude oil preheating is shown in Figure 8.8.

Example 8.6

Considering a shell and tube heat exchanger, as in Exercise 8.5, determine the number of heat exchangers to be connected in a series if the number of tube passes is 1 instead of 4.

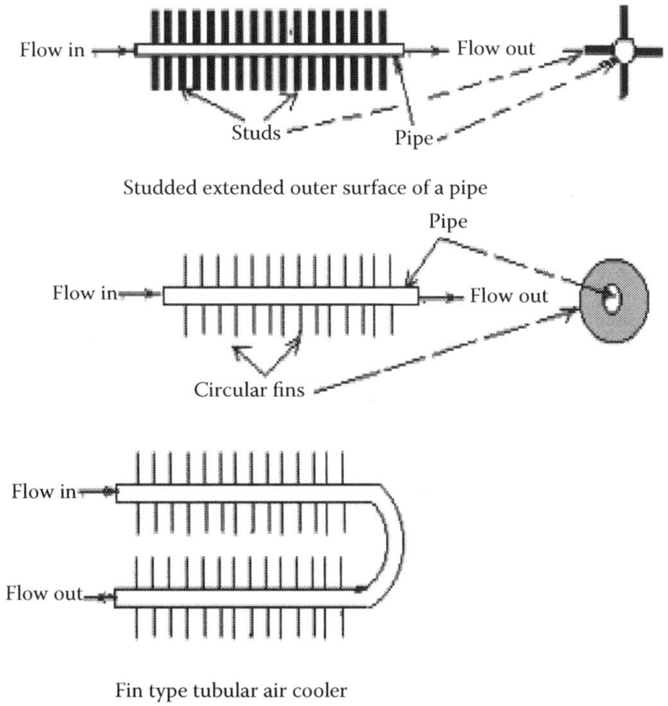

FIGURE 8.6 Fin type and stud type extended heat exchangers.

Solution

Proceeding to calculate the required heat transfer area in a similar fashion as done in Exercise 8.5, taking a 1-1 heat exchanger, the area required becomes 103.274 m^2 (or 1111.632 ft^2), whereas the available area is 61.45 m^2 (or 661.49 ft^2). Hence, the number of heat exchangers to be taken in a series is 102.274/61.45 = 1.67

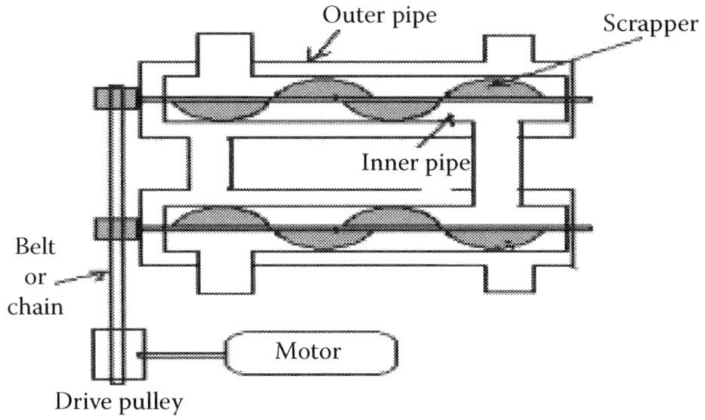

FIGURE 8.7 Scraped pipe chillers in a series.

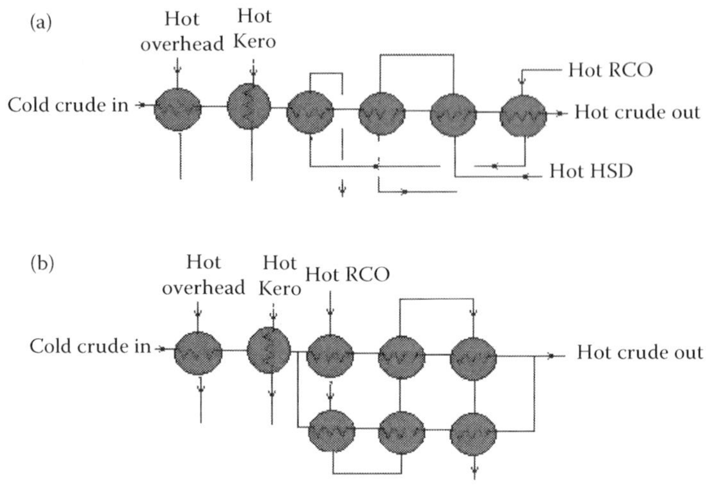

FIGURE 8.8 Typical heat exchanger trains for crude preheating.

or 2. The pressure drop in the tube passes will be 4.44 kPa instead of 7.74 kPa. Hence, two heat exchangers in a series will be equivalent to a 1-4 heat exchanger and with a lower pressure drop.

8.8 PIPE-STILL FURNACE

Tube-still heaters are used in petroleum and petrochemical plants for various purposes, such as crude distillation (atmospheric and vacuum), reboiling, and preheating of oil stocks, where heating is carried out for vaporisation purposes only. Tube-still furnaces are also used as reactors, as vis-breaking furnaces, coking furnaces, pyrolysis heaters in olefin production, steam reformers for hydrogen production, etc. These are low temperature furnaces (as compared to furnaces used in metallurgical, ceramic, or cement plants), Where heat duty of burners vary from 0.5 to 100 million kcal/hr and temperature goes upto 1200°C. The pressure may be allowed as high as 10 MPa (100 atm, approximately). These furnaces are commonly housed in a cylindrical or rectangular box type refractory-lined steel casing supported on a steel structure above ground. The heater has a radiant section and a convection section. Feed flowing through the pipes in the radiant section receives heat from the flames of the burning fuel from the burners. In the convection section, feed entering the furnace recovers heat from the hot flue gas leaving the radiant section. Various designs of such furnaces are available, such as helical coil type, vertical cylindrical, box type, twin cell type, terrace wall type, multisection pyrolysis type, etc. Typical tube-still furnaces are shown in Figures 8.9 through 8.12. A small duty furnace (<2 million kcal/h) uses a pipe arranged in the form of a helical coil having only a radiant section. These furnaces are low efficiency furnaces as no convection zone is present. However, they are easy to clean and maintain, but are costly and occupy a large space. Vertical cylindrical

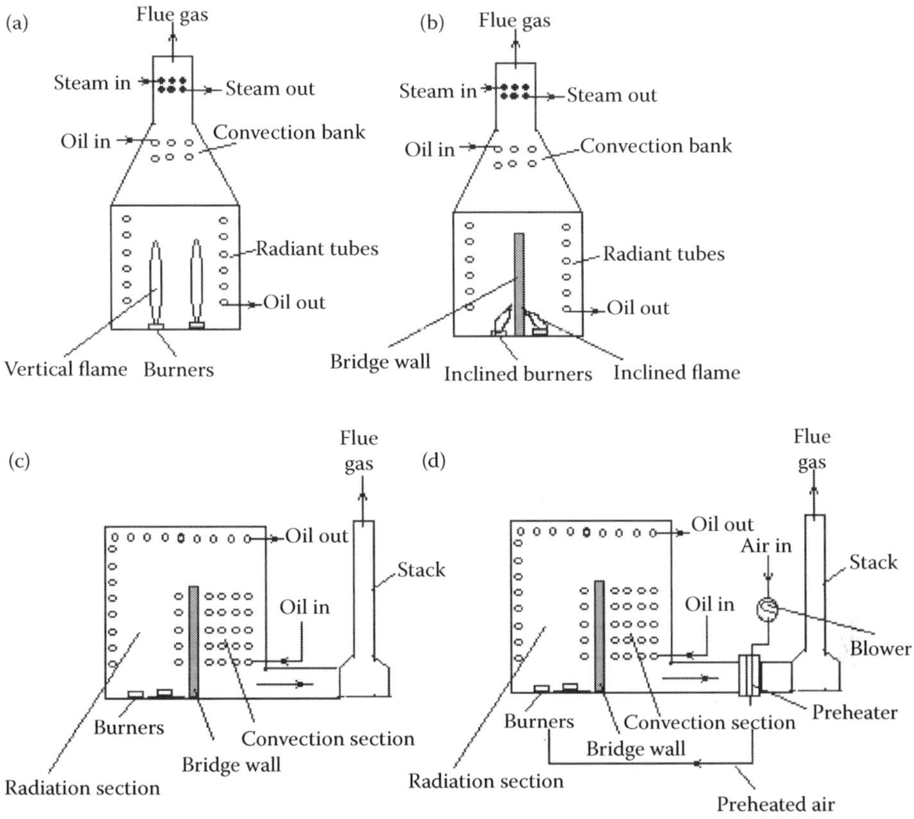

FIGURE 8.9 (a) Pipe-still furnaces, (b) natural draught circular furnaces, (c) natural draught box furnace, and (d) box furnace with force draught with air preheater (APH).

furnaces are built with a convection section at the top and a large radiant section at the bottom. Vertical box type furnaces also have the convection and radiation sections, but the cross section is rectangular instead of circular. In both types of vertical furnaces, burners are placed on the bottom of the furnace cross sections. Twin cell furnaces are designed to have two radiant boxes with a common convection section. Burners are inclined to heat the radiant walls such that overheating of the tubes is avoided for high temperature services. These are widely used as mild pyrolysis furnaces used in vis-breaking and coking. Terrace wall furnaces are used for steam reformer to generate hydrogen from methane or naphtha. Inclined burners are used to heat the refractory walls such that catalyst-filled tubes through which reactant hydrocarbons flow are uniformly heated at the highest temperature possible. Multizone furnaces are designed to have multiple pyrolysis sections to render cracking of the hydrocarbon feedstocks at the highest possible temperature in a very short time. These are used in olefin production from ethane, naphtha, and even heavier fractions.

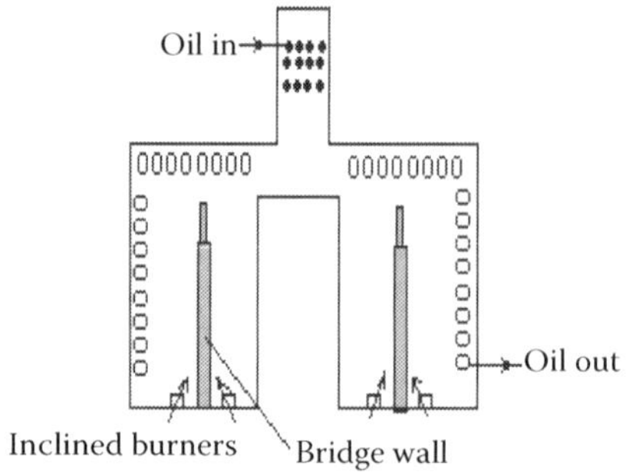

FIGURE 8.10 Twin cell furnace.

8.9 PIPE-STILL FURNACE ELEMENTS

The common elements of pipe furnaces consist of pipes or tubes, supports of the pipes, refractories, burners, still shell and structural elements. Section wise it consists of stack, convection, and radiant sections.

8.9.1 Heater Pipes or Tubes

Steel pipes are used for normal heating operations with nominal bore diameters varying from 2 to 6 in and a thickness varying in the range of 40–80 schedule. For certain cases, the diameter may be as high as 8 in or more, especially for vacuum services. Tubes are vertically arranged in single or multiple rows surrounding an equally spaced burner on the bottom cross section of the furnace. The material of construction of the tubes is selected based on the flue gas composition, the nature of the feed hydrocarbons, and the temperature and pressure of the operation. Alloys of steel containing chromium, molybdenum, and nickel are widely used as the material for tubes

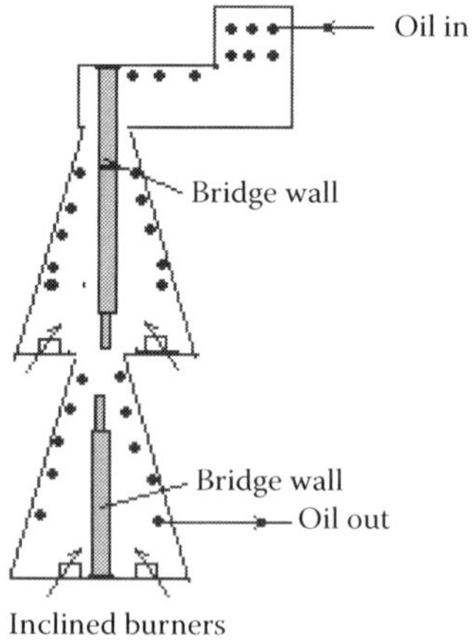

FIGURE 8.11 Terrace wall furnace.

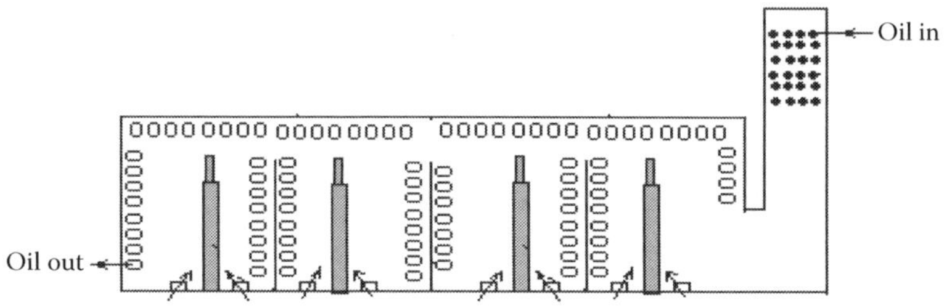

FIGURE 8.12 Multizone pyrolysis furnace.

in the furnace. The presence of chromium resists corrosion from hydrogen sulfide, mercaptans, organic acids, high temperature oxide corrosion, etc. Since chromium reduces the creep strength of the metal, a small amount of molybdenum is added to increase the strength. A typical distillation furnace in a refinery contains 5% chromium and 0.5% molybdenum in steel. High chromium steel becomes brittle at high temperatures in the absence of nickel, therefore, a small amount of nickel is also allowed in the alloy. Nickel acts as the toughening component and a small amount of silicon is also desirable in the alloy to resist scale formation due to oxidation.

Long pipes are connected to one another by bends welded to the pipe ends. These welded bends are kept outside the combustion chamber in a separate box, known as a header box, to avoid the joints inside the combustion chamber, as these joints are prone to leaking due to thermal shocks. This type of arrangement is required for easy cleaning of the tubes and also for repair of the bends.

8.9.2 Refractories

The surface of the inside walls of the furnace is covered with refractory lining to reflect most of the radiant heat generated by the flames. That is why these surfaces are called re-radiating surfaces. High alumina tiles or refractory lining containing as high as 97% alumina are commonly employed as re-radiating walls backed by high silica-alumina refractories. Insulating bricks with high porosity surround the refractory walls and protect the steel shell (which holds the refractories and the pipes' structures and burners) from the high temperature of the furnace and loss of heat to the atmosphere.

8.9.3 Burners

Burners are devices by which the fuel and air mixture is brought in contact with a spark or ignition source and allow continuous combustion at a high temperature at its mouth. It contains a nozzle through which the fuel gas or vaporised liquid fuel emerges at a high velocity and causes air (primary air) to be entrained with the jet. Continuous combustion is sustained by the supply of secondary air in the combustion chamber. Heavy liquid fuel, which is not easily vaporisable, is atomised with dry steam by passing it through the nozzle of the burner. Liquid fuel and primary air are thus simultaneously entrained with the steam jet. Construction of gas burners is different from oil burners due to the abovementioned different mechanisms of firing. The flame temperature depends on the heating value of the fuel and the rate of firing. A high turn down ratio (as the ratio of maximum firing rate to the minimum firing rate) of firing must be maintained for a clean burner nozzle tip with a uniform flame of the desired diameter and length. (Turndown ratio is defined as the ratio of maximum firing rate to the minimum firing rate.) In the natural draught furnaces, primary and secondary air flow due to the difference of pressure of atmosphere and the hot gas within the furnace. This difference in pressure occurs owing to the difference between the atmospheric temperature and the furnace gas temperature. Burners of natural draught furnaces give longer flames and have the capacity to release heat in the range of 3–4 million kcal/h. The excess air requirement for these furnaces varies from 15%–20% with <70% thermal efficiency. In a forced draught furnace, air is mechanically forced through the burners with the help of suitable air blowers and the draught does not depend on the temperature difference of the atmosphere and furnace gas. Burners of these furnaces are capable of releasing a heat of 20–30 million kcal/h with shorter flame lengths. In modern furnaces, air is preheated with the hot flue gas leaving the stack. The thermal efficiency is as high 90% with a reduced requirement for excess air, which may be as low as 5%.

8.9.4 Convection Zone

In this section, feedstock is heated by the hot flue gas leaving the radiant section below. A bank of large numbers of tubes are placed either countercurrently or cross-currently to the upgoing flue gas stream. The tubes are connected with bends or in the form of coils in this section. Heat transfers from hot flue gas to the colder feedstock entering the convection zone. The diameter of the pipes is usually smaller than the radiant tubes. The outer tube surfaces have an extended surface with finned or studded type designs to promote convective heat transfer. Cold or preheated feed enters a bank or coils of tubes in the convection section. In this section, heat transfer between the feed flowing through the tubes and the hot flue gas passing outside the tubes takes place mainly by convection. The film heat transfer coefficients of the tube side fluid and flue gas outside the tubes play the main role in heat transfer. These film coefficients vary with the flow rates and the properties of the fluids. The higher the flow rate or velocity either inside or outside the tubes, the more the heat transfer, and higher will be the temperature of feed stock exiting from the convection section. The conductivity, thickness, diameter, and surface area of the metallic tubes also influence the overall heat transfer coefficient. Calculations for the design of this section are similar to the design calculation for heat exchangers.

8.9.5 Radiant Section

This section is the combustion chamber of the fuel in the presence of air and the heat is transferred to the tubes from the flames by radiation. An air and fuel mixture is burnt by a number of burners distributed over the cross section of the chamber to allow uniform heating without touching the surfaces of the tubes (i.e., without impinging the tube walls). Since the flame temperature is higher than 1000°C, the metal of the tube may be damaged due to impingement, such as local melting, oxidation or change in mechanical properties due to thermal shock, etc., and finally give way. The maximum allowable temperature of the surface of metallic pipes is restricted by the composition of the alloy used for construction. This temperature is also known as the allowable skin temperature of the metallic pipes. For very high-duty heat transfer and temperature, it is necessary to use burners to heat the walls of the furnace, which then heat the tubes. Such indirect heat transfer is essential to avoid impingement of the flames on the tubes.

8.9.6 Stack or Chimney

The stack or chimney is an exhausting duct for the flue gas to the atmosphere. Combustion gases or flue gas contains carbon dioxide, carbon mono oxide, sulfur dioxide, water vapour, unburnt hydrocarbons, nitrogen, oxygen, etc. The temperature of the flue gas of most of the furnaces is maintained above 100°C so that no moisture is liquified to cause acidic corrosion due to the presence of acid gases, such as sulfur dioxide, hydrogen sulfide, mercaptans, etc., which are highly corrosive in the aqueous phase. The usual stack gas temperature is 120°C–150°C for the majority of furnaces, though it can be reduced by recovering heat in the convection section and also by preheating the combustion air in the forced draught furnaces. For such preheaters, the excess air requirement is reduced to 5% and, as a result, overall efficiency may be as high as 95%–98%. Depending on the ease of the contacting pattern of hot flue gas with the feedstock, combustion air, pressure, and the flow rate of flue gas, the stack may be placed on the top of the furnace or it may be on the ground or there may be single or multiple stacks. Dampers are also provided in the stack to adjust the draught of the furnace.

8.10 OPERATION OF A FURNACE

Furnace housing is placed above grade (i.e., above ground level) to provide enough space to facilitate lighting of burners, observation of flames through sight glasses, air suction registers for the ingress of combustion air, steam purging and inert gas purging of the radiant box, maintenance facilities for the burners, etc. An operating platform is provided above grade to allow opening and closing of the fuel and steam valves. Sight glasses are also provided to view inside the radiant section to monitor and adjust the flame on the burner tips. Dampers can also be operated from this platform. The steel shell wall temperature of the furnace must be maintained near ambient temperature to avoid thermal shocks to the operating personnel. Usually, the outside surface of the furnace need not be insulated as this may cause a rise in the steel shell wall temperature, which may damage the shell. Judicious insulation inside the shell by the bricks of low thermal conductivity encircling the refractory walls is sufficient to bring down the steel shell temperature. Purging with low-pressure steam (snuffing steam) is essential before lighting the burners to avoid explosion due to the accumulated unburnt fuels from unsuccessful lighting events. Soot consisting of fine carbon particles, coke and ash accumulate in the banks of tubes in the convection zone, thereby reducing the heat transfer. Hence, periodic soot blowing is essential for the convection zone. Steam or an air jet are blown over the tube banks to remove the soot by various means, e.g., manual or mechanical means.

8.11 DRAUGHT IN A FURNACE

The air column outside the furnace at ambient temperature has a higher pressure than the hot gas column inside the furnace chamber and, as a result of this pressure difference, air enters the combustion chamber through the air registers. This is called a natural draught and such a draught is maintained without any mechanical blowers. For forced draught furnaces, air is forced into the combustion chamber by blowers. A draught can also be made by suction of the flue gas by blowers and such a draught is called an induced draught. Both induced and forced draught systems may be used simultaneously in many furnaces to have good control over the draught and temperature of preheated air. Natural draught furnaces are cheaper, but yield poor efficiency, high excess air requirement, and are prone to back firing during light up owing to poor control over the draught.

8.12 FURNACE DESIGN BY THE WILSON, LOBO AND HOTTEL METHOD

In this method, heat transfer is considered as combined radiation and convection. The overall relation for heat transfer is given as, according to the Wilson, Lobo, and Hottel equation, fraction (R) of heat generated by combustion that is absorbed by the tubes is given as

$$R = \frac{1}{\left\{1 + \dfrac{G\sqrt{(Q/\alpha A_{cp})}}{4200}\right\}}, \tag{8.20}$$

where:

Q: total heat absorbed in the cold plane surface of the radiation section, Btu/h
A_{cp}: equivalent total refractory cold plane surface = $L \times N \times C$ ft^2, where L, N, and C are the length, number, and clearance of spacing of the tubes, respectively
α: fraction of cold plane surface covered by tubes as a function of the number of rows and the ratio of clearance to o.d. as obtained from Figure 8.13
G: air/fuel ratio, wt/wt
Q: total heat generated by fuel, Btu/h = firing rate × net heating value

Exercise 8.7

A petroleum stock having the following specification is to be heated in a tube-still furnace at a rate of 1200 bbl/h. The coil inlet temperature is 220°C. Determine the

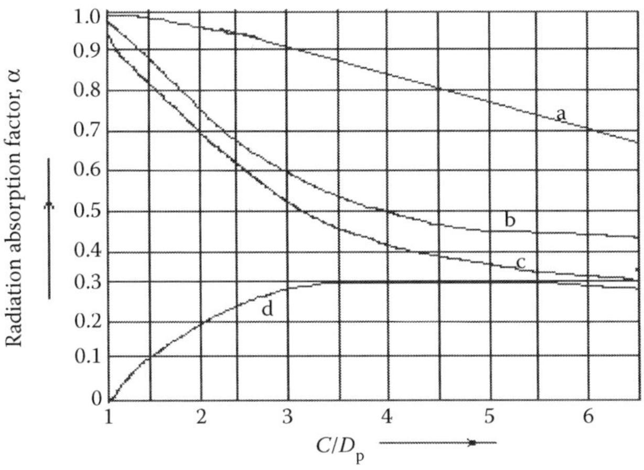

FIGURE 8.13 Radiation absorption factor as a function of tube spacing. Radiation absorption factor, α [where (a) total to two rows when two are present, (b) total to one row when only one is present, (c) total to first row when two are present, (d) total to second row when two are present and C/D_p is the ratio of centre-centre distance to the pipe diameter; ref: Hottel, *Mech Engg (year could not be ascertained)*].

coil outlet temperature if (1) vaporisation of oil is neglected and (2) vaporisation of oil is taken into account.

Feed data: The API of oil is 34.5, specific heat is 2.268 kJ/kg °C; average latent heat of vaporisation is 100 Btu/lb.

Fuel data: Net heating value is 47.46×10^3 kJ/kg.

Rate of firing is 3500 kg/h.

Air/fuel ratio is 25.

Furnace data: Assume that the heater has only a radiant section having 120 tubes each of 10.5 cm o.d. and 12 m long and clearance between the tube centres is twice the o.d. Assume only a single row of tubes surrounding the burners that take a = 0.88.

Conversion factors:

1 bbl = 42 US gal = 5.6 cft = 0.158 m^3 = 158 L

1 g/cc = 62.5 lb/cft

1 Btu = 252 cal; 1 kcal/kg °C = 1 Btu/lb °F

Stephan–Boltzmann constant = 1.72×10^{-9} Btu/ft^2 h $(R)^4$

Solution

According to the Wilson, Lobo, and Hottel equation, fraction (R) of heat generated by combustion that is absorbed by the tubes is given as

$$R = \frac{1}{\left\{1 + \frac{G\sqrt{(Q/\alpha A_{cp})}}{4200}\right\}},$$

where

G = air/fuel ratio = 25 wt/wt

Q = total heat generated by fuel, Btu/h = firing rate × net heating value = 3500 × 47.46 × 1000/4.2 kcal/h = 3500 × 47.46 × 1000/(4.2 × 0.252) = 1.56944×10^8 Btu/h

A_{cp} = area of furnace wall that has tubes mounted on it, ft^2; = $L \times N \times C/12$ = (12 × 3.28) × 120 × (2 × 10.5/2.54)/12 = 3254.1732 ft^2

L = 12 m = 12 × 3.28 ft; N = number of tubes

o.d. = 10.5 cm = 10.5/2.54 in

aA_{cp} = equivalent cold plane surface for a single row; = 0.88 × 3254.1732 ft^2 = 2863.6724 ft^2

So

$$R = \frac{1}{\left\{1 + \frac{25\sqrt{1.56944 \times 10^8 / 2863.6724)}}{4200}\right\}} = 0.417,$$

Heat transferred to feedstock = 0.417 × 1.56944×10^8 = 6.557×10^7 Btu/h.

$$\text{API} = \frac{141.5}{\text{spg}} - 131.5,$$

so

$$\text{spg} = \frac{141.5}{(\text{API} + 131.5)} = 141.5/166 = 0.852.$$

Density = 0.852 × 62.5 = 53.25 lb/cft.

Mass flow rate of feedstock = 1,200 × 5.6 × 53.25 = 357,840 lb/h.

Specific heat of stock = 2.268 kJ/kg °C = 2.268/4.2 kcal/kg °C = 0.54 kcal/kg °C = 0.54 Btu/lb °F.

1. Neglecting vaporisation of feed

Heat absorbed by feedstock = mass flow rate × specific heat × temperature rise = 357,840 × 0.54 × ΔT Btu/h,

so

$$357{,}840 \times 0.54 \times \Delta T = 6.557 \times 10^7,$$

or

$$\Delta T = 339.33°\text{F} = 188.52°\text{C}.$$

So the coil outlet temperature = 220 + 188.52 = 408.52°C.
Ans.

2. Vaporisation not neglected

Heat absorbed by feedstock = mass flow rate × (specific heat × temperature rise + latent heat of vaporisation) = 357,840 × (0.54 × ΔT + 100) Btu/h = 6.557 × 10^7,

or

$$\Delta T = 154°\text{F} = 86°\text{C}.$$

So, the coil outlet temperature = 220 + 86 = 306°C.
Ans.

Exercise 8.8

Determine the flame temperature and flue gas temperature leaving the furnace in Exercise 8.7. Given that the maximum allowable tube metal skin temperature is 800°C.

Solution

If the burnt gas temperature (flame temperature) is T_g in Rankine, the heat radiated by this flame is

$$Q_r = \delta(T_f^4 - T_w^4)A,$$

where A is the total surface area of the furnace wall containing the tubes encircling the flames and δ is the Stephan–Boltzmann constant. If heat loss to the surroundings from the furnace wall is neglected, then this is also equal to the heat generated by the fuel.

From the previous calculation, $Q = 1.56944 \times 10^8$ Btu/h

$$\delta = 1.72 \times 10^{-9}\ \text{Btu/ft}^2\text{h(R)}^4,$$

$$A = A_{cp} = 3254.1732\ \text{ft}^2,$$

Skin temperature, T_w = 800°C = 1472°F = 1472 + 460 = 1932 R,

so

$$1.56944 \times 10^8 = 1.72 \times 10^{-9} \times 3254.1732 \times (T_f^4 - 1932^4),$$

or

$$T_f = \{(2.80398 \times 10^{13} + 1932^4)\}^{1/4} = 2545.3105R = 2085.3105°\text{F} = 1140.728°\text{C}.$$

As already calculated in the previous problem that 41.7% of the radiated heat is absorbed, so the flue gas temperature is given as

$$(1 - 0.417) \times 1.56944 \times 10^8 = \text{mass flow rate of gas} \times \text{enthalpy of gas}$$
$$= (3500 \times 26 \times 2.2) \times 0.276 \times (T_g - 0),$$

{where the enthalpy of the flue gas leaving, taking absolute zero as the base temperature of enthalpy = $0.276 \times (T_g - 0)$ Btu/lb},
so

$$T_g = 1655.97°\text{F} = 902°\text{C}.$$

Hence, the flue gas temperature is very high where around 60% of the heat is lost with the stack gas.

Exercise 8.9

In a modification of the abovementioned furnace, a convection section is provided above the radiation section where heat is recovered from the flue gas leaving the radiation section by the cold feedstock entering the furnace. In this section, assume that the tubes are placed vertically such that the incoming cold feedstock and the outgoing flue gas are contacted exactly countercurrently. Since flue gas has a poor heat transfer coefficient, the tubes surface in this section may be provided with fins to extend the surface to enhance heat transfer. Determine the coil outlet temperature and flue gas temperature leaving the convection section. Details of the convection section data are given below. Also, evaluate the possibility of raising superheated steam from saturated steam at 200°C.

Convection section: One pass parallel 50 tubes of the following specifications are used.

o.d. (D_o) = 10.5 cm; i.d. (D_i) = 9.5 cm; length = 5 m
Thermal conductivity of metal (k_m) = 180 Btu/ft °F h
Properties of feedstock in the tube:
Specific heat (Cp_i) = 0.667 Btu/lb °F
Thermal conductivity (k_i) = 0.066 Btu/ft °F h
Viscosity (μ_i) = 0.605 lb/ft h
Density (ρ_i) = 53.25 lb/cft

Flue gas: Heat transfer coefficient due to a combined convection and radiation effect is given by Monrad's equation:

$$h_c = \frac{1.6G^{0.667}T^{0.3}}{D^{0.33}} + (0.0025T - 0.5),$$

where

h_c: combined heat transfer coefficient of flue gas, Btu/ft² °F sec
G: mass velocity of flue gas at minimum cross section, lb/sec ft²
T: average flue gas temperature °F absolute, i.e., in Rankine
D = o.d. of the tube, in

Solution

Let us designate the mass flow rate of feed as w_c, the inlet temperature as T_{c1}, and the exit temperature as T_{c2}. The mass flow rate of flue gas is w_g, and the flue gas inlet and exit temperatures of the convection section are T_{h1} and T_{h2}, respectively. Then the heat absorbed by the liquid in the tubes is

$$U_i A_i \text{LMTD} = w_c cp_c (T_{c2} - T_{c1}). \tag{1}$$

Also

$$w_c cp_c (T_{c2} - T_{c1}) = w_g cp_g (T_{h1} - T_{h2}) \tag{2}$$

A_o = outside surface area for heat transfer = $3.14 \times D_o \times L \times n = 3.14 \times 10.5/(2.54 \times 12) \times 5 \times 3.2 \times 50 = 865\ \text{ft}^2$

A_i = inside surface area for heat transfer = $3.14 \times D_i \times L \times n = 3.14 \times 9.5/(2.54 \times 12) \times 5 \times 3.2 \times 50 = 782.5\ \text{ft}^2$

$$\text{LMTD} = (\Delta T_2 - \Delta T_1) / \ln(\Delta T_2 / \Delta T_1), \tag{3}$$

where

$\Delta T_2 = T_{h2} - T_{c1}$,
$\Delta T_1 = T_{h1} - T_{c2}$,
$T_{c1} = 200°\text{C} = 428°\text{F}$,
$T_{h1} = 1655.9°\text{F}$,
T_{h2} = unknown,
T_{c2} = unknown,
and
U is obtained as

$$\frac{1}{U_i A_i} = \frac{1}{h_i A_i} + \frac{x_m}{k_m A_m} + \frac{1}{h_o A_o},$$

also

$$U_i A_i = U_o A_o.$$

Calculation of h_i (tube side):

A_{ix} = area of cross section of a tube or area for fluid flow per tube $= 3.14 \times D_i^2/4 = 3.14 \times \{9.5/(2.54 \times 12)\}^2/4 = 0.0763\ \text{ft}^2$

using the Dittus–Boeltier relation in turbulent flow condition in the tube side,

$\text{Nu} = 0.023\ \text{Re}^{0.8}\ \text{Pr}^{0.33}$,

also

Nusselt number:

$$\text{Nu} = \frac{h_i D_i}{k_i} = \frac{h_i 9.5/(2.54 \times 12) \times 3600}{0.066} = 17000.7 h_i,$$

where h_i in Btu/ft^2 °F sec.

Reynolds number:

$$\mathrm{Re} = \frac{D_i V_i \rho}{\mu_l} = \frac{9.5/(2.54 \times 12) \times 24.5}{(0.605/3{,}600)} = 45438.277 > 10{,}000,$$

(as $V_i = 1200 \times 5.6/(3600 \times A_i) = 1.87/0.0763 = 24.5$ ft/sec).
Hence, the flow is turbulent and the Dittus–Boeltier relation is applicable.
Prandtl number:

$$\mathrm{Pr} = \frac{c p_i \mu_i}{k_l} = \frac{0.667 \times 0.605}{0.066} = 6.11,$$

$$\mathrm{Nu} = 0.023(45438.277)^{0.8}\,(6.11)^{0.33}$$

$$= 0.023 \times 5320.3167 \times 1.82 = 222.36,$$

or

$$17000.7 h_i = 222.36,$$

or

$$h_i = 222.36/17000.7 = 0.01308 \text{ Btu/ft}^2\ \text{°F sec} = 47.00 \text{ Btu/ft}^2\ \text{°F h}.$$

Flue gas side:

$$h_c = \frac{1.6 G^{0.667} T^{0.3}}{D^{0.33}} + (0.0025T - 0.5).$$

Assume that the exit temperature of flue gas is 350°C (662°F). The average temperature of flue gas = (1655.9 + 662)/2 = 1158.95°F.
Calculation of the mass velocity of flue gas, G lb/sec ft^2:
Mass flow rate of flue gas = 3,500 × (1 + 25) kg/h = 91,000 × 2.2/3,600 lb/sec = 55.611 lb/sec

Minimum area of cross section is at the stack, maximum allowable velocity allowable from practice = 20 ft/sec (Nelson, W.L, *Petroleum Refinery Engineering*, 4th ed, p. 613, 1958.)

So, the minimum area for flue gas flow = volumetric flow rate/velocity
Considering flue gas, the ideal gas density is given as

$$\rho_o = PM/RT,$$

where

P = pressure in the furnace ≈ 1 atm (natural draught furnace)
M = average molecular weight, which is taken as that of air (as air is present in large amounts in the flue gas) = 29
R = 0.0832 lit-atm/gm mol °C)

$T = 1158.95°\text{F} = 626°\text{C} = 899\ \text{K}$

so

$$\rho_o = 1 \times 29/(0.0832 \times 899) = 0.3877\ \text{g/L} = 0.3877 \times 10^{-3}\ \text{g/cc} = 0.02423\ \text{lb/cft}$$

Volumetric flue gas flow = 55.611/0.02423 = 2294.869 cft/sec
Area of cross section of stack = 2294.869/20 = 114 ft^2
Hence, the diameter of the duct is 12 ft.
$G = 55.611/114 = 0.487$ lb/sec ft^2

so

$$h_o = \frac{1.6(0.487)^{0.667}(1158.95+460)^{0.3}}{4.133^{0.33}} + (0.0025 \times 1158.95 - 0.5)$$

$$= 5.69 + 2.397 = 8.08,$$

(as $D_o = 10.5/2.54 = 4.133$ in),

so

$h_o = 8.08$Btu/ft^2 °F sec.

U for each tube based on inside and outside surfaces per tube:

Metal of the tube:

$k_m = 180$ Btu/ft °F h = 0.05 Btu/ft °F sec
Thickness of tube wall $(x_m) = (10.5 - 9.5)/2 = 0.5$ cm = 0.196 in
A_m = average metal surface area = $(A_o - A_i)/\ln(A_o/A_i) = (17.3 - 15.65)/\ln(10.5/9.5) = 16.48$ ft^2
$h_o = 8.08$ Btu/ft^2 °F sec
$h_i = 0.01308$ Btu/ft^2 °F sec
$k_m = 0.05$ Btu/ft °F sec
$x_m = 0.196$ in = 0.01633 ft
$A_m = 16.48$ ft^2
A_o = outside surface area for heat transfer per tube = 17.3 ft^2
A_i = inside surface area for heat transfer per tube = 15.65 ft^2

$$\frac{1}{U_iA_i} = \frac{1}{0.01308 \times 15.65} + \frac{0.01633}{0.05 \times 16.48} + \frac{1}{8.08 \times 17.3}$$

$$= 4.885 + 0.0198 + 0.007154 = 4.9119,$$

so

$$U_iA_i = 1/4.9119 = 0.2035,$$

or

$$U_i = 0.2035/15.65 = 0.013\ \text{Btu/ft}^2\ °\text{F sec} = 46.83\ \text{Btu/ft}^2\ °\text{F h}.$$

Heat balance in the convection section is therefore checked with the assumed exit flue gas temperature of 350°C (the assumed temperature should be greater than the feed inlet temperature) as follows

$$Q = U_i \times (n \times A_i)\text{LMTD}, \quad (1)$$

$$Q = w_c cp_c (T_{c2} - T_{c1}) = w_g cp_g (T_{h1} - T_{h2}). \quad (2)$$

Now,

$w_c = 357{,}840$ lb/h
$cp_c = 0.54$ Btu/lb °F
$T_{c1} = 220°C = 428°F$
$w_g = 55.611$ lb/sec $= 200199.60$ lb/h
$cp_g = 8$ Btu/lb mol °F $= 0.276$ Btu/lb °F
$T_{h1} = 902°F = 1655.9°F$
$T_{h2} = 350°C = 662°F$

Evaluation of T_{h2} and T_{c2} to check the assumed values used above:
From Equation 2:

$$\text{Heat released by flue gas} = 200199.60 \times 0.276 \times (1655.9 - T_{h2}), \quad (a)$$

$$\text{Heat released by feed stock} = 357{,}840 \times 0.54 \times (T_{c2} - 428). \quad (b)$$

Equating (a) and (b)

$$T_{h2} = 3152.616 - 3.497 T_{c2}. \quad (c)$$

Also,

$$357{,}840 \times 0.54 \times (T_{c2} - 428) = U_i(A_i \times n)\text{LMTD},$$

or

$$357{,}840 \times 0.54 \times (T_{c2} - 428) = 46.83 \times 15.65 \times 50 \times \text{LMTD},$$

or

$$T_{c2} = 0.18963 \times \text{LMTD} + 428, \quad (d)$$

$$\Delta T_2 = T_{h2} - T_{c1} = T_{h2} - 428, \quad (e)$$

$$\Delta T_1 = T_{h1} - T_{c2} = 1655.9 - T_{c2}, \quad (f)$$

$$\text{LMTD} = \{(T_{h2} - 428) - (1655.9 - T_{c2})\} / \ln\{(T_{h2} - 428)/(1655.9 - T_{c2})\}. \quad (g)$$

Assume T_{c2}, determine T_{h2} from (c), determine LMTD and redetermine Tc2 from (g). If the assumed and the redetermined value of T_{c2} are equal, then the correct T_{c2} has been evaluated. This has been found to be $T_{c2} = 1092.394°F = 589.1°C$ and $T_{h2} = 589.13°F = 309.5°C$. (However, because of this new gas temperature, the properties of the gas and liquid should have been re-evaluated to determine the film heat transfer coefficients. This has been left as an additional exercise for the students.)

The COT will increase as feedstock leaving the convection section is at 309.5°C. Considering vaporisation of the feedstock, the coil outlet temperature will be COT = (309 + 86)°C = 395°C.

Rate of superheated steam generation:
Since the gas temperature is still high, a separate additional bundle or coil of tubes could be provided at the top of the furnace to preheat the steam or air as required to reduce the exit gas temperature of around 140°C. If saturated steam at 200°C enters this section, then it can be superheated to 400°C at a rate of

200199.60 * 0.276 * (1655.9 – 1092.394) / (0.5 × 200 × 1.8) = 172980.97 lb/h = 78.6 ton/h.

Exercise 8.10

If the flue gas analysis of the furnace in the previous problem is obtained by Orsat analysis as

	vol%
CO_2	11.8
O_2	6.2

determine the percentage of excess air used and the hydrogen to carbon ratio in the fuel.

Solution

Gas	Moles	C-atoms	Moles O_2
CO_2	11.8	11.8	11.8
O_2	6.2	–	6.2
N_2	82.0	–	–
Total	100.0	11.8	18.0

$$\text{Oxygen supplied from air} = \frac{82 \times 21}{79}\ \text{mol} = 21.79\ \text{mol},$$

(taking air as a mixture of 21% oxygen and 79% nitrogen by volume).

Hence, oxygen unaccounted for in the above analysis = 21.79 – 18 = 3.7974 mol, which must have been consumed to produce water not measured.

$$H_2 + \frac{1}{2}O_2 = H_2O.$$

Hence,

Moles of water produced = 2 × 3.7974 = 7.5948 mol,

Moles of hydrogen reacted = 7.5948 mol,

so

Hydrogen present in the fuel was 7.5948 mol.

Also, total oxygen reacted to produce CO_2 and H_2O = 11.8 + 3.7974 = 15.5974 mol.

So

$$\text{Excess oxygen supplied} = \frac{(21.79 - 15.5974) \times 100}{15.5974} = \frac{6.2 \times 100}{15.5974}$$

$$= 39.75 = 40\% \text{ (approx).}$$

So, excess air supplied is 40% by volume.

$$\text{Hydrogen to carbon ratio in fuel} = 7.5948 / 11.8 = 0.6436 \text{ mol/mol}$$

$$= \frac{(7.5948 \times 2)}{(11.8 \times 12)} = \frac{15.1896}{141.6}$$

$$= 0.107 \text{ kg/kg.}$$

Ans.

Exercise 8.11

For the above furnace, if a natural draught is maintained, determine the draught available in millimetres of water. Assume the ambient air temperature is 30°C and the height of the furnace is 20 m.

Solution

From the previous calculation, the gas temperature in the radiation box is 1140.728°C (better take an average temperature) and the air temperature is 30°C. Assuming that the flue gas is rich with air composition, then using the ideal gas law, the densities of ambient air and flue gas are given as

$$\rho_{air} = \frac{1 \times 29}{0.0832 \times (273 + 30)} = 1.1503 \text{ g/L,}$$

$$\rho_{flue} = \frac{1 \times 29}{0.0832 \times (273 + 1140.728)} = 0.2465 \text{ g/L,}$$

$$\text{Pressure difference} = H \times (\rho_{air} - \rho_{flue}) \times g = H \times (1.1503 - 0.2465) \times 0.001$$

$$\times 981 \text{ g/cm}^2$$

$$= 0.88657H \times 0.001 \text{ kg/cm}^2$$

$$= 8.86576 \times 10^{-4} H \text{ kg/cm}^2,$$

where H cm is the height of the furnace.

Draught in centimetres of air column is

$$\frac{\Delta p}{\rho_{air}} = \frac{8.86576 \times 10^4}{1.1503 \times 0.001} H = 0.77 H \text{cm of air}$$

Hence, $\Delta pw = 0.77H \times 1.1503 \times 0.001/1 = 7.7 \times 10^{-4} H$ cm of water col $= 7.7 \times 10^{-4} H \times 10$ mm of water $= 0.0077$ H mm of water.

If the height of the furnace up to the stack is taken as 20 m, then the maximum draught will be $0.0077 \times 20 \times 100 = 15.4$ mm of water.

Ans.

8.12.1 Furnace Design by the Lobo and Evans Method

The design of equipment can be done in two ways: (1) using fundamental theoretical rules and desired constraints or (2) studying existing designed equipment and checking for fulfilment of the desired requirements. The first one is more rigorous and difficult, whereas the second one is less rigorous and a comparatively easy method. The Lobo and Evans method falls into the second category. The steps of design in this method are summarised below.

1. Assume a furnace of the circular or box type and given the design data, i.e., number of tubes in a row, number of rows, tubes spacing, diameter of tube, and length of tube.
2. Determine the effective cold plane surface area as described earlier (αA_{cp}) and the total wall area (A_T), and calculate the effective radiation area ($A_R = A_T - \alpha A_{cp}$).
3. Determine the *mean beam length*, which is defined as the average length of the radiant beam of the flames and is evaluated from the dimension of the furnace as

$$\frac{2}{3}\sqrt[3]{\text{furnace volume}}$$

4. Determine the total pressure of CO_2 and water vapour (P) from the C/H ratio of the fuel and percentage of excess air used.
5. Determine the *effective flame emissivity* (ε), which is obtained from the Values of product of pressure (P) and beam length (L), i.e (PL), assumed flame temperature, and tube skin temperature.
6. Determine the overall exchange factor (φ), which determines the radiation and convection heat exchange and is obtained from the values of (ε) and the ratio of $A_R/\alpha A_{cp}$.

7. Determine the heat generation rate (Q) due to fuel firing and determine the fraction absorbed (R) in the radiation zone using the relation given in Equation 8.20.
8. Determine the rate of heat absorption factor as,

$$\frac{RQ}{\alpha A_{cp} \phi}$$

9. Check the flame temperature from the rate of heat absorption factor evaluated in step 8 and at a permissible tube skin temperature. If the flame temperature evaluated at this step does not match with the assumed temperature in step 5, then repeat the steps from the beginning, i.e., assuming a new furnace design at step 1.

QUESTIONS

1. What are the basic differences between a heat exchanger and a furnace?
2. Define the terms "recuperator," "regenerator," and "contact heat exchanger."
3. Distinguish between a cooler and a condenser.
4. Describe the various elements of construction of a pipe-still heater.
5. Why are there two separate zones of convection and radiation in a furnace?
6. A furnace has to be used to heat crude oil before feeding into a distillation tower. Determine the coil outlet temperature of the crude oil from the following information.
   Crude throughput: 400 m^3/h, API = 36, coil inlet temperature of 260°C at an average specific heat of 2.13 kJ/kg K. The number of pipes in the radiation zone is 180, the pipe o.d. is 2 inch, the length of 40 ft and placed with a clearance of 2.5 in from the centre of each pipe. The pipes are arranged in a single row circularly and vertically surrounding the burners distributed over the floor of the furnace. The fuel gas has a net heating value of 42,000 kJ/kg, the air to fuel ratio is 20 kg/kg, and the fuel firing rate is 2500 kg/h. Neglect the heat absorption in the convection zone.

9 Distillation and Stripping

9.1 PROCESSES OF DISTILLATION AND STRIPPING

Distillation is a process in which more volatile substances in a mixture are separated from less volatile substances. This separation can be carried out either by vaporising a liquid mixture or by condensing a vapour mixture. In the actual practice of distillation, liquid containing the volatile substances is heated to generate vapour which is separated from the mixture followed by condensation of the vapours as the distillate. The residue is rich with the less volatile components. Distillation may be carried out for both batch and continuous processes. Removal of low boiling components in small amounts by heating or by steam or by gas is called *stripping*. This is also similar to distillation and is used to remove dissolved gases from naphtha in a stabiliser column similar to a distillation column; to remove hydrogen sulfide from a solvent (e.g., diethyl amine) used up in a hydrogen sulfide absorber from a hydrocarbon gas mixture; and to remove undesired light fractions from straight run fractions for adjusting the flash point, etc.

9.2 BATCH DISTILLATION

Batch distillation is carried out for testing and analysis of small-scale separation processes. In refineries and petrochemical plants, distillation is carried out in a continuous manner for large-scale separation. As already explained in Chapter 2, crude oil and its products are analysed by ASTM and TBP distillation analysis, which are batch processes. In the ASTM distillation process, a certain amount of sample is gradually heated in a retort to separate the vapour, which is then condensed and collected as the fractions. In this process, fractions boiling with increasing boiling points (bubble points) are collected as the temperature of the retort increases. Thus, separation takes place solely by heat transfer. In a TBP distillation process, the sample is heated in the retort similarly as in the ASTM, but in a long column packed with packing materials. The vapour separated from the liquid is condensed and partly returned (refluxed) to the retort and partly collected as the fraction. In this case, separation is enhanced by the reflux by simultaneous heat and mass transfer of the components. As the vapour temperature rising from the retort is at a higher temperature than the temperature of the refluxed liquid, the less volatile components are condensed from the vapour and join the reflux stream, whereas the more volatile components in the reflux stream are vaporised and join the vapour stream. This enhances the separation solely by heat transfer because of the temperature and phase differences. In addition to this mechanism, separation is also enhanced because of the concentration differences of the components in the vapour and liquid reflux streams. As the vapour rising from the retort is rich with less volatile components and the reflux is rich with more volatile components, there is a transfer of more volatile components from the liquid to the vapour stream and less volatile components are transferred from the vapour to the liquid stream. Because of this simultaneous heat and mass transfer process, the vapour stream is enriched with more volatile components and the reflux stream is enriched with less volatile components. Thus, fractions collected are nearly the true representative of the fractions originally present in the sample. Hence, such a distillation is called true boiling point (TBP) distillation. However, the contacting pattern of the liquid and vapour streams plays the role of separation. The more intimate the contact between the vapour and liquid streams, the greater the separation. The highest possible separation occurs with such an intimate contact when the composition of the components in the vapour and liquid reaches equilibrium. Such a state is known as vapour-liquid equilibrium (VLE). To promote the equilibrium state, contact between the liquid and vapour streams is carried out in a plate or a packed tower. A plate is a circular disk equal to the diameter of the column, with perforations or nozzles or valves through which vapour rises through the liquid hold up (liquid level) over the plate. The liquid falls from the plate through separate larger holes (known as downcomers) provided on the plate surface at suitable positions, usually at the sides of the plate. The more the number of plates, the greater will be the separation of more volatiles from the less volatiles; hence, separation is also expressed in terms of equivalent plates. In a standard TBP distillation column, packing is used that provides separation achieved by 20 equivalent plates.

9.3 BOILING POINT AND EQUILIBRIUM DIAGRAMS

The theory of distillation is most easily understood from binary (two components) component analysis, i.e., the binary distillation process. Consider a binary mixture of two volatile components, A and B, where A is more volatile than B. If the vapour and liquid are in equilibrium, then the composition of A in the vapour and liquid phases will be functions of temperature, T, at a constant total pressure. A plot of the temperature and composition of A in the vapour and liquid phases is shown in Figure 9.1. This is called the boiling point diagram of the components, where y and x are the mole fractions of A in the vapour and liquid phases, respectively. A plot of y vs. x is known as the equilibrium diagram (Figure 9.2). Vapour composition (y) and liquid composition (x) are related by relative volatility (α) as follows

$$y = \alpha x/\{1 + (1 - \alpha)x\}. \tag{9.1}$$

9.4 THEORY OF DISTILLATION

Consider a vapour mixture of two volatile components, A and B, having boiling points 50°C and 100°C, respectively, which is brought in intimate contact with its liquid mixture (reflux) falling on the plate as shown in Figure 9.3. If the vapour temperature and liquid temperature are 120°C and 60°C, respectively, B will condense from the vapour stream as the liquid temperature is lower than its boiling point and will join the liquid stream. Whereas, A from the liquid stream will vaporise as the vapour temperature is greater than its boiling point and will join the vapour stream. Thus, because of this simultaneous vaporisation of A and the condensation of B, the

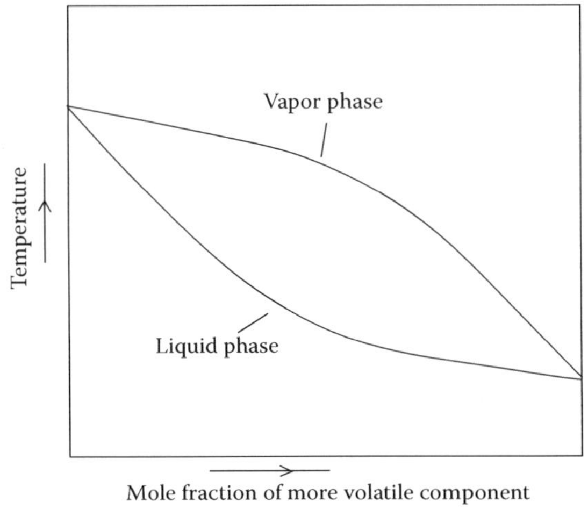

FIGURE 9.1 Boiling point diagram.

liquid stream will be B rich and the vapour stream will be A rich. This is qualitatively shown in Figure 9.3a. Again, since the liquid stream (reflux) is rich with A as compared to the vapour stream, assume A and B are 90% and 10%, respectively, in the reflux and 80% and 20% in the vapour. Because of the concentration difference, the diffusion of A from the liquid stream to vapour and B from vapour to the liquid stream will take place simultaneously. This is explained in Figure 9.3b. Hence, the simultaneous effects of both heat and mass transfer will enrich the vapour stream with A and the liquid stream with B during distillation with reflux. This is explained in Figure 9.3c. It can also be said that in the absence of reflux, separation will take

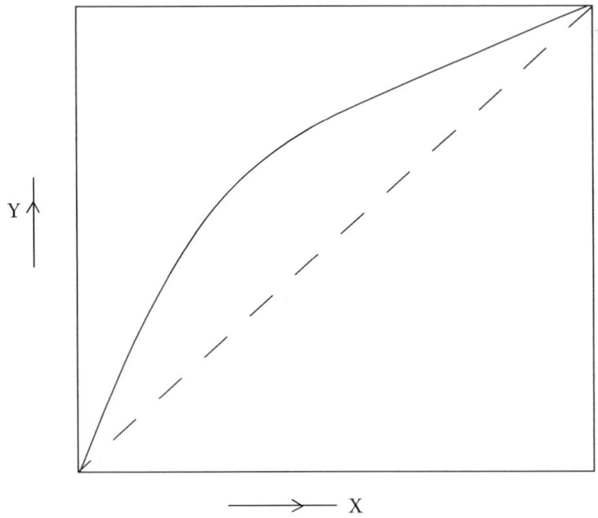

FIGURE 9.2 Vapour-liquid equilibrium diagram.

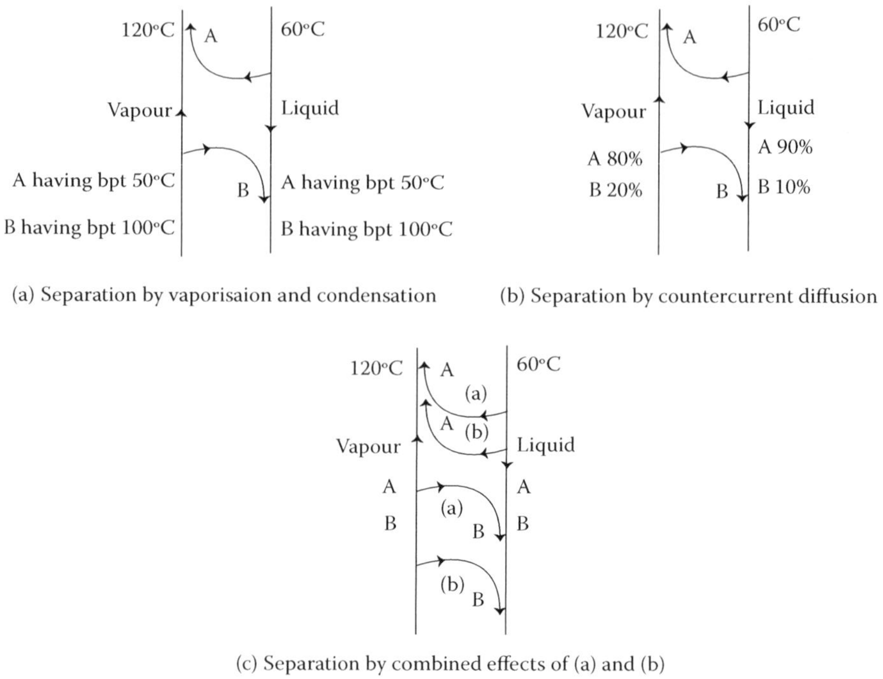

FIGURE 9.3 Separation of a more volatile component A from a mixture of A and B by simultaneous heat and mass transfer processes in a distillation with reflux.

place only because of the heat transfer effect, therefore the separation will be poorer than the distillation with reflux.

9.5 CONTINUOUS DISTILLATION

Continuous distillation is usually carried out in plated columns. Packed columns, combinations of packed sections, and plated sections are also used. Consider a plated column, as shown in Figure 9.4, which contains a number of plates (n_t), each bringing vapour and liquid streams in equilibrium such that the vapour leaving the plate and the liquid leaving the plate are in equilibrium. The column plates are provided with bubble caps for vapour rising and downcomers are provided for liquid to flow down. If the number of plates is counted from the top, the top plate is numbered 1 and the bottom plate is n_t. The feed plate on which the feed mixture is introduced is the n_fth plate. Any plate located in the column may be designated as the nth plate, plates above and below this plate are designated as $n - 1$th and $n + 1$th plates, respectively. Feed plate location divides the column into two distinct sections, namely, *rectification* and *stripping* sections, which are above and below the feed plate, respectively. Vapour from the top plate is condensed by a condenser (usually water cooled) and condensate is collected in a vessel, known as the reflux drum, from which a part of the condensate is refluxed back to the top of the column. Condensate drawn from the reflux drum as product is called the distillate or top product. Residue liquid from the bottom of the column enters

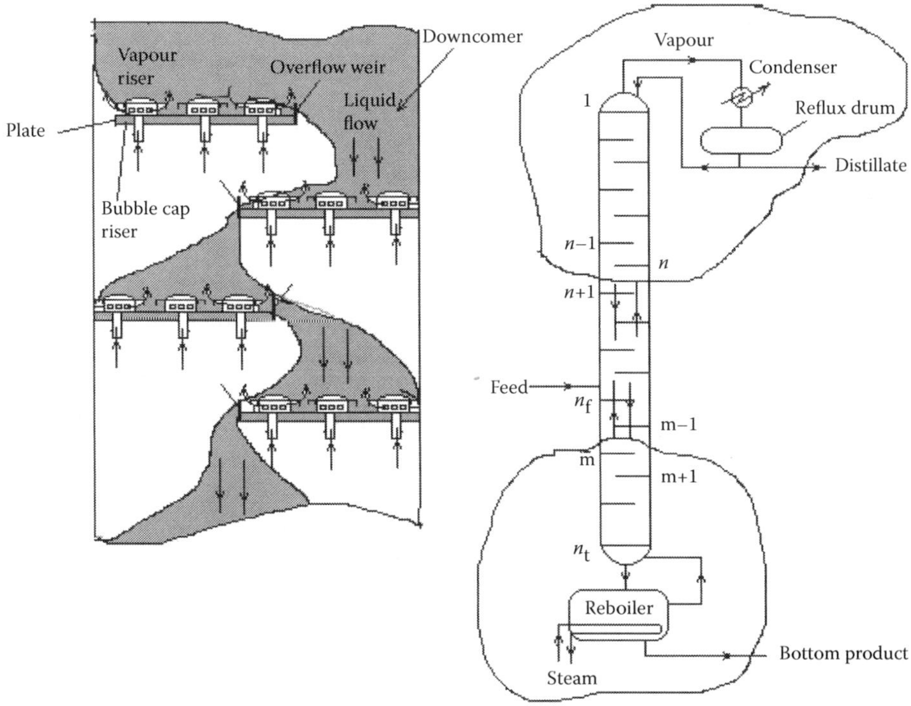

FIGURE 9.4 A distillation column with plates and sections.

a reboiler, which partly reboils or vaporises the liquid back to the bottom plate. Liquid drawn from the reboiler is known as the bottom product or residue. The top product will be rich with a more volatile component and the bottom product will be rich with a less volatile component. Consider a feed of a binary mixture of A and B entering the feed plate of the column at a flow rate of F mol/time, the feed containing the more volatile component A as the z_f mole fraction. The top and bottom products are withdrawn from the reflux drum and the reboiler at the rates of D and B mol/time with the desirable compositions of A in the distillate as x_d and of B in the bottom as x_b mole fractions. Let us analyse the column in sections, assuming that the steady state operation is maintained such that no accumulation or depletion of any component, vapour or liquid streams occur in the plates, reflux drum, reboiler, or any part of the column.

9.5.1 Top Reflux Drum

Assume that the vapour from the top plate is totally condensed to liquid, i.e., the condenser is a *total condenser*. If the rate of vapour leaving the top plate is V mol/time, then

$$V = L + D, \tag{9.1}$$

where L is the reflux rate in mole per time and is expressed as a fraction of the distillate drawn as RD, where R is the reflux ratio, which is the ratio of L to D, i.e.,

$R = L/D$. Thus, the above overall material balance equation for the vapour and liquid streams entering and leaving the reflux drum is rewritten as

$$V = RD + D = (R + 1)D. \tag{9.2}$$

If the vapour composition of A from the top plate is y_1, then the material balance of component A is given as

$$Vy_1 = (R + 1)x_d. \tag{9.3}$$

So

$$y_1 = x_d. \tag{9.4}$$

9.5.2 Rectification Section

Consider the envelope over the *n*th plate in the rectification section, as shown in Figure 9.4, the vapour and liquid streams and their corresponding compositions in mole fractions are given below.

9.5.2.1 Streams Leaving the Envelope under Study

Overhead distillate is leaving at the rate of D moles/time, with x_d mole fraction of more volatile component and liquid is leaving the enveloped section from the *n*th plate at a rate of L_n mole/time with x_n mole fraction of the more volatile component.

Thus, the overall outgoing streams leaving the section are $(L_n + D)$ mol/time and the moles of more volatile components are $(L_n x_n + Dx_d)$.

9.5.2.2 Streams Entering the Envelope under Study

Vapour is entering the *n*th plate from the $(n + 1)$th plate at a rate of V_n mole/time with y_{n+1} mole fraction of the more volatile component.

Thus, the total input stream to the section is V_{n+1} mol/time and a more volatile component entering is $V_{n+1}\, y_{n+1}$ mol/time.

At the steady state condition, overall material balance: the total moles entering and leaving the enveloped section must be same, and component material balance: the moles of the more volatile component entering and leaving the enveloped section must be same. Similarly mole balances is also true for the less volatile components, i.e.,

$$V_{n+1} = (L_n + D), \tag{9.5}$$

and

$$V_{n+1}\, Y_{n+1} = L_n x_n + Dx_d, \tag{9.6}$$

or

$$Y_{n+1} = (L_n x_n + Dx_d)/V_{n+1}. \tag{9.7}$$

This is the general relation of a plate applicable for a rectification zone for the plates from the top plate to the plate above the feed plate ($n = 1$ to $n_f - 1$). The vapour and liquid rates in the plates are actually determined from an enthalpy balance relation over the section. However, the calculation becomes simpler according to the assumption (of McCabe-Thiele) of equal molal heat of vaporisation of both the components, therefore, it can be proved that the vapour rates and liquid rates in the rectification section are unchanged. The same is also true for the vapour and liquid rates in the stripping section.

Thus,

$$V_1 = (R + 1)D = V_2 \ldots = V_n, \tag{9.8}$$

and

$$L_n = RD = L_{n+1} = \ldots = L_1. \tag{9.9}$$

Thus, Equation 9.7 becomes,

$$\begin{aligned} y_{n+1} &= (RDx_n + Dx_d)/(RD + D) \\ &= (R/R + 1)x_n + (1/R + 1)x_d. \end{aligned} \tag{9.10}$$

9.5.2.3 Reboiler

Reboiling is carried out in various methods. For example bottom liquid is boiled by a submerged steam coil or by heating through a furnace. If the vapour generated from the reboiler has a rate and composition of V_b and y_b, respectively, and the rate and composition of the liquid drawn are B and x_b, respectively, then the overall (total mole balance) and component material balance relations are given as

$$L_{nt} = V_b + B, \tag{9.11}$$

and

$$L_{nt}x_{nt} = V_b y_b + Bx_b. \tag{9.12}$$

9.5.2.4 Stripping Section

This section contains the feed plate and the plates below it down to the bottom of the column. Consider an mth plate (counted from the top plate) in the envelope containing the section below the feed plate including the reboiler. Designate the rates and composition of the entering and leaving streams as below.

V_m = molal rate of vapour leaving the mth plate and the section under study with the mole fraction of the more volatile component, y_m, and

L_{m-1} = molal rate of liquid leaving the $(m - 1)$th plate and entering the mth plate and the section under study with a mole fraction of the more volatile component, x_{m-1}.

Where $m = n_{i+1}$ to n_t.
Equating the overall molal flows of the streams,

$$V_m + B = L_{m-1}, \tag{9.13}$$

$$V_m y_m + Bx_d = L_{m-1} x_{m-1}. \tag{9.14}$$

According to the McCabe-Thiele assumption, the vapour and liquid rates are taken as constant in this section.

$$V_m = V_{m-1} \dots = V_{nt} = V_b, \tag{9.15}$$

and

$$L_m = L_{m-1} = \dots = L_{nt}. \tag{9.16}$$

Thus, the relationship for the plate compositions in this section is given as

$$\begin{aligned} x_{m-1} &= (V_b y_m + Bx_b)/L_m \\ &= \{(L_m - B)y_m + Bx_b\}/L_m. \end{aligned} \tag{9.17}$$

9.5.2.5 Feed Plate or Flash Zone

This is the plate on which feed is introduced. This plate cannot establish equilibrium separation, rather the vapour and liquid are in a turbulent condition due to the interaction of the vapour and liquid loads with the reboiled vapour and the reflux liquid. Usually, in order to allow vapour separation, a larger space is provided between the feed plate and the plate above to accommodate all the vapour and liquid loads. Hot crude is fed in the distillation tower and a large amount of vaporisation takes place due to flashing, hence the name 'flash zone' is more appropriate for the feed plate of such a tower.

Considering the envelope containing the n_{nf-1}th plate and the rectification section up to the distillate, the component material balance is given by the rectification equation as,

$$(R + 1)Dy_{nf} = RDx_{nf-1} + Dx_d, \tag{9.18}$$

or

$$Vy_{nf} = RDx_{nf-1} + Dx_d, \tag{9.19}$$

and considering the envelope containing the feed plate and the reboiler, as shown in Figure 9.4, the component material balance equation is given by the stripping section as

$$(L_{nt} - B)y_{nf} = x_{nf-1} L_{n/t} - Bx_b, \tag{9.20}$$

or

$$V_b y_{nf} = x_{nf-1} L_{n/t} - B x_b. \tag{9.21}$$

Subtracting Equation 9.21 from Equation 9.19,

$$(V - V_b) y_{nf} = (RD - L_{n/t}) x_{nf-1} + D x_d + B x_b.$$

Considering the feed is flashed over the feed plate and vapour is formed as f fraction (i.e., $f = V/F$) and the liquid is $(1 - f)$ fraction of the feed molal rate, F, where f depends on the feed condition whether sub-cooled, saturated liquid, saturated vapour, superheated vapour, partial vapour, liquid, etc.

Thus,

$$V = V_b + fF, \tag{9.22}$$

and

$$L_{nt} = RD + (1 - f)F. \tag{9.23}$$

Also, the overall material balance of the column for the component is

$$F.z_f = D x_d + B x_b. \tag{9.24}$$

Hence, replacing the relations 9.23 and 9.24 with f in Equation 9.22, we get the feed plate equation as

$$fF y_{nf} = -(1 - f) F x_{nf-1} + F.z_f, \tag{9.25}$$

or

$$y_{nf} = -\{(1 - f)/f\} x_{nf-1} + z_f / f. \tag{9.26}$$

The slope of the line is $-(1 - f)/f$ and is dependent on the quality of the feed. It is evaluated by enthalpy balance as mentioned next.

9.5.2.6 Evaluation of Fraction Vaporised (f) from the Quality of the Feed

Consider a feed of enthalpy h_f, which is flashed into vapour and liquid at boiling point (bubble point). Thus, the enthalpy of vapour and liquid at boiling point (T_b) are H_v and h_l, respectively, and

$$H_v = h_l + \lambda = cp_l T_b + \lambda, \tag{9.27}$$

taking 0 K as the base temperature, where λ is the molal heat of vaporisation of the mixture.

The enthalpy balance for the feed, vapour, and liquid at a steady state is then obtained as

$$Fh_f = H_v V + h_l L, \tag{9.28}$$

or

$$h_f = fH_v + (1 - f)\, h_l, \tag{9.29}$$

so

$$f = (h_f - h_l)/(H_v - h_l) = (h_f - h_l)/\lambda. \tag{9.30}$$

The enthalpy of the feed will vary with its quality as evaluated below:

1. Considering the feed as a liquid below its boiling point, then its enthalpy is given as

$$h_f = cp_l T_f, \tag{9.31}$$

so from Equation 9.30,

$$f = (h_f - h_l)/\lambda = (cp_l T_f - cp_l T_b)/\lambda = cp_l(T_f - T_b)/\lambda. \tag{9.32}$$

Since $T_f < T_b$, so $f < 0$ or negative.

2. Considering the feed as a liquid at its boiling point, i.e., it is a saturated liquid, then its enthalpy is given as

$$h_f = cp_l T_b, \tag{9.33}$$

so from Equation 9.30,

$$f = (h_f - h_l)/\lambda = (cp_l T_b - cp_l T_b)/\lambda = 0. \tag{9.34}$$

3. Considering the feed as a mixture of liquid and vapour at its boiling point, then its enthalpy is given as

$$h_f = (cp_l T_b + \lambda)\varphi + cp_l T_b(1 - \varphi), \tag{9.35}$$

where φ is the fraction of vapour in the feed mixture. It can be said straightforward that $f = \varphi$. It can also be seen that from Equation 9.30,

$$f = (h_f - h_l)/\lambda = \{(cp_l T_b + \lambda)\varphi + cp_l T_b(1 - \varphi) - cp_l T_b\}/\lambda = \varphi\lambda/\lambda = \varphi. \tag{9.36}$$

4. Considering the feed as a vapour at its boiling point, then its enthalpy is given as

$$h_f = cp_l T_b + \lambda, \tag{9.37}$$

so from Equation 9.30,

$$f = (h_f - h_1)/\lambda = \{(cp_1 T_b - cp_1 T_b) + \lambda\}/\lambda = (cp_1 T_b - cp_1 T_b)/\lambda + 1. \quad (9.38)$$

Hence, for this case $f = 1$ and if the feed is superheated vapour, $f = (C_{pf}\, T_f + \lambda)/\lambda > 1$.

9.6 MCCABE–THIELE METHOD

A graphical solution is made stepwise using the McCabe-Thiele assumptions. For the construction of an operating line for the plates in the rectification section: In a distillation column as discussed in the foregoing section vapour and liquid composition leaving each plate can be found by graphical technique. Draw an equilibrium diagram on graph paper as shown in Figure 9.5, where the vapour mole fraction is along the y-axis and the liquid mole fraction is along the x-axis. Each of the axes starts from 0.00 mole fraction and ends at 1.00 mole fraction. Also, join the origin with the last point (1,1), such that each point on this line indicates $y = x$.

Drawing of operating lines:

Locate the top distillate composition ($y = x_d$, $x = x_d$,) on the $y = x$ line. Now using the material balance relation in the rectification section as

$$y_{n+1} = (Rx_n + x_d)/(R + 1), \quad (9.10)$$

replacing x_n with x_d, we get $y_{n+1} = x_d$.

This means that the above relation for a straight line passes through the point (x_d, x_d). Also at $x_n = 0$, the line will have an intercept with the y-axis equal to $x_d/(R + 1)$. Joining this intercept and the point (x_d, x_d), we get the operating line for the rectification section. For the top plate composition, (y_1, x_1) is obtained by drawing

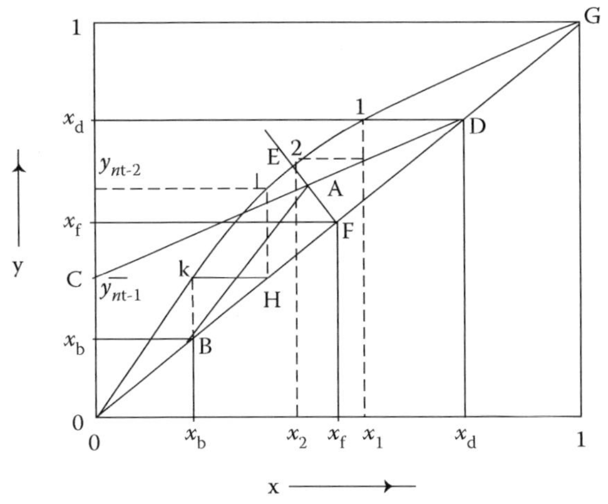

FIGURE 9.5 OEG, the equilibrium curve; DAC, the operating line for the rectification section; EAF, the feed line; AB, the operating line for the stripping section; OBFDG, the $y = x$ line.

a horizontal line parallel to the x-axis from the point (x_d, x_d) and intersecting the equilibrium curve. Drop a vertical line from (y_1, x_1) to the operating line to get (y_2, x_1). To get the composition on the next plate (y_2, x_2), again draw a horizontal line until it cuts the equilibrium curve and repeat this procedure until the composition reaches the plate composition (y_{nf-1}, x_{nf-1}), which is the composition of the plate above the feed plate. It is to be noted that the longer the length of the horizontal or vertical lines drawn in locating the next plate composition on the operating line, the greater the vapour liquid separation and the less the number of plates required for the desired separation.

9.6.1 Operating Line for the Feed Section or Feed Line

The feed line is obtained from Equation 9.26 as,

$$y_{nf} = -\{(1 - f)/f\}x_{nf-1} + z_f /f. \tag{9.26}$$

It is seen that this line passes through the point (z_f, z_f), i.e., at $x_{nf-1} = z_f$, $y_{nf} = z_f$. This point is located on the $y = x$ line and using the slope of the feed line, $-(1 - f)/f$ as mentioned above, the feed line (FE) is drawn. The point of intersection with the operating line for the rectification section already drawn is located at point "A".

9.6.2 Operating Line and Plates for the Stripping Section

$$x_{m-1} = \{(L_{nt} - B)y_m + Bx_b\}/L_{n/t}. \tag{9.17}$$

This equation shows the at $x_{m-1} = x_b$, $y_m = x_b$, i.e., it passes through the point (x_b, x_b).

Since this will also pass through the point of intersection of the feed line and the rectification line obtained, by joining these two points A and B, the operating line for the stripping section is obtained. This could also be drawn with knowledge of the vapour and liquid rates in this section from the above equation. Starting from the (x_b, x_b) point, a horizontal line is drawn to intersect the equilibrium line at K to get the vapour composition for the upper plate ($nt - 1$) and this is repeated until the plate's liquid composition reaches the feed composition or crosses the common point of intersection with the feed line. Thus, the number of plates can be evaluated. The greater the reflux rate, the less the number of plates required for the given separation.

Thus, the number of plates will be minimum at total reflux, i.e., when $D = 0$. This can be verified from the operating lines while $D = 0$, i.e., $R = \infty$, then the slope becomes unity, i.e., it will coincide with the $y = x$ line.

Similarly at $R = \infty$, $L_{nt} = RD + fF = \infty$, i.e., slope = $(L_{nt} - B)/L_{n/t}$ = unity. Hence, both the operating lines of the rectification and stripping sections will coincide on the $y = x$ line. The number of plates will be given by the steps drawn as described above. Note that the difference between y and x is greater as compared to the corresponding values at a reflux rate less than the total. The minimum number of plates can also be determined using the Fenske relation.

According to this relation,

$$N_{min} = \frac{\text{In}[x_D(1-x_B)/x_B(1-x_D)]}{\text{In}\alpha_{AB}} - 1, \quad (9.39)$$

where In α_{AB} is the relative volatility of A with respect to B, given by the relation,

$$\alpha_{AB} = (y_A/x_A)/(y_B/x_B), \quad (9.40)$$

i.e.,

$$y_A = \alpha_{AB}x_A/\{1 + (\alpha_{AB} - 1)x_A\}, \quad (9.41)$$

Again the number of plates will be infinite when $R = 0$, and this can be proved by observing the slopes to be zero. If the operating lines intersect or are tangential to the equilibrium curve, the number of plates will become infinitely large well before reflux reaches zero. In reality a distillation column cannot be operated at zero or total reflux, but at a certain reflux ratio determined by the operating cost. At a minimum cost of operation, the reflux ratio is called the optimum reflux ratio at which the distillation column should operate. The cost of the operation includes the pumping cost of reflux and the fixed cost of the column. The pumping cost increases with reflux rate and the fixed cost increases with the number of plates. Hence, at a high reflux rate, as the number of plates decreases, the fixed cost (which is directly proportional to the number of plates) comes down. Thus, initially when the reflux rate increases from the minimum (i.e., when the number of plates is infinite), the cost of pumping increases but the fixed cost decreases, as a result, the total cost decreases and reaches a minimum and then starts increasing with an increasing reflux ratio as the cost of pumping becomes larger than the reduction in fixed cost. The optimum reflux ratio is determined as the total minimum cost.

9.7 ENTHALPY BALANCE METHOD

Distillation column calculations using enthalpy and material balances are more appropriate than the McCabe-Thiele method. Here, the rates of liquid and vapour from each plate are determined along with the plate temperatures.

9.7.1 Reflux Drum

Material balances:
Overall:

$$V_1 = (R + 1)D. \quad (9.42)$$

Component:

$$V_1y_1 = (R + 1)Dx_D. \quad (9.43)$$

Enthalpy balance:

$$V_1H_1 = (R+1)Dh_D + q_c, \tag{9.44}$$

where q_c = condenser duty required = latent heat of vapour × rate of vapour from the top plate = λV_1 (if a saturated liquid reflux is used).
Assuming total condenser, $y_1 = x_D$ and

$$H_1 = h_D + q_c/(R+1)D, \tag{9.45}$$

where h_d is the enthalpy of distillate drawn as the drum temperature is known and obtain T_1 from H_1. Assume L_1 and get V_2 from Equation 9.46 and determine y_2 from Equation 9.47 and finally establish L_1 with additional trials with the help of Equation 9.48. Solution can be done repeatedly until assumed values converge with calculated values. A small computer program can be used for this purpose.

9.7.2 Top Plate

From the value of H_1, the top temperature and equilibrium liquid composition (x_1) corresponding to $y_1(=x_D)$ and also the liquid enthalpy on the plate (h_1) are obtained.

Considering the section from the top plate up to the distillate draw.
Material balances:
Overall:

$$V_2 = L_1 + D. \tag{9.46}$$

Component:

$$V_2y_2 = L_1x_1 + Dx_D. \tag{9.47}$$

Enthalpy balance:

$$V_2H_2 = L_1h_1 + Dh_D + q_c. \tag{9.48}$$

Assume plate temperature, T_2, obtain equilibrium composition (y_2, x_2) by trial and error and corresponding H_2, h_2. Solve simultaneously for V_2 and L_1 from Equations 9.47 and 9.48.

Rectification section:
Continue the above steps for the calculations of plate temperature, composition, and enthalpies of both vapour and liquid streams.

Feed plate:
As done before, if f is the fraction of feed that is vaporised, then

at rectification and feed section:

$$V_f = L_{f-1} + D - fF, \tag{9.49}$$

at the stripping and feed section:

$$V_f = L_{f-1} + (1 - f)F - B, \tag{9.50}$$

This, in fact, is the overall material balance for the column

$$F = D + B \tag{9.51}$$

The component balances are

$$V_f y_f = L_{f-1} x_{f-1} + D x_D - fF z_f, \tag{9.52}$$

$$V_f y_f = L_{f-1} x_{f-1} + (1 - f)F z_f - B h_B. \tag{9.53}$$

Heat balance equations:

$$V_f H_f = L_{f-1} h_{f-1} + D h_D - fF h_f, \tag{9.54}$$

$$V_f H_f = L_{f-1} h_{f-1} + (1 - f)F h z_f - B h_B, \tag{9.55}$$

where x_{f-1}, h_{f-1} are known from the previous plate.

Assume T_f and obtain H_f and h_f, then solve for V_f and L_{f-1} simultaneously from Equations 9.54 and 9.55. To check for T_f, use Equations 9.52 and 9.53.

9.7.3 Reboiler

Material balances:
Overall:

$$L_n = V_B + B. \tag{9.56}$$

Component:

$$L_n x_n = V_B x_B + B x_B. \tag{9.57}$$

Heat balance:

$$L_n h_n = V_B H_B + B h_B - q_B. \tag{9.58}$$

Determine the bubble point temperature and find the corresponding equilibrium composition from the liquid composition, x_B, where

$$q_B = D h_D + B h_B + q_C - F h_f, \tag{9.59}$$

by the overall heat balance of the column.

Exercise 9.1

A continuous distillation column has to separate benzene and toluene from a mixture containing 44% benzene and 56% toluene in mole percentage. If the feed rate is 350 mol/h and the required benzene content in the top and bottom products are 97.4% and 2.35% by mole, respectively, determine the number of plates required in the column. Assume McCabe-Thiele assumptions of equimolal overflow. Given that the density and specific heat of the feed are 440 kg/m^3 and 1.85 kj/kg K, respectively. The latent heat of vaporisation is 376 kj/kg. The relative volatility of benzene with respect to toluene is 2.434. Take the reflux ratio as 3.5 mol/mol of distillate. Assume that the feed is a mixture of two-thirds vapour and one-third liquid.

Solution

Draw the equilibrium curve by using the relation of relative volatility,

$$y = \alpha x/\{1 + (1 - \alpha)x\} = 2.434x/\{1 - 1.434x\}$$

Draw the $y = x$ line.

Locate the distillate composition (0.974, 0.974) and the bottom composition (0.0235, 0.0235), as shown in Figure 9.6. Locate the intercept on the y-axis at $y = 0.974/(3.5 + 1) = 0.2164$ and draw the operating line for the rectification section. Also, locate the feed point as (0.44, 0.44) and draw a line with a slope −(1−f)/f, where $f = 2/3$, i.e., slope = − 0.5. Locate the point of intersection of the operating line of the rectification section with this feed line. Connect the point of intersection with a straight line through the bottom composition point, which yields the operating line of the stripping section. Next, draw horizontal lines starting from the point of the distillate composition located on the operating line of the rectification section and extend the line to intersect with the equilibrium line to get the equilibrium liquid composition. Drop a vertical line from this point of intersection

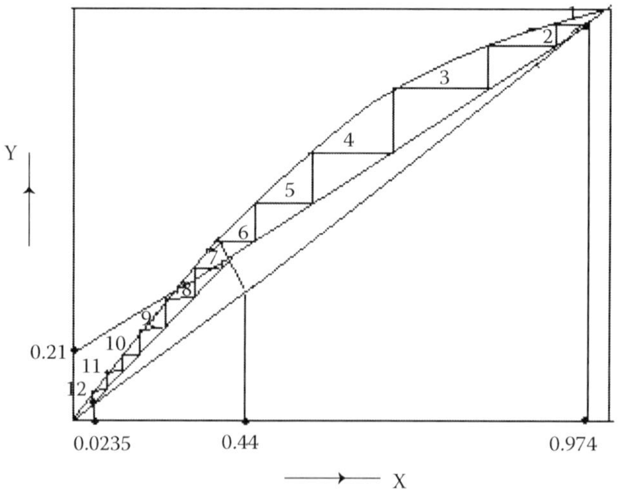

FIGURE 9.6 Graphic solution to Exercise 9.1.

to the operating line to get the vapour composition for the next plate. Repeat the process by drawing horizontal lines for the next plate until the last horizontal line crosses the feed line intersection. Count the numbers of horizontal lines drawn, which will be equal to the number of theoretical plates in the rectification column.

Next, start from the bottom composition located on the $y = x$ line and draw a vertical line to intersect the equilibrium line to get the vapour composition. From this point of intersection, draw a horizontal line to intersect the operating line of the stripping section to get the liquid composition, and continue to repeat the process by drawing vertical and horizontal lines until the last horizontal line crosses the feed line. Count the horizontal lines as the number of plates required for the stripping section. The total number of theoretical plates is the sum of the plates in the rectification and the stripping section. The number of plates counted is 12.

9.7.4 Numerical Solution

For total condenser, $y_1 = x_d = 0.974$.
Top plate liquid composition in equilibrium is

$$x_1 = y_1/\{\alpha + (1-\alpha)y_1\} = 0.974/\{2.434 - 1.434 \times 0.974\} = 0.949.$$

The vapour composition leaving the second plate is obtained from the material balance Equation 9.07,

$$y_2 = (L_m x_m + D x_d)/V = (3.5 \times 0.949 + 153.4 \times 0.974)/(4.5 \times 153.4) = 0.900.$$

Equilibrium $x_2 = 0.900/\{2.434 - 1.434 \times 0.900\} = 0.879$.

Proceed to the third and other plates, computing in a similar fashion until the point of intersection with the feed line and the rectification operating line is reached. This point of intersection is obtained by solving the following relations,

$$y = (Rx + x_d)/(R+1), \quad \text{(a)}$$

and

$$y = -\{(1-f)/f\}x + z_f/f, \quad \text{(b)}$$

solving

$$\begin{aligned} x &= \{z_f/f - x_d/(R+1)\}/\{R/(R+1) + \{(1-f)/f\} \\ &= (0.44/0.67 - 0.974/4.5)/(3.5/4.5 + 0.44/2) = 0.347. \end{aligned}$$

The number of plates is counted until the liquid composition reaches 0.347 mole fraction. Similarly for the stripping section, calculation is started from the reboiler where $x_b = 0.0235$ and the equilibrium vapour composition is $y_b = \alpha x_b/\{1 + (\alpha - 1)x_b\} = 2.434 \times 0.0235/\{1 + 2.434 \times 0.0235\} = 0.0553$. The liquid composition in the bottom plate is obtained from the stripping section operating line, $X_{nt} = \{(L_{nt} - B)y_b + Bx_b\}/L_m = \{(654.267 - 196.6) \times 0.0553 + 196.6 \times 0.0235\}/654.267 = 0.0457$ (where $L_{nt} = R \times D + (1-f) \times F = 3.5 \times 153.4 + 0.33 \times 350 = 654.267$). The equilibrium vapour composition of this plate is $y_{nt} = 2.434 \times 0.0457/$

$\{1 + 2.434 \times 0.0457\} = 0.1748$. Proceed to the next upper plates similarly until the point of intersection with the feed line for the liquid composition (0.347) is reached. This is shown in the following table. From the computations, it is seen that the seventh plate from the top or the sixth plate from the bottom nearly meets the feed line intersection liquid composition. Hence, the feed plate is the seventh plate from the top, there are 13 plates including the feed plate required in the column; 7 in the rectification and 6 in the stripping section.

	Rectification section	
Plate No.	**Vapour mole fraction, *y***	**Liquid mole fraction, *x***
Reflux drum		0.974
1	0.974	0.939
2	0.9468	0.8796
3	0.9006	0.7882
4	0.8295	0.6666
5	0.7349	0.5325
6	0.6306	0.4122
7	0.5371	0.3228*
	Stripping section	
Reboiler	0.0553	0.0235
Bottom plate, *nt*	0.1045	0.0457
nt – 1	0.1748	0.0801
nt – 2	0.2654	0.1292
nt – 3	0.3692	0.1925
nt – 4	0.4656	0.2636
nt – 5	0.5479	0.3324*

*Nearest to the liquid composition of 0.347, the point of intersection between the feed line and the operating lines.

Exercise 9.2

The above problem can also be solved for feeds of different quality, i.e., when the feed is saturated liquid ($f = 0$), sub-cooled liquid at 20°C ($f = -0.37$), and saturated vapour ($f = 1$). Where f for the sub-cooled liquid, $f = cp(T_f - T_s)/\lambda$, where cp = 1.85 kj/kg k, $T_f = 20$, $T_s = 95$ and $\lambda = 376$ kj/kg, so, $f = 1.85 \times (20 - 95)/376$ or $f = -0.36$.

The solution is given below.

		Rectification section	
Feed quality, *f* = 0	**Plate No.**	**Vapour mole fraction, *y***	**Liquid mole fraction, *x***
	Reflux drum		0.974
	1	0.974	0.939
	2	0.9468	0.8796

		Rectification section	
Feed quality, $f = 0$	**Plate No.**	**Vapour mole fraction, y**	**Liquid mole fraction, x**
	3	0.9006	0.7882
	4	0.8295	0.6666
	5	0.7349	0.5325
	6	0.6306	0.4122*
		Stripping section	
	Reboiler	0.0553	0.0235
	Bottom plate, nt	0.1098	0.0480
	$nt - 1$	0.1950	0.0900
	$nt - 2$	0.3120	0.1570
	$nt - 3$	0.4450	0.2480
	$nt - 4$	0.5690	0.3520
	$nt - 5$	0.5479	0.4480*

*Nearest to the liquid composition of 0.440, the point of intersection between the feed line and the operating lines.

For $f = 0$, the total number of plates is 10, 6 in the rectification section including the feed plate (sixth from the top) and 4 in the stripping section. Also, note that the point of intersection with the feed line is 0.44 for the liquid composition.

		Rectification section	
Feed quality, $f = 1$	**Plate No.**	**Vapour mole fraction, y**	**Liquid mole fraction, x**
	Reflux drum		0.974
	1	0.974	0.939
	2	0.9468	0.8796
	3	0.9006	0.7882
	4	0.8295	0.6666
	5	0.7349	0.5325
	6	0.6306	0.4122
	7	0.5371	0.3228
	8	0.4675	0.2651*
		Stripping section	
	Reboiler	0.0553	0.0235
	Bottom plate, nt	0.1000	0.0437
	$nt - 1$	0.1589	0.0720
	$nt - 2$	0.2300	0.1093
	$nt - 3$	0.3077	0.1544
	$nt - 4$	0.4502	0.2517*
	$nt - 5$	0.5033	0.2940*

*Nearest to the liquid composition of 0.280, the point of intersection between the feed line and the operating lines.

For $f = 1$, the number of plates required is 14 including the feed plate (eighth from the top), 8 in the rectification section, and 6 in the stripping section, where the point of intersection is 0.280 for the liquid composition.

Rectification section

Feed quality, $f = -0.36$	Plate No.	Vapour mole fraction, y	Liquid mole fraction, x
	Reflux drum		0.974
	1	0.974	0.939
	2	0.9468	0.8796
	3	0.9006	0.7882
	4	0.8295	0.6666
	5	0.7349	0.5325*
	Stripping section		
	Reboiler	0.0553	0.0235
	Bottom plate, nt	0.1098	0.04920
	$nt - 1$	0.1950	0.0946
	$nt - 2$	0.3120	0.1680
	$nt - 3$	0.4450	0.2701
	$nt - 4$	0.5690	0.3865
	$nt - 5$	0.5479	0.4924
	$nt - 6$	0.7639	0.5707*

*Nearest to the liquid composition of 0.52, the point of intersection between the feed line and the operating lines.

For $f = -0.36$, the number of plates required is 12 including the feed plate (seventh from the top), 5 plates in the rectification section, and 7 in the stripping section, where the point of intersection is 0.52 for the liquid composition.

Example 9.3

Determine the number of plates for a distillation column when the reflux is maximum with zero draws of products.

Solution

The number of plates will be minimum and can be found theoretically for the following conditions: (a) the condensation is total; (b) the McCabe-Thiele assumption holds good; (c) constant relative volatility; (d) reboiler is an equilibrium vaporiser, i.e., ideal plate; and (d) applicable for a binary mixture only.

If the number of plates is n counted from the top of the column, then $n + 1$th plate is the reboiler.

The top liquid composition in the reflux drum for A and B in mole fractions are x_{AD} and x_{BD}, respectively, and in the reboiler liquid these are $x_{Aw}(x_{An+1})$ and $x_{Bw}(x_{Bn+1})$, respectively.

The operating lines' equations are $y_{n+1} = x_n$, i.e.,

$$y_1 = x_d,\ y_2 = x_1,\ y_2 = x_2 \ldots y_{n+1} = x_n. \tag{1}$$

From the definition of relative volatility of A, with regard to B is

$$\alpha_{AB} = (y_{An} / x_{An})/(y_{Bn} / x_{Bn}). \tag{2}$$

So,

$$\alpha_{AB} = (y_{A1} / x_{A1})/(y_{B1} / x_{B1}) = (x_{AD} / x_{A1})/(x_{BD} / x_{B1}). \tag{3}$$

Also,

$$\alpha_{AB} = (y_{A2} / x_{A2})/(y_{B2} / x_{B2}) = (x_{A1} / x_{A2})/(x_{B1} / x_{B2}). \tag{4}$$

Similarly,

$$\alpha_{AB} = (x_{A2} / x_{A3})/(x_{B2} / x_{B3}). \tag{5}$$

Thus,

$$\begin{aligned}(x_{AD}/x_{A1})/(x_{BD}/x_{B1}) \times (x_{A1}/x_{A2})/(x_{B1}/x_{B2}) \times (x_{A2}/x_{A3})/(x_{B2}/x_{B3}) \ldots (x_{An}/x_{An+1})/(x_{Bn}/x_{Bn+1}) \\ = (x_{AD}/x_{BD})/(x_{Aw}/x_{Bw}) \\ = (x_{AD}x_{Bw}/x_{BD}x_{Aw}),\end{aligned} \tag{6}$$

or

$$(\alpha_{AB})^n = (x_{AD}x_{BW} / x_{BD}x_{AW}). \tag{7}$$

Taking log,

$$n = \log\{(x_{AD}x_{BW} / x_{BD}x_{AW})\}/\log(\alpha_{AB}), \tag{8}$$

where n includes the reboiler as a plate, hence the number of plates required in the distillation column is,

$$N_{min} = \log\{(x_{AD}x_{BW} / x_{BD}x_{AW})\}/\log(\alpha_{AB}) - 1. \tag{9}$$

Equation 9 is the Fenske equation already mentioned in Equation 9.39.

It is to be noted that with an increase in either the overhead reflux rate or the reboiler heating rate, the number of plates required for a desired product decreases. This can be easily verified by observing the slopes of the operating lines, while the slope approaches the $y = x$ line, the number of plates decreases.

Exercise 9.4

Repeat Exercise 9.1 applying the enthalpy balance method.

Solution

Graphically, the solution can be found using the Savarit-Ponchon method. This requires an enthalpy concentration diagram for the benzene-toluene system to be constructed. The ordinates represent enthalpies (H, h) and the abscissa mole fractions (x, y). Vapour phase enthalpy (H) with the liquid phase enthalpy (h) are plotted along with the equilibrium vapour mole fraction (y) and the liquid mole fraction (x) in the same plot. The feed, distillate, and bottom compositions are located in the diagram followed by drawing of the operating lines. The points of the solution are determined with the help of equilibrium lines. The detailed method will not be presented here. With the availability of computers, the numerical method is more appropriate and will be discussed here. The following steps for the solution are presented below.

Reflux drum: Assume that the condenser totally condenses all the vapours, then the vapour mole fraction of benzene is 0.974. The vapour leaving the top plate at a rate of $(R+1)D$ mole/time. The liquid temperature is taken as 50°C which is around the boiling point of benzene. The enthalpy of $h_D = c_p T_D$ and the cooler duty $q_c = (R+1)D\lambda$ is evaluated where $\lambda = 376$ kj/kg, $= 376/80$ kj/mol, taking the average molecular weight of the reflux to be 80.

Top plate: The liquid composition of this plate is obtained by assuming the top temperature, say 80°C, and then equilibrium (x) for both benzene and toluene are obtained using Raoult's law and it is checked that the sum of these liquid compositions is unity, otherwise a new guess is made for the temperature, and calculations are repeated. In this process of evaluating the temperature and liquid compositions, enthalpies for both the liquid and vapour are determined.

Rectification plates from the second plate downward: From the following relations,

$$V_2 = L_1 + D, \tag{9.46}$$

$$V_2 y_2 = L_1 x_1 + D x_D, \tag{9.47}$$

$$V_2 H_2 = L_1 h_1 + D h_D + q_C. \tag{9.48}$$

Next, the equilibrium liquid mole fraction is determined as before by assuming the plate temperature. The temperature is checked by using the enthalpy balance equation. The vapour and the liquid leaving rates are evaluated next. This process of calculations is repeated until the vapour composition from the plate reaches the feed plate composition, which is given as

$$y_{nf} = -\{(1-f)/f\} x_{nf-1} + z_f/f. \tag{9.26}$$

At this point, the feed plate is identified.

Reboiler: From the given reboiler duty ($q_B = 260{,}000$ kcal/h), the rate of vapour generation is evaluated as $V = q_B/\lambda$, and assuming the temperature of the reboiler, the vapour composition (y_B) is evaluated and checked that the sum of these compositions is unity. After knowing the temperature and compositions, the enthalpies of both the vapour and liquid drawn are obtained.

Bottom plate: The liquid rate, $L = B + V$, is determined, and the liquid composition is given as

$$x_m = Vy_B + Bx_B.$$

Next, the bottom plate temperature is assumed followed by evaluation of the equilibrium vapour composition, and checking the sum to unity, the temperature is confirmed. The vapour and liquid rates are evaluated next. This is continued for the plates above the bottom plate until the vapour composition reaches the feed line composition as per Equation 9.26.

A small program suffices to solve the above problem and can be used to see the effects of changes in feed rate, composition, thermal quality of feed, reflux ratio, and reboiler duty. Some of the results are given below.

9.7.5 Types of Refluxes

External refluxes are the part of the vapour or liquid streams that are separated from the column, cooled (by external coolers or condensers), and returned to the column. Any reflux returned from outside the column is external reflux. Vapour streams leaving the top plate of the column are condensed usually by a water cooler and collected in a reflux drum. Condensate is drawn as the top product or distillate and a part of this, known as the *overhead or top reflux*, is returned back to the column top. Therefore, the overhead reflux is an external reflux. If the vapour is just condensed to its saturated liquid (or bubble point temperature), the overhead reflux is then a saturated liquid reflux. If the condensate is further cooled below the saturated liquid temperature, the reflux is a sub-cooled reflux. In fact, the overhead reflux is a sub-cooled reflux in reality. The overhead reflux cools the uprising vapour from which the components that are at boiling point below the temperature of the overhead reflux are condensed and separated from the vapour stream leaving the column top. The lower the temperature of the overhead reflux, the greater this separation, causing a reduced amount of vapour flow to leave the column. Too low a temperature and too high an overhead reflux flow rate will cause too much condensation when too little vapour will leave the column and too large a flow rate of liquid will flow down the column plates. Liquid may also weep through the risers and trickle down to the lower plates allowing no vapour to flow up (or pass through), causing little or no separation of the components. On the other hand, when the reflux rate is low as compared to the vapour rising, separation will be poor due to incomplete contact of vapour with the liquid and the liquid flow rate is affected by the vapour flow rate, i.e., *loading* is initiated and when the vapour flow rises further resisting liquid to flow down. Foaming may also be generated over the plate surrounding the risers. This situation is called *flooding*, which, in fact, is the phenomenon limiting the maximum vapour velocity in the column. Hence, the maximum flow rates of both the liquid and vapour in a distillation column must be taken care of as far as the quality of the products is concerned. A circulating reflux or pumparound are other types of external reflux where the liquid stream separated from the column is cooled in an external cooler either by a water cooler or by other cold streams and returned back to the column. External refluxes keep the heat load in the column under control.

9.7.6 Internal Reflux

A part of the volatile components in the vapour stream gets condensed as it rises up the column as the temperature in the upper plates is lower. This condensate then falls to the lower plate, only to be vaporised again and condensed in the upper plates. This process is continuously repeated and the stream, which takes part in the repeated condensation and vaporisation without leaving the column, is known as the *internal reflux*. The internal reflux flow cannot be physically measured, but it can be evaluated by making an enthalpy balance calculation over the plates.

9.7.7 Minimum Reflux

If the reflux increases, the distillates become more enriched with more volatile components for a column of a specified number of plates. The higher the reflux, the more the pumping cost, the less the distillate rate, the more heater duty, etc., causing the cost of the operation to increase directly with the increase in the reflux rate. On the other hand, if the reflux is reduced, the number of plates in the column required to maintain the same degree of enrichment of more volatile components in the top product (or enrichment of less volatile components in the bottom product) will have to be increased, leading to high cost of equipment. In order to reduce the cost of the operation, the reflux rate is minimized so that total cost which is the sum of the cost of operation and the fixed cost becomes minimum. This reflux is called the optimum reflux rate. Thus, the actual operating reflux rate is maintained near the optimum rate.

9.8 GAP AND OVERLAP

In petroleum distillation, the separated products from the distillation column are, in fact, a mixture of hydrocarbons, and the use of VLE constants (k) are not exact, though their average values are available for certain light fractions. Empirical methods are widely used in the absence of VLE information for boiling fractions. The degree of separation and the quality of the products are conveniently expressed in terms of *gap* or *overlap* of the separated streams. Gap is defined as the difference between the 5% and the 95% points of the heavier and the lighter fractions, respectively. For instance, if a 5% point of kerosene is 145°C and a 95% point of naphtha is 140°C, then the gap between them is + 5°C. This indicates that the naphtha is completely free from the kerosene hydrocarbons and the kerosene is also free from the naphtha hydrocarbons, i.e., the ASTM distillation curves for these products are not intersecting anywhere, hence no mix-up. On the other hand, if the 5% and 95% points were 140°C and 145°C, respectively, for kerosene and naphtha, then there would be an overlap, i.e., gap = –5°C or overlap = 5°C. In this case of overlap, some of the naphtha components (heavier parts) will be present in the kerosene and some of the kerosene components (lighter parts) will be found in the naphtha. Thus, this 5% and 95% point difference is positive for the gap and negative for the overlap. Gap or overlap are monitored to adjust the initial boiling point (IBP) or flash point of the adjacent streams from a column. A higher reflux rate and a larger number of plates maintain a good degree of separation or gap, but at high cost of refining. Designers, therefore, prefer to compromise on overlap and gap. In fact, internal reflux between the side stream drawing sections can be varied to a certain extent in a specified column and hence gap/overlap can be manipulated. The presence of steam also enhances gap/overlap between the side products.

9.9 PACKIE'S CORRELATION

Packie has developed correlations between the number of plates and the internal reflux ratio between the side streams as a function of gap/overlap. Some of this information is presented in Table 9.1.

Example 9.5

In a crude distillation column, the following data have been collected in a test run of a refinery. Determine the internal reflux and the number of plates required for separating a 270°C–340°C cut and a 340°C–365°C cut as a function of gap/overlap.

Data for the Draw Plates for 270°C–340°C and 340°C–365°C Cut Side Streams

Entering streams/phase	Tons/h	Draw temperature, °C, 340°C–365°C cut
IBP-140°C cut/vap	47.61	328
Top cir. reflux to HSD pool/vap	6.15	328
140°C–270°C cut/vap	49.06	328
270°C–340°C cut/vap	53.40	328
340°C–365°C cut/vap	12.79	328
Leaving streams	**Tons/h**	**Draw temperature, °C, 270°C–340°C cut**
IBP-140°C cut/vap	47.61	292
Top cir. reflux to HSD pool/vap	6.15	292
140°C–270°C cut/vap	49.06	292
270°C–340°C cut/vap	53.40	292
340°C–365°C cut/liq	12.79	292

The streams entering and leaving the column are schematically given in Figure 9.7.

TABLE 9.1
Typical Packie's Correlations

Gap/overlap, °F	*Δ50%, 100°F	*Δ50%, 150°F	*Δ50%, 200°F	*Δ50%, 250°F
	$N \times$ IR**	$N \times$ IR**	$N \times$ IR**	$N \times$ IR**
(A) Without Using Steam				
−50	8			
−40	10			
−30	12			
−20	15	8		
−10	18	9		
0	25	12		
+10	45	15	8	
+20		21	10	
+30		35	13	
+40			17	9
+50			30	13
+60				18
+70				35
+80				
+90				
(B) Using Steam				
−50				
−40				
−30	8			
−20	10			
−10	13			
0	16	8		
+10	25	10		
+20	50	13		
+30		20	8	
+40		35	12	
+50			17	9
+60			30	13
+70				20
+80				50
+90				

*Δ50% indicates the difference between the 50% points of the side streams.
**$N \times$ IR indicates N, the number of plates, and IR, the internal reflux ratio.

Solution

The enthalpies of the vapour and liquid streams are obtained from an enthalpy chart for petroleum fractions (*Petroleum Refiner*, Supplement to Vol. 24, No. 4, April 1945) and a heat balance is carried out to determine the internal reflux rate.

Heat balance of the streams

Entering streams/ phase	Tons/h	Draw temperature, °C for 340°C–365°C cut	Enthalpy, kcal/kg	kcal/h × 1000
IBP-140°C cut/vap	47.61	328	270	12854.7
Top cir. reflux to HSD pool/vap	6.15	328	260	1599.0
140°C–270°C cut/vap	49.06	328	254	12467.24
270°C–340°C cut/vap	53.40	328	246	13136.4
340°C–365°C cut/vap	12.79	328	241	3082.39
Total	169.01			43133.73
Leaving streams	**Tons/h**	**Draw temperature, °C, 270°C–340°C cut**		
IBP-140°C cut/vap	47.61	292	244	11616.84
Top cir. reflux to HSD pool/vap	6.15	292	235	1445.25
140°C–270°C cut/vap	49.06	292	230	11283.80
270°C–340°C cut/vap	53.40	292	225	12015.0
340°C–365°C cut/liq	12.79	292	195	2494.05
Total	169.01			38854.94

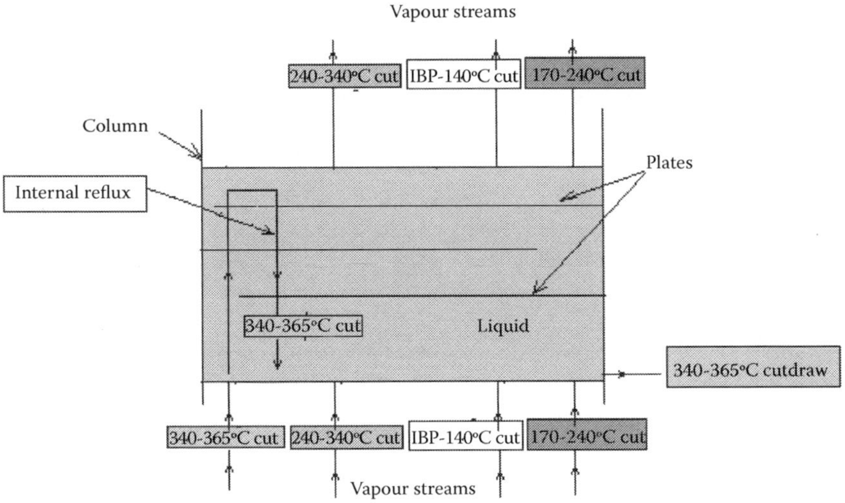

FIGURE 9.7 Input and output streams in a plated section for 270°C–340°C and 340°C–365°C cut side streams in Example 9.5.

Hence, the heat absorbed by the internal reflux by condensation is (433,133.73 − 38,854.94) × 1,000 = 4,278.79 × 1,000 kcal/h.

Therefore, the internal reflux rate is 4,278.79 × 1,000/(241 − 195) = 93.8 × 1,000 = 93,880 kg/h.

The reflux ratio is 93,880/169,010=0.554.

Taking the 50% point of 340°C–365°C cut as 360°C or 680°F and the 50% point of 270°C–340°C cut as 30.5°C or 580°F, i.e., the 50% point difference between these cuts is 100°F.

From Packie's correlation table without steam, the number of plates by the IR ratio is listed below for various gaps/overlaps.

Gap/overlap, °F	No. of plates × IR ratio	No. of plates, calculated
−50	8	14
−40	10	18
−30	12	22
−20	15	27
−10	18	33
0	25	45
10	45	81

The following results are obtained with steam. If superheated steam is introduced at a rate of 5,000 kg/h at the bottom, the amount of heat to be balanced is 4,278.79 × 1,000 + 5,000 × (328 − 248) = 4,678,790 kcal/h and the IR is 4,678,790/(241 − 195) = 101,712.83 kg/h. The IR ratio is 101,712.8/(169.01 × 1,000) = 0.60. From Packie's correlations with steam, the following results are obtained.

Gap/overlap, °F	No. of plates × IR ratio	No. of plates, calculated
−30	8	13
−20	10	17
−10	13	22
0	16	27

QUESTION

Prove that using open steam helps in vaporising the hydrocarbon components. Hints: Total pressure = vapour pressure of hydrocarbon oil + vapour pressure of steam (note that HC are insoluble in water will exerts its vapour pressure not p.p and steam will also at its vapour pressure, both of them are dependent on temperature but not on the compositions). The addition of steam will reduce the vapour pressure of hydrocarbon oil (= total pressure − pressure of steam), as a result hydrocarbon oil will vaporize. The use of superheated steam helps reduce the steam consumption, which would have been more if the steam was saturated. Hydrocarbon exert its vapour pressure at the prevailing temperature (which is higher than the saturation temperature of steam) of the superheated steam. The use of steam below atmospheric pressure (vacuum) causes a further increase in the vaporisation of hydrocarbons and a reduced amount of steam consumption.

10 Extraction

10.1 EXTRACTION PRINCIPLE

If two liquids, a solvent and a solute, are in a homogeneous solution and a third liquid is introduced into the solution, there will be two liquid layers separated into two homogeneous solutions as shown in Figure 10.1. This type of phenomenon will occur only for certain selected third liquids, which solubilise any one or both of the components, which were formerly in the homogeneous solution. Phase separation occurs due to the difference in density of the resulting solutions. This phenomenon is utilised for separating two miscible liquid components from a solution and such a method is called extraction. If the solute (A), the carrier liquid (B) in the feed mixture, and the pure solvent (C) are brought in intimate contact by thorough mixing, the solute will transfer to the solvent due to the concentration gradient set up between the feed and the solvent. This transfer (mass transfer) process will continue until the equilibrium concentration of the solute between the feed and the solvent is achieved, as long as the feed and the solvent are in intimate contact.

Extraction is applicable for separating components that are difficult to separate by distillation, settling, or other means. For instance, if the boiling points of the components are very close or the desired components are too sensitive to high temperature,

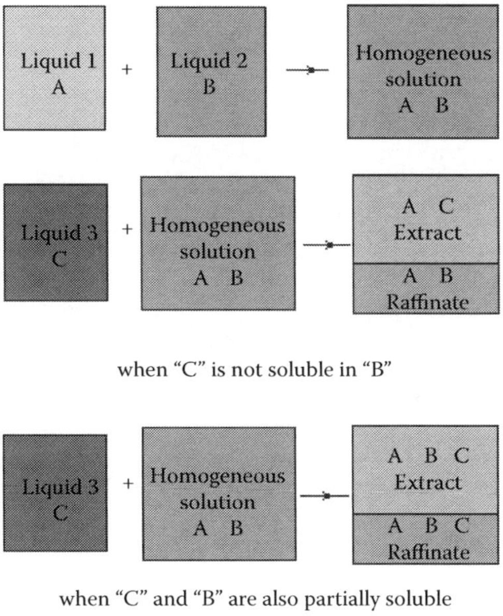

FIGURE 10.1 Extraction phenomenon.

extraction should be used as the best method for separation. When all the components involved in extraction are in liquid phases, the process is known as liquid extraction. While the feed is a solid mixture containing the solute, extraction by a liquid solvent is carried out and the process is known as leaching. For example, separation of aromatic hydrocarbons from vacuum distillates by furfural (solvent) is an example of liquid extraction, whereas removal of paraffinic hydrocarbons from waxy distillates by ketone (solvent) is an example of leaching. In refineries and petrochemical plants, extraction processes also involve more than one solute and more than one solvent.

10.2 EXTRACTION PROCESS

Any extraction process requires three operating steps, i.e., mixing, separating, and solvent recovery. The solvent and the feed must be mixed intimately to transfer the solute to the solvent. After mixing, the final mixture is allowed to settle, while two phases containing the solvent rich phase or the extract and the solvent lean phase or raffinate are separated. These phases will separate by gravity or by centrifuge settler. Usually, the solvent is also partially soluble in the feed and vice versa and as a result both the extract and raffinate phases will contain solvent, solute, and the carrier components. The next step will be to recover the solvent from these phases usually by distillation. Finally, the extract phase will yield solute and the raffinate phase will be the feed liquids nearly free from solutes. The success of any extraction process depends on various factors as listed below. The steps for an extraction process are schematically explained in Figure 10.2.

10.3 DEFINITION OF TERMS RELATED TO EXTRACTION

10.3.1 Partition Coefficient

If the amount of solute in the extract is x and that in raffinate is y, then the ratio y/x is known as the partition or distribution coefficient or equilibrium constant. The greater this value, the greater the separation of the solute from the original mixture.

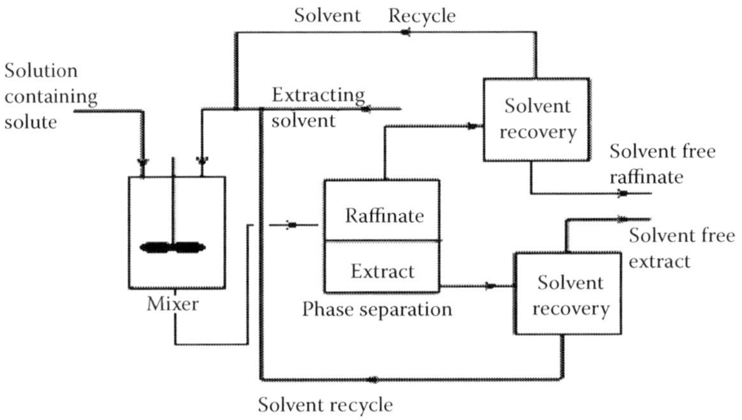

FIGURE 10.2 Extraction process.

10.3.2 Partial Solubility

If the solvent (C) solubilises the solute components (A) alone without interfering the carrier components (B), then the original mixture will be separated from the extract containing no solvent. But many of the solvents also solubilise the carrier (B) components, causing ternary mixture of A, B, and C in both the extract and raffinate phases.

10.3.3 Solvent to Feed Ratio

If the amount of solvent and feed is W_c and W_f, then the ratio W_c/W_f is the solvent to feed ratio. If the solubility of the desired solute in the solvent is high, a lower solvent to feed ratio will be required. If the solvent is also partially soluble in the carrier liquid, a higher solvent to feed ratio will be desirable.

10.3.4 Solvent Recovery

Solvents must be recovered from the extract and the raffinate as:

1. Solvents are not desirable in the final products
2. Solvents are costly
3. Solvents cannot be disposed of as they will pollute the environment

A solvent is commonly recovered by distillation or stripping. Care must be taken that the quality of the solvent should not degrade during this recovery process nor should it contain a high concentration of residual solute, which will otherwise make it unfit for reuse.

10.3.5 Separation of Phases

Usually, the greater the density difference between the extract and raffinate phases, the quicker the separation by gravity and the cheaper it will be. For small density difference, a costlier separation, e.g., centrifuging, is essential to speed up the process. The viscosity of the phases also plays an important role during separation. A high viscosity of one or both of the phases reduces the speed of separation. Other factors, such as selectivity, solvent power, and critical solution temperature (CST), must also be taken into consideration for the selection of the solvent.

10.3.6 Selectivity

Selectivity is the ratio of the distribution coefficient of the solute to that of the solute free carrier. For instance, if paraffin and aromatic hydrocarbons are present in feed, then the selectivity (β) of the solvent will be defined by the ratio of distribution coefficients of aromatic hydrocarbon to paraffin hydrocarbon.

Thus,

$$\beta = (Y_A/X_A)/(Y_B/X_B), \tag{10.1}$$

where Y_A and Y_B are the mass or mole fraction of the solute and solute carrier, respectively, in the extract phase, and X_A and X_B are the mass or mole fractions of the solute and carrier, respectively, in the raffinate phase.

10.3.7 Solvent Power

Solvent power is defined as the ratio of mass of solvent free extract to the mass of solvent used. Thus, the greater the dissolving power of the solvent, the greater this ratio will be. However, high solvent power also dissolves undesired feed components and thus reduces the yield of raffinate. High solvent power also makes less impure extract. Therefore, a compromise must be made between the solvent power and selectivity while the solvent is selected.

10.3.8 Critical Solution Temperature

When a solvent is mixed with the feed hydrocarbons, phase separation occurs in a certain range of temperatures. As the temperature is raised, the separated phases start remixing and yield a homogeneous solution at a certain temperature, known as the critical solution temperature (CST). Hence, a solvent of high CST must be selected so that phase separation will occur in a wide range of temperatures. The CST for the same solvent is also affected by the nature of the components involved, e.g., straight chain or paraffins have a higher CST as compared to branched chain or ring compounds and aromatics. Thus, the CST increases as the feed (containing paraffins and aromatics) is gradually freed from aromatics. Hence, the temperature gradient in an extractor plays an important role in extraction.

10.4 PHASE EQUILIBRIUM IN THE EXTRACTION PROCESS

For a mutually soluble solvent in a binary feed of solute and carrier, the resulting raffinate and extract phases each become a ternary mixture of solute, solvent, and the carrier feed solvent. Theoretically, if infinite time of separation of the phases is allowed, the phases will attain equilibrium. Thus, the composition of each component in the raffinate and extract will be related by their corresponding distribution constants. A ternary mixture is best represented graphically by a triangular diagram as shown in Figure 10.3, where points E and R are the locations of the equilibrium compositions of the components A, B, and C, as Y and X, respectively. The line ER in the diagram connects the equilibrium compositions in the extract and raffinate phases. Alternatively, phase relationships are also conveniently handled in rectangular diagrams as shown in Figure 10.4, where the solvent concentrations (extract and raffinate phases, ordinates) per unit of solvent free mixture are plotted against the solvent free solute compositions in the raffinate and extract phases along the x-axis. The equilibrium points, y_a and x_a, are located on this phase diagram with the help of the lower plot, where y_a and x_a are the solvent free solute compositions in the extract and raffinate phases, respectively. At the plait point, y_a and x_a are equal and hence no separation is possible when this point is reached.

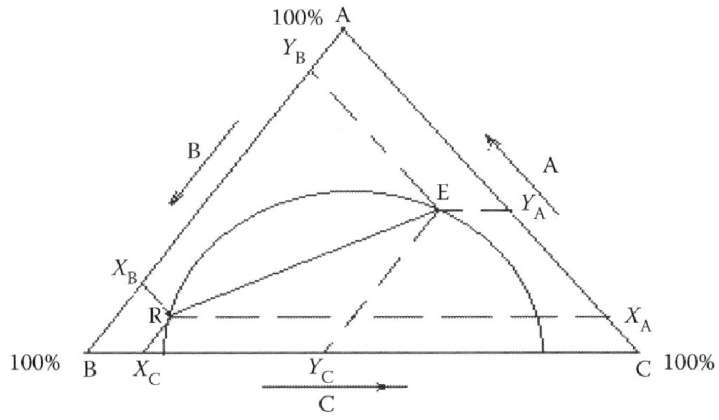

FIGURE 10.3 Phase diagram in triangular co-ordinates.

10.5 BATCH EXTRACTION

In the batch process, the solvent and feed are mixed in the desired proportions in a stirred vessel for some time and allowed to settle. Extract and raffinate phases are then transferred to separate vessels where the solvent is driven out by steaming or heating. The solvents are collected and reused. A multistage batch extraction process is shown in Figure 10.5, where fresh solvents are added to each batch and the extracts are collected separately. Feed is introduced in the first stage and the raffinate from each batch is further processed in the subsequent batches and emerges

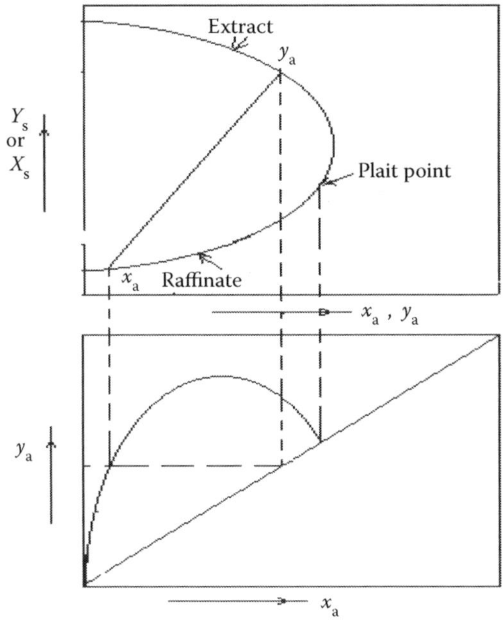

FIGURE 10.4 Phase diagram using rectangular co-ordinates.

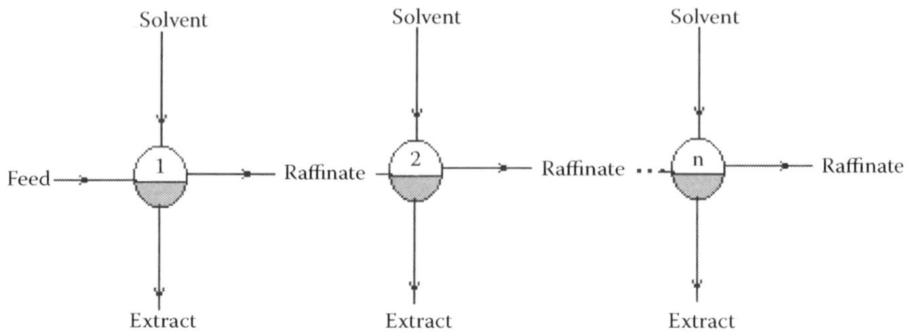

FIGURE 10.5 A schematic arrangement of a multistage batch extraction process.

from the last (nth) batch. Solvent recovery (not shown in the diagram) is carried out from the extracts and the final raffinate solutions. Batch processes are not suitable for large-scale extraction, such as filling, mixing, separation, etc., are time consuming and labour intensive, although these may be preferred in high-valued products where profit is justifiably high. Large-scale commercial extractions are carried out in continuous extractors as described next.

10.6 CONTINUOUS EXTRACTION

Continuous extraction is carried out in a tall plated tower where the feed and solvent are treated countercurrently. Simultaneous mixing and continuous phase separation occur while one phase is drawn from the top and the other from the bottom of the tower. Each plate provides intimate contact between the phases and enhances the enrichment of the extract and raffinate phases leaving the plates. A countercurrent plated tower extractor is shown in Figure 10.6, where the feed is introduced near the bottom of the tower and the solvent from the top. Usually, the solvent has a higher density than the feed and hence the extract becomes heavier than the raffinate. Therefore, the extract is drawn from the bottom and the raffinate from the top of the tower. The temperature gradient along the tower plays an important role for both in the extraction and phase separation. Each plate provides intimate mixing between the extract falling from the top plate with the raffinate rising from the plate below it. Theoretically, the compositions of the extract and the raffinate leaving a plate are in equilibrium. However, in practice, equilibrium compositions are not achieved because of inadequate mixing and the time of separation. This is compensated by using a higher number of plates than the ideal number of plates to achieve maximum extraction.

10.6.1 Computation of Number of Plates

Consider an extraction process involving a feed containing solute A in a carrier solvent (raffinate solvent) B and an extracting solvent C which are countercurrently treated in a multiplated column. If the number of plates in a column is n as measured from the bottom as the first plate, the feed entry section, the following equations can

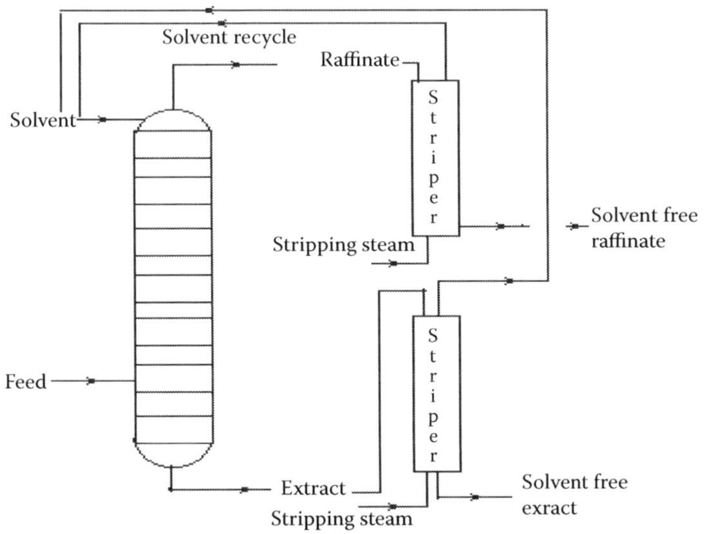

FIGURE 10.6 A continuous multi-plated extraction unit.

be set up. Designate the raffinate and the extract mass flow rates entering and leaving plates as R_n and E_n, and the corresponding compositions are $Y_{An,}$ Y_{Bn}, Y_{Cn} and X_{An}, X_{Bn}, X_{Cn}. For the rest of the plates, the corresponding flow rates and compositions are obtained by replacing n with 1, 2, 3,.... With reference to Figure 10.7, in section 1, the overall material and component balances are written as

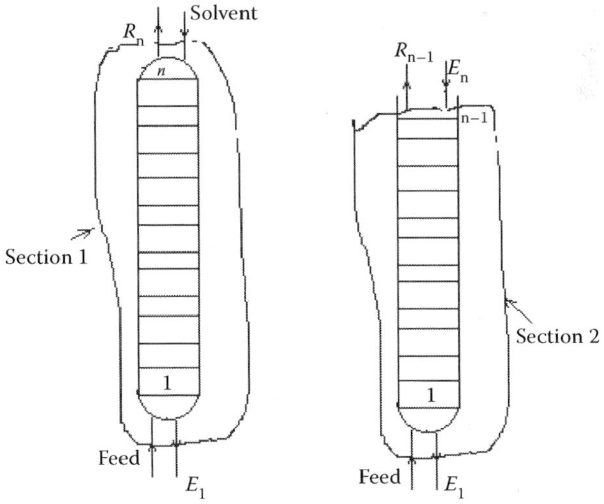

FIGURE 10.7 Sectional studies for flow rates and compositions.

$$R_n + E_1 = F + S, \tag{10.2}$$

$$R_n X_{An} + E_1 Y_{A1} = F X_{Af}, \tag{10.3}$$

$$R_n X_{Bn} + E_1 Y_{B1} = F X_{Bf}, \tag{10.4}$$

$$R_n X_{Cn} + E_1 Y_{C1} = F X_{Cf}, \tag{10.5}$$

where the feed compositions for A and B are X_{Af} and X_{Bf}, respectively, and the solvent is pure containing only C. Usually, the feed rate, feed composition, solvent rate, and the desired solute content in the final extract, Y_{A1}, are specified. The number of plates can be calculated numerically by the following steps:

1. From the equilibrium relation, obtain the corresponding compositions of A, B, and C in the extract and the raffinate as Y_{B1}, Y_{C1}, X_{A1}, X_{B1}, and X_{C1} at the given value of Y_{A1}.
2. Next, assume Y_{An} and obtain the other equilibrium compositions as step 1, where N is the top plate of the column.
3. Solve Equations 10.3 and 10.4 to evaluate R_n and E_1.
4. Evaluate S from Equation 10.2.
5. Check whether Equation 10.5 is satisfied or not. Otherwise, repeat from step 2 with another guess of Y_{An}.

Proceed to the next section as explained below.

With reference to Figure 10.7, considering section 2, for the n–1th plate and the first plate, the material balances are given as

$$R_{n-1} + E_1 = F + E_n, \tag{10.6}$$

$$R_{n-1} X_{An-1} + E_1 Y_{A1} = F X_{Af} + E_n Y_{An}, \tag{10.7}$$

$$R_{n-1} X_{Bn-1} + E_1 Y_{B1} = F X_{Bf} + E_n Y_{Bn}, \tag{10.8}$$

$$R_{n-1} X_{Cn-1} + E_1 Y_{C1} = E_n Y_{Cn}. \tag{10.9}$$

6. Assume Y_{An-1} and get the corresponding values of the other components as before from the equilibrium relation.
7. Solve for R_{n-1}, E_n, and E_1 from Equations 10.7 through 10.9.
8. Check for Equation 10.6.

Proceed for the calculations for the next plates from $n = n - 2$, n − 3,…2.

The rates and compositions are evaluated repeatedly until the raffinate compositions tally with the equilibrium composition (X_{A1}, X_{B1}, and X_{C1}) with the desired extract composition, i.e., Y_{A1}, Y_{B1}, and Y_{C1}.

The above steps can also be conducted graphically, which is available from the book by RE Traybal entitled *Mass Transfer Operations*.

Example 10.1

Determine the number of plates required for the extraction of a solute in feed containing 25% by mass of solute. The feed rate is 100 kg/h and the desired solute content in the extract is 40% by mass. Pure solvent is fed in a countercurrent at the top of the tower. Assume that the solvent does not dissolve the raffinate solvent (carrier) of the feed. The equilibrium of solute distribution is given below.

Raffinate phase		Extract phase	
Mass fraction of solute		Mass fraction of solute	
Solute A	Carrier B	Solute A	Solvent C
0.1214	0.8786	0.40	0.60
0.2708	0.7292	0.6369	0.3631
0.1479	0.8521	0.5090	0.491
0.1408	0.8592	0.4412	0.5588
0.038	0.9620	0.1005	0.8995

Solution

Here, $F = 100$, $X_{Af} = 0.25$, and $Y_{A1} = 0.40$, where n is the number of stages required.

Taking a single plate or stage, $n = 1$, the material balances are given as

$$R_1 + E_1 - S = F,$$

$$R_1 X_{A1} + E_1 Y_{A1} = F * X_{Af},$$

$$R_1 X_{B1} + E_1 0.0 = F * X_{Bf},$$

$$R_1 0.0 + E_1 Y_{C1} = S.$$

From the table of equilibrium data, $X_{A1} = 0.1214$ corresponding to $Y_{A1} = 0.40$, $X_{B1} = 1 - 0.1214 = 0.8786$, and $Y_{C1} = 1 - 0.40 = 0.60$.

The equations are rewritten as

$$R_1 + E_1 - S = F,$$

$$R_1 0.1214 + E_1 0.4 - S.0.0 = 25,$$

$$R_1 0.8786 + E_1 0.0 - S.0.0 = 75,$$

$$R_1 0.0 + E_1 0.60 - S = 0.$$

$$\begin{bmatrix} R_1 \\ E_1 \\ S \end{bmatrix} \begin{bmatrix} 0.1214 & 0.4 & 0.0 \\ 0.8786 & 0.0 & 0.0 \\ 0.0 & 0.60 & -1.0 \end{bmatrix} = \begin{bmatrix} 25 \\ 75 \\ 0.0 \end{bmatrix}.$$

Solving $R_1 = 85.36$, $E_1 = 36.5$, and $S = 21.955$.

Hence, the solvent required is about 22 kg/h in a single stage and percentage extraction is

$$36.5 * 0.40 / 25 \times 100 = 58.4\%.$$

For two-stage extraction:

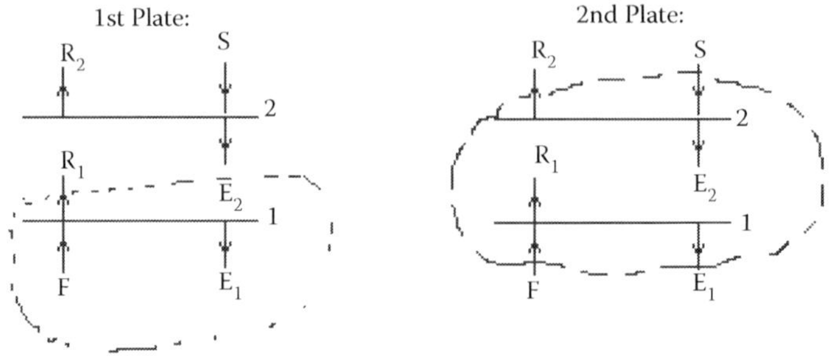

Considering the envelope for the first plate, the material balance equations are given as,

$$R_1 + E_1 - E_2 = F,$$

$$R_1X_{A1} + E_1Y_{A1} = E_2Y_{A2} = FX_{AF},$$

$$R_1X_{B1} + E_10.0 = E_20.0 = FX_{BF},$$

$$R_10.0 + E_1Y_{C1} = E_2Y_{C2} = 0.0.$$

$$\begin{bmatrix} R_1 \\ E_1 \\ E_2 \end{bmatrix} \begin{bmatrix} 0.1214 & 0.4 & -0.25 \\ 0.8786 & 0.0 & 0.0 \\ 0.0 & 0.60 & -0.75 \end{bmatrix} = \begin{bmatrix} 25 \\ 75 \\ 0.0 \end{bmatrix},$$

where the intermediate extract composition, Y_{A2}, is assumed to be 0.25 as the first trial.

The corresponding values of Y_{C2}, X_{A2}, and X_{B2} are obtained from the equilibrium data table by interpolation as 0.75, 0.0797, and 0.9203, respectively.

Solving $R_1 = 85.36$, $E_1 = 73.18$, and $E_2 = 58.547$.

Considering the envelope for the second plate,

$$R_2 + E_1 - S = F,$$

$$R_2X_{A2} + E_1Y_{A1} - S0.0 = FX_{Af},$$

$$R_2X_{B2} + E_10.0 - S0.0 = FX_{Bf},$$

$$R_20.0 + E_1Y_{C1} - S = 0.0,$$

$$\begin{bmatrix} R_2 \\ E_1 \\ S \end{bmatrix} \begin{bmatrix} 0.0797 & 0.40 & 0.00 \\ 0.9203 & 0.00 & 0.00 \\ 0.00 & 0.60 & -1.00 \end{bmatrix} = \begin{bmatrix} 25 \\ 75 \\ 0.00 \end{bmatrix},$$

where X_{A2} and X_{B2} are 0.0797 and 0.9203, respectively, at $Y_{A2} = 0.25$, solving, $R_2 = 81.49$, $E_1 = 46.26$, and $S = 27.75$.

Hence for more accurate results, a few more trials may be made. The following results are obtained for a two-plate extractor as

Plate	Y_A	Y_C	X_A	X_B	E, kg/h	R, kg/h	S, kg/h	% Extraction
1	0.4	0.6	0.1214	0.8786	50.00	85.36	35.00	70
2	0.15	0.85	0.0519	0.9481	31.00	79.00		

The above results are confirmed for the trial value of $Y_{A2} = 0.15$ for which both the plate material balances yield nearly the same value of $E_1 \approx 50.0$ kg/h. The percentage extraction is, therefore, $50*0.40/25 \times 100 = 80\%$ and the solvent required is 35.0 kg/h.

Example 10.2

Determine the solvent rate required to obtain an extract containing solute of 40% by weight, leaving the bottom of the column for (a) single stage and (b) two plates. Assume that the feed contains 25% by weight of solute in a mixture of its carrier (raffinate solvent) and a pure solvent is used in the extractor. Equilibrium data are given in Table 10.1.

Solution

(a) For a single stage extractor, from the table, at $Y_{A1} = 0.40$, the other compositions in the extract and the raffinate phases are obtained by interpolation as

$$Y_{B1} = 0.2103,\ Y_{C1} = 0.3897,\ X_{A1} = 0.2143,\ X_{B1} = 0.7171,\ \text{and}\ X_{C1} = 0.0686.$$

The solution is carried out for the bottom plate as

$$R_1 + E_1 - S = F,$$

$$R_1X_{A1} + E_1Y_{A1} - S\,0.0 = FX_{AF},$$

$$R_1X_{B1} + E_1Y_{B1} - S\,0.0 = FX_{BF},$$

$$R_1X_{C1} + E_1Y_{C1} - S\,1.0 = 0.0.$$

$$\begin{bmatrix} R_1 \\ E_1 \\ S \end{bmatrix} \begin{bmatrix} 0.2143 & 0.40 & 0.00 \\ 0.7171 & 0.2103 & 0.00 \\ 0.0686 & 0.3897 & -1.0 \end{bmatrix} = \begin{bmatrix} 25 \\ 75 \\ 0.0 \end{bmatrix}.$$

Solving $E_1 = 7.67$, $R_1 = 102.36$, and $S = 10.02$.

TABLE 10.1
Equilibrium data

Raffinate phase			Extract phase		
Solute	**Carrier**	**Solvent**	**Solute**	**Carrier**	**Solvent**
0	0.95	0.05	0	0.10	0.90
0.025	0.925	0.0505	0.079	0.101	0.82
0.050	0.899	0.051	0.150	0.108	0.742
0.075	0.873	0.052	0.210	0.115	0.675
0.100	0.846	0.054	0.262	0.127	0.611
0.125	0.819	0.056	0.300	0.142	0.558
0.150	0.791	0.059	0.338	0.159	0.503
0.175	0.763	0.062	0.365	0.178	0.457
0.200	0.734	0.066	0.390	0.196	0.414
0.250	0.675	0.075	0.425	0.246	0.329
0.300	0.611	0.089	0.445	0.280	0.275
0.350	0.544	0.106	0.450	0.333	0.217
0.400	0.466	0.134	0.430	0.405	0.165
0.416	0.434	0.150	0.416	0.434	0.150

So, the solvent rate is 10.02 kg/h and the percentage extraction is $7.67 \times 0.40/25 \times 100 = 12.27\%$.

(b) For the two-plated column, for the top plate, i.e., the second plate as shown in Problem 10.1, assume $Y_{A2} = 0.36$, the corresponding values for Y_{B2}, Y_{C2}, X_{A2}, X_{B2}, and X_{C2} are 0.1774, 0.4585, 0.1742, 0.7639, and 0.0619, respectively, by interpolation. Hence, the material balance equations for the components for the first plate are given as

$$\begin{bmatrix} R_1 \\ E_1 \\ E_2 \end{bmatrix} \begin{bmatrix} 0.2143 & 0.4 & -0.36 \\ 0.7171 & 0.2103 & -0.1774 \\ 0.0686 & 0.3897 & -0.4585 \end{bmatrix} = \begin{bmatrix} 25 \\ 75 \\ 0.0 \end{bmatrix}.$$

Solving, we get

$$E_1 = 99.56,\ R_1 = 100.44,\ \text{and}\ E_2 = 95.32.$$

The material balance equations for the components for the second plate, i.e., the bottom plate, are given as

$$\begin{bmatrix} R_2 \\ E_1 \\ S \end{bmatrix} \begin{bmatrix} 0.1742 & 0.4 & 0.0 \\ 0.7639 & 0.2103 & 0.0 \\ 0.0619 & 0.3897 & -1.0 \end{bmatrix} = \begin{bmatrix} 25 \\ 75 \\ 0.0 \end{bmatrix}.$$

Solving, we obtain

$$E_1 = 22.43,\ R_2 = 92.00, \text{ and } S = 14.44.$$

Since E_1 values are widely different, a new trial value of $Y_{A2} = 0.30$ is attempted next.

For $Y_{A2} = 0.30$, the corresponding values for Y_{B2}, Y_{C2}, X_{A2}, X_{B2}, and X_{C2} are 0.142, 0.558, 0.125, 0.819, and 0.056, respectively. With these values the equations are rewritten as

$$\begin{bmatrix} R_1 \\ E_1 \\ E_2 \end{bmatrix} \begin{bmatrix} 0.2143 & 0.40 & -0.30 \\ 0.7171 & 0.2103 & -0.142 \\ 0.0686 & 0.3897 & -0.555 \end{bmatrix} = \begin{bmatrix} 25 \\ 75 \\ 0.0 \end{bmatrix}.$$

For the first plate, solving

$$E_1 = 36.85,\ R_1 = 101.34, \text{ and } E_2 = 38.19.$$

$$\begin{bmatrix} R_2 \\ E_1 \\ S \end{bmatrix} \begin{bmatrix} 0.125 & 0.40 & 0 \\ 0.819 & 0.2103 & 0 \\ 0.056 & 0.3897 & -1 \end{bmatrix} = \begin{bmatrix} 25 \\ 75 \\ 0.0 \end{bmatrix},$$

and for the second plate:

$$E_1 = 36.83,\ R_2 = 82.15, \text{ and } S = 18.95.$$

Hence, the second trial is accepted as E_1 is found to be almost equal.

Also, the overall balances are checked as

$$R_1 + E_1 - E_2 = 101.34 + 36.85 - 38.19 = 100,$$

and

$$R_2 + E_1 - S = 82.15 + 36.83 - 18.95 = 100.03.$$

Hence, the solvent consumption should be 18.95 kg/h and the intermediate solute concentration in the extract is 0.30 mass fraction.

For more than two plates, computation will be time consuming and becomes difficult by manual calculations. Small computer programs are sufficient to evaluate the solvent requirement and the composition and the rates of all the streams for any number of plates.

QUESTIONS

1. What are the basic differences between extraction and distillation?
2. Define the terms solute, solvent, extract, and raffinate.
3. Differentiate the extraction processes between the operations in a plated tower and the operations in a series of mixer/separator tanks.

11 Reactor Calculations

11.1 REACTORS IN REFINERIES AND PETROCHEMICAL PLANTS

A reactor is a piece of equipment in which reactions are carried out to manufacture product materials chemically and physically different from the raw materials (reactants) put into it. These are vessels made of specially selected materials to withstand the operating temperature, pressure, and the chemical corrosion resulting from the reacting fluids present within the reactors. Common reactions encountered in petroleum refining are thermal and catalytic cracking, hydrogenation, desulfurisation, isomerisation, dehydrogenation, etc., which are fast reactions as compared to polymerisation reactions involved in resin or plastic manufacturing plants. The reactors may be either batch or continuous types depending on the reaction type. Usually, slow reactions are carried out in batch reactors whereas fast reactions are carried out in continuous flow reactors. Catalytic reactors using a solid catalyst are predominantly packed bed flow reactors. Fluidised bed reactors use a circulating catalyst with a steam or inert gas fluidising medium coupled with simultaneous regeneration facilities. Reactors with a circulating catalyst without fluidisation are also in use. A circulating catalyst through a liquid medium falls in the category of slurry reactors.

11.2 REACTION STOICHIOMETRY, MECHANISM, AND PATHWAYS

If two chemical species, A and B, react to form a product, C and D, which have different chemical and physical properties than those of the reactants A and B, the total mass of the system will remain unchanged according to the Law of Conservation of Matter. This can be represented by a simple mathematical relation as

$$aA + bB = cC + dD. \tag{11.1}$$

This relation is known as the stoichiometric balance or equation, where a moles of A and b moles of B react to form c moles of C and d moles of D. If reactant A is present in excess over its stoichiometric amount a, while B is present in its stoichiometric amount b, then A is called the *excess reactant* and B is called the *limiting reactant*. The reverse is also true while B becomes excess and A becomes limiting. In addition to these excess or limiting reactants, there may be incomplete reaction due to a shorter reaction time and/or maintenance of improper operating conditions, such as temperature and pressure, in the reactor. If none of the species, whether reactants or products, are removed during or after the reaction, the total mass of the contents in the reactor will be unchanged. Analysis of the composition of the product stream will reveal unreacted reactants A or B or both along with the products C and D, therefore, stoichiometry of the said reaction may be determined. In case of multiple reactions multiple pathways with different stoichiometry will be present. For instance, if Equation 11.1 is rewritten as equations 11.2 and 11.3 as, which are two reaction paths instead of one.

$$a_1\mathrm{A} + b_1\mathrm{B} = c_1\mathrm{C} + d_1\mathrm{D}_1, \tag{11.2}$$

$$a_2\mathrm{A} + b_2\mathrm{B} = c_2\mathrm{C} + d_2\mathrm{D}_2. \tag{11.3}$$

These are two parallel reactions, both producing product C, but the degree of completion may be different. Hence, if the pathways are correctly selected, stoichiometric relations will hold good. In fact, pathways are imagined and tracking of the composition of the resulting mixture is carried out to establish the pathways. Stoichiometric relations based on the composition analysis after the reaction may differ from the stoichiometry during the reaction, which may involve more complicated active states of the species. Thus, the formation of active states of the species from its inert state and thereby formation of a new species is understood by the ionic or electronic or radical species involved. Thus, the pathways determined by this activated state to completion are called the *mechanism* of reaction. Pathways established on the mechanism of reaction can only lead to correct stoichiometry of the reaction. Further discussion is out of the scope of this book. Varieties of reaction pathways as established through certain mechanisms are single, series, parallel, or series-parallel multiple, irreversible, or equilibrium reactions.

11.3 RATE OF REACTION AND KINETIC EQUATIONS

The rate of reaction is defined as the rate at which the moles or mass of reactants are depleted or products formed with respect to time per unit volume (or per unit weight or surface of catalyst in the case of solid catalysed reactions) of the reactor. For instance, if the remaining concentration of A during the reaction is C_A at a time of t seconds, then the rate of reaction or disappearance of A is given as

$$r_\mathrm{A} = \frac{1}{V}\frac{dN_\mathrm{A}}{dt} = \frac{dC_\mathrm{A}}{dt}, \tag{11.4}$$

where N_A is the moles of A in the reactor at anytime t (sec) and V is the volume of the reactor. From the concentration of reactant A data taken for different reaction times, the rate can be evaluated from the above relation. The rate is also conveniently expressed as a function of C_A. Thus,

$$r_\mathrm{A} = kC_\mathrm{A}^n, \tag{11.5}$$

where n is the order of the reaction that may assume any value, positive or negative, integer or floating point specific for the reaction. Although many irreversible reactions behave like this nth order rate, which may be first order ($n = 1$), second order ($n = 2$), or third order ($n = 3$), fraction order (e.g. $n = 0.5$, 0.2, 1.5, etc.), etc., many reactions also manifest more complicated relations like those given below, especially for multiple or reversible reactions.

$$r_A = \frac{kC_A^n}{k_1 C_A^m + k_2}, \tag{11.6}$$

where k, k_1, and k_2 are the kinetic parameters for the reactions.

In Equation 11.5, k is called the reaction rate constant, which influences the rate of reaction and is a function of temperature. According to the Arrhenius relation of temperature dependence, the specific rate constant is related as

$$k = \mathrm{A}e^{-E/RT}, \tag{11.7}$$

where T is the reaction temperature in absolute scale, E is the energy of activation required for the reaction, and A is the frequency factor of the reaction rate. The greater the temperature, the greater the rate constant and hence the rate of reaction. In case of a reversible reaction, as shown below,

$$a\mathrm{A} + b\mathrm{B} \underset{k_2}{\overset{k_1}{\rightleftharpoons}} c\mathrm{C} + d\mathrm{D} \tag{11.8}$$

A and B are both converted in the forward reaction to C and D which are converted back to A and B in the reverse reaction. Here, k_1 and k_2 are the reaction rate constants in the forward and reverse reactions, respectively. If the rate of reactions of the reactant A in the forward and backward directions are assumed as forward:

$$r_A = k_1 c_A c_B, \tag{11.9}$$

and backward:

$$r_A = k_2 c_C c_D, \tag{11.10}$$

then, the net rate of disappearance of A, is given by

$$r_A = k_1 c_A c_B - k_2 c_C c_D. \tag{11.11}$$

In order to theoretically determine the composition at any given operating temperature at equilibrium, i.e., when the rate of reactions in both the forward and backward direction are the same, then, $r_A = 0$. This is also true for r_B, r_C, and r_D. Substituting $r_A = 0$ in Equation 11.11, we get

$$k_1 c_A c_B - k_2 c_C c_D = 0, \tag{11.12}$$

or

$$k_1/k_2 = c_C c_D / c_A c_B. \tag{11.13}$$

The ratio, $k_1/k_2 = K$, is known as the equilibrium constant. If the value of K is greater than unity, it indicates that the forward reaction is favoured more than the reverse reaction. The equilibrium rate constant is a strong function of the temperature, like the specific rate constants. An increase in reaction rate will enhance forward and backward reactions equally. Catalytic activity will also enhance the reactions equally in both directions. The compositions in Equation 11.13 are the equilibrium compositions and will vary with operating temperature. The rate of reaction is also a function of operating pressure, while any gaseous component is present as the reactant.

11.4 BATCH, CONTINUOUS STIRRED TANK REACTOR, AND PLUG FLOW REACTOR CONCEPTS

A batch reactor is a reaction vessel where reactants, liquid or gas or both (with or without a catalyst as the case may be), are filled in the vessel initially and closed. Then, proper operating conditions, e.g., temperature and pressure, are raised to carry out a reaction and left for a certain period for the reaction to continue. Thus, the reaction is a time-varying process in such a reactor. The reaction time may vary from reaction to reaction. The reactor is then cooled to release the products. The amount of product obtained is called the batch quantity. The reactor is then cleaned and refilled with the reactants and the reaction is repeated. This type of operation involves much manual activity and a large amount of time is lost to filling and discharging, raising the operating temperature and pressure, and finally cooling for withdrawal of the products. A good amount of heat, especially for a high temperature reaction, is lost every time during cooling before product withdrawal. A large production rate batch reaction process is not economical where a flow or continuous reactor must be used. However, high value products in small quantities are manufactured in batch reactors.

Theoretically, the rate equation and composition of reactants and products can be related in a constant volume batch reactor as

Input – output – disappearance of reactant "A" = rate of depletion of reactant within the reactor

$$\text{i.e } 0 - 0 - r_A V = \frac{d}{dt}\left(VC_A\right), \tag{11.14}$$

or

$$r_A = \frac{d}{dt} C_A, \tag{11.15}$$

where V is the constant volume of the reactor, C_A is the concentration of reactant A in moles/volume, and r_A is the rate of disappearance of A in moles per unit time per unit volume of the reactor. Conversion (X_A) of A is expressed as moles of A disappeared per mole of A initially present (C_{A0}) and is given as

$$X_A = (VC_{A0} - VC_A)/(VC_{A0}) = (C_{A0} - C_A)/C_{A0}, \tag{11.16}$$

or

$$C_A = C_{A0}(1 - X_A). \tag{11.17}$$

Example 11.1

Ethane gas (A) cracking reaction takes place in a batch laboratory reactor with a volume of 1 m^3 at a pressure of 20 atm and a temperature of 680°C. The endothermic heat of the reaction (average) is 1,000 kcal/kmol. If the average molal heat capacity is 30 kcal/kmol °C, calculate the relation between the conversion and the time under,

1. Adiabatic condition
2. Isothermal condition

The variation of the first order reaction rate constant with temperature is given as

T °C	600	610	620	630	640	650	680
K, h^{-1}	1.2	1.68	2.33	3.28	4.61	7.2	9.41

Solution

1. Designating V, the constant reactor volume occupied by the gaseous species, m^3
   C_{A0}, concentration of reactant in the feed, kmol/m3
   C_A, concentration of reactant, kmol/m3
   X_A, moles of A converted per mole of A in the feed
   N, total moles of reactant and products in the reactor
   C, average molar specific heat of the reactant and products in the reactor, kcal/kmol °C
   T_0, initial feed temperature, °C
   P, pressure in atm
   ΔH_R, endothermic heat of reaction, kcal/kmol of A converted
   R_A, rate of disappearance of A, kmol/m^3h

   First order reaction rate equation,

$$R_A = KC_A, \tag{1}$$

component material balance of the reactant for the batch reactor,

$$-R_A = dC_A/dt, \tag{2}$$

where

$$C_A = C_{A0}(1 - X_A). \tag{3}$$

Hence, Equation 2 is rewritten as

$$R_A = dX_A/dt. \tag{4}$$

The heat balance over the reactor for the adiabatic condition, i.e., the reactor is well insulated so that no heat is transferred from the reactor surface, is given as

$$R_A \Delta H_R V = NC dT/dt, \tag{5}$$

Now combining Equations 3 and 4, and Equations 3 and 5,

$$\int \frac{dX_A/(1-X_A)}{K} = t. \tag{6}$$

Integration is done either graphically or numerically as K varies with temperature and hence is also a function of X_A as obvious from the following relation,

$$dT/dt = -R_A \Delta H_R (V/NC) \text{ or } dT/dt = -\Delta H_R (V/NC) C_{A0} dX_A/dt,$$

or

$$T = T_0 - \Delta H_R (V/NC) C_{A0} X_A. \tag{7}$$

Here,
$V = 1$ m^3
$P = 20$ atm
$T_0 = 680$°C

$C_{A0} = 1 \times 20 \times 273/(680 + 273)$ kmol/lit (assuming ideal gas law) = 0.2557 kmol/m^3

$N = 1 \times 20 \times 273/(680 + 273)/22.4 = 0.2557$ kmole (assumed unchanged)

$C = 30$ kcal/kmole°C

$\Delta H_R = 1{,}000$ kcal/kmole of A converted

Using the above data and Equations 6 and 7, conversion and temperature are evaluated as presented below,
where the rate constant (K) is obtained from the table and by interpolation at the intermediate temperatures.

Time, h	Temp, °C	X_A, %
0	680	0
0.1	654	19.4
0.2	639	31.09
0.3	629	38.7
0.4	620	45.2
0.5	616	48.7
0.6	613	51.2
0.7	610	53.4
0.8	607	55.3

Time, h	Temp, °C	X_A,%
0.9	605	57.0
1.00	603	58.5
1.1	601	59.9
1.2	600	61.0

2. For isothermal condition at a constant temperature of 680°C, by using Equation 6, as T is constant, integration is given as

$$\ln(1 - X_A) = -Kt.$$

The following result is obtained as

Time, h	X_A,%
0	0
0.1	21.4
0.2	38.2
0.3	51.2
0.4	61.9
0.5	70.0
0.7	81.5
0.8	85.5
0.9	88.60
1.00	91.0
1.1	92.0
1.2	94.5

This indicates that the maintenance of a constant highest possible reaction temperature increases the conversion manifold as compared to an adiabatic reactor. This helps in reducing the required volume of the reactor as compared to an adiabatic one.

A *flow reactor* is a reactor where reactants and products enter and leave simultaneously and continuously. Steady state operating conditions, such as temperature, pressure, feed flow, and composition, are required to be maintained to have uniform product quality. In this flow process, the operating conditions have to be raised at the start to the desired level before the reactants enter. After this is complete, the reactants are introduced. At the beginning, the product quality varies with time to yield a uniform composition. This period required for raising steady state operating conditions and uniform product quality is known as the *start-up period*. Production is then continued for a long time and is stopped only when repair or maintenance is required or when the supply of reactants or demand for products falls short. Usually, a continuous run period varies from 6 months to 1 year while maintenance or repair is made for the entire plant. There are continuous operations without any *shut down* where maintenance is done online, e.g., fluidised catalytic cracking (FCC), circulating catalytic reformer, and catalytic reformer with a *swing reactor*. Reactors are either tubular or stirred tank type. Thermal reactions are carried out in tubular type reactors, solid catalysed reactions are carried out in tubular packed bed reactors, and polymerisation and certain reactions are carried out in a continuous stirred tank reactor (CSTR).

Theoretically a tubular reactor is a long tube or pipe where reactants are converted to products as they pass through the tube from the entrance to the exit. Consider such a tube as shown in Figure 11.1 where a first order reaction with respect to reactant A is carried out.

$$\mathrm{A} \xrightarrow{k} \mathrm{B}$$

Let us take that the first order reaction is taking place and the rate is

$$r_A = -kC_{A,} \tag{11.18}$$

Consider an element of the tube of infinitely small thickness of ΔZ and at a distance Z from the feed entry point, the following component material balance equation can be written as

$$FC_A\Big|_{\text{at } Z} - FC_A\Big|_{\text{at } Z+\Delta Z} - r_A a \Delta Z = \frac{\Delta(aZC_A)}{\Delta t}, \tag{11.19}$$

where F is the volumetric flow rate, r_A is the rate of reaction in mole per unit volume per unit time, C_A is the concentration of A in moles/volume, a is the surface area per unit volume of the reactor, and ΔZ and Δt are the infinitely small thickness and time, respectively. Equation 11.19 is the unsteady state general equation

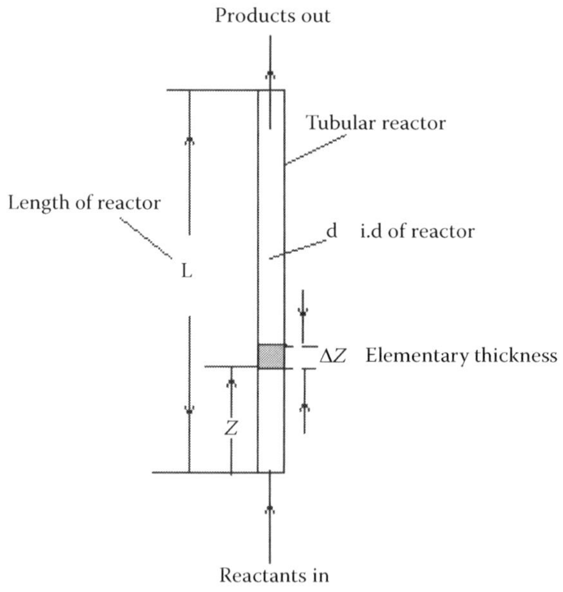

FIGURE 11.1 An ideal tubular reactor.

applicable. However, during steady state operation, the time gradient becomes zero and the resulting equation is given as

$$FC_A\Big|_{\text{at } Z} - FC_A\Big|_{\text{at } Z+\Delta Z} - r_A a\Delta Z = 0, \qquad (11.20)$$

Dividing both sides of Equation 11.20 by ΔZ and taking the limits as $\Delta Z \to 0$, we get the following differential equation for the reactor,

$$\lim_{\Delta Z \to 0} \frac{FC_A\Big|_{\text{at } Z} - FC_A\Big|_{\text{at } Z+\Delta Z}}{\Delta Z} - r_A a = 0, \qquad (11.21)$$

or

$$F\frac{dC_A}{dZ} = -r_A a, \qquad (11.22)$$

The equation is valid for any steady state tubular reactor. Now, substituting the rate (r_A) by the first order relation as given in Equation 11.18, the above relation becomes,

$$F\frac{dC_A}{dZ} = -kC_A a, \qquad (11.23)$$

$$\frac{dC_A}{dV} = -\frac{kC_A}{F}, \qquad (11.24)$$

Finally, the volume of the tubular reactor can be found as

$$\frac{V}{F} = -\int_{C_{Ai}}^{C_{Af}} \frac{dC_A}{kC_A}, \qquad (11.25)$$

where V is the volume of the reactor, and C_{Ai} and C_{Af} are the concentration of A at the entrance and at the exit of the reactor. After integration of the above equation,

$$V = -\frac{1}{k}\ln\left(\frac{C_{Af}}{C_{Ai}}\right), \qquad (11.26)$$

Similarly, the volume can be evaluated for reactions of a different order. However, if analytical integration is not possible then graphical or numerical integration has to be done for a complicated rate equation. In the above relation, k is a function of temperature and hence Equation 11.26 is applicable for an isothermal reactor. It is also to be noted that the velocity of the fluid through the tube is assumed to be the same at any cross section at all radial positions, i.e., a *plug flow* is assumed. Hence, such a model as given by Equation 11.25 is also known as the plug flow reactor model. In case the heat effect is substantial in exothermic or endothermic reactions, the temperature will vary with the reaction along the length of the tube. If the reaction has a heat of reaction ΔH_r defined as the heat unit evolved (in exothermic reaction) or absorbed (in endothermic reaction) per unit of A reacted, then the heat balance equation can be set up under adiabatic conditions as shown in Equation (11.27) for an endothermic reaction,

$$F\delta C_p T\Big|_{\text{at } Z} - F\delta C_p T\Big|_{\text{at } Z+\Delta Z} - r_A a \Delta Z \Delta H_r = \frac{\Delta\left(aZ\delta C_p T\right)}{\Delta t}, \qquad (11.27)$$

where δ and C_p are the average density and specific heat of the fluid mixture flowing in the reactor. This is valid for unsteady state conditions. For steady operating conditions, the time derivative will be zero and the equation is reduced to

$$F\delta CpT\Big|_{\text{at } Z} - F\delta CpT\Big|_{\text{at } Z+\Delta Z} - r_A a \Delta Z \Delta H_r = 0, \qquad (11.28)$$

Taking the limits we get

$$\lim_{\Delta Z \to 0} \frac{F\delta CpT\Big|_{\text{at } Z} - F\delta CpT\Big|_{\text{at } Z+\Delta Z}}{\Delta Z} - r_A a \Delta H_r = 0, \qquad (11.29)$$

or

$$\frac{dT}{dZ} = -\frac{r_A a \Delta H_r}{F\delta C_p}, \qquad (11.30)$$

Thus, for an endothermic reaction there will be a drop in temperature along the length of the reactor. In the case of an exothermic reaction, there will be a rise in temperature and the equation will be

$$\frac{dT}{dZ} = \frac{r_A a \Delta H_r}{F\delta C_p}, \qquad (11.31)$$

It is noted that Equations 11.30 and 11.31 are valid for steady state and adiabatic conditions, i.e., when no heat is lost from the reactor to the surroundings. This is true for perfectly insulated reactors. Simultaneous solution of material balance equation (11.22) coupled with Equations 11.30 or 11.31 as the case may be, both the composition C_A and temperature T can be evaluated for a given rate equation.

Example 11.2

Repeat the Example (11.1) cracking pure ethane in a tubular reactor (of volume 100 m^3) at a rate of 100 m^3/h at 680°C.

Solution

The component material balance of the reactant for the flow reactor is obtained by considering an elemental reactor volume (ΔV) from its feed entry,

$$FC_A|_V - FC_A|_{V+\Delta V} - R_A \Delta V = 0,$$

(where F is the flow rate in cubic meters per hour, assumed to be constant for a negligible change in the density of the fluids) or

$$FdC_A/dV = R_A \text{ or } FdX_A/dV = K(1 - X_A),$$

or

$$dX_A/d\tau = K(1 - X_A), \tag{8}$$

The heat balance over the reactor for the adiabatic condition is given as

$$FC\rho T|_V - FC\rho T|_{V+\Delta V} - R_A \Delta V \Delta H_R = 0,$$

or

$$FC\rho dT/dV = -R_A \Delta H_R \text{ or } dT/dV = -R_A \Delta H_R/(F\rho N),$$

or

$$dT/dV = -\Delta H_R/(FC\rho) dX_A/dV,$$

i.e.,

$$T = T_0 - \Delta H_R/(FC\rho) X_A, \tag{9}$$

where $\tau = V/F$ is the space time for the flow reactor and ρ is the average molar density of the gases which is

$$\rho = PM/RT = 20 \times 30/\{0.0832 \times (273 + 640)\} = 7.89 \text{ kmol/m}^3,$$

at an average temperature of 640°C and the average molecular weight is taken as 30.

The following results are obtained using Equations 8 and 9 under adiabatic condition,

τ, h	Temperature °C	X_A, %
0	680	0
0.1	679	21
0.20	678	38
0.3	677	51
0.4	676	61
0.5	676	69
0.6	676	75
0.7	676	80
0.8	676	85
0.9	676	88
1.00	676	90.5

The above result indicates that the temperature drop is negligible but conversion varies widely with space time and conversion achieved at the end of the reactor is 90.5%.

3. For isothermal operation of the reactor at 680°C, the component material balance is simplified as

$$\ln(1 - X_A) = -k\tau = -9.41\tau.$$

The results are given as

τ	X_A
0	0
0.1	68
0.2	84.7
0.30	94.0

This also indicates that at an isothermal operation at the highest possible temperature, conversion is greater than that obtained in an adiabatic reactor. The above result shows that only one-third of the reactor volume is required to get 94% conversion.

Example 11.3

If the above problem is carried out in a stirred vessel reactor of the same volume and throughput, the solution will be as shown below. For this case, the reactor is taken as a stirred tank reactor as shown in Figure 11.2.

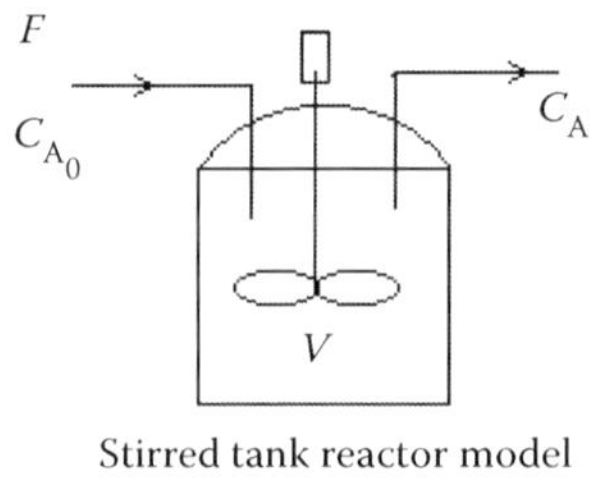

FIGURE 11.2 Reactor in the Example 11.3.

The component material balance is given as

$$FC_{A0} - FC_A - R_AV = 0, \tag{10}$$

at steady state condition.

Substituting $R_A = KC_A$ in Equation 10,

$$C_A = C_{A0}/(1 + KV/F) = C_{A0}/(1 + K\tau), \tag{11}$$

where τ is V/F, the space time for the reactor.

The heat balance is given as

$$F\rho C(T_0 - T) - R_AV\Delta H_R = 0,$$

or

$$\begin{aligned} T &= T_0 - R_AV\Delta H_R/(F\rho C) \\ &= T_0 - KC_AV\Delta H_R/(F\rho C). \end{aligned} \tag{12}$$

where C is the average specific heat of the reactant and products.

i. For an adiabatic reactor, the results are given as, at the given flow rate of 100 m^3/h, i.e., space time, $\tau = V/F = 100/100 = 1$ h, conversion, $X_A = 90\%$, and exit temperature, $T = 678°C$.

For other values of flow rates, the conversion decreases as the flow increases as shown below, where the temperature drops marginally.

τ, h	**X_A, %**	**Exit temperature, °C**
0.05	32	672
0.10	48	674
0.20	65	676
0.50	82	678
1.00	90	678.9
5.00	99	679
10.0	98	679.8

The conversion and exit temperatures will also change as the starting temperature is changed from 680°C.

ii. For isothermal operation at 680°C, $X_A = 90\%$.

For other operating temperatures under isothermal operations, the following conversions are obtained.

Temperature, °C	**X_A, %**
680	98.9
620	69.9
610	62.6

Example 11.4

If ethane cracking is carried out in a tubular reactor and in a stirred tank reactor separately under isothermal and steady state continuous conditions, then compare the conversions achieved from these reactors under identical operating conditions.

Solution

As already evaluated in the previous problems, under isothermal condition:

At 680°C,

Space time	Stirred tank	Tubular
τ, h	X_A, %	X_A, %
	$X_A = k\tau/(1 + k\tau)$	$X_A = 1 - \exp(-k\tau)$
10.0	98	98
1.0	90	90
0.5	82.5	82.5
0.2	65.3	84.7

Exit conversion in the tubular reactor (assuming a plug flow or ideal reactor) is found to be higher than those in the stirred tank reactor as revealed from the above results.

Example 11.5

Catalytic dehydrogenation of ethylbenzene, represented by the reaction,

$$C_6H_5C_2H_5 \Leftrightarrow C_6H_5CH = CH_2 + H_2,$$

$$\text{Ethyl benzere (E)} \Leftrightarrow \text{Styrene (S)} + \text{Hydrogen (H)}$$

is carried out in a tubular (of 1 m inner diameter) reactor packed with catalyst pellets.

The global rate is

$$R_E = K(P_E - P_S P_H/K_{eq}),$$

where:

R_E, kmole ethyl benzene converted per hour per atmosphere per cubic meter of catalyst

K, forward rate constant and given by the relation

$$K = \exp(-48/T + 4.0),\ h^{-1},$$

P_E, P_S, and P_H are the partial pressure of ethyl benzene (E), styrene (S), and hydrogen (H), respectively in atmosphere

K_{eq}, is the equilibrium constant with the following information:

K_{eq}	0.002	0.03	0.22	1.5
Temperature,	673	773	873	973

Assume that there is no heat exchange between the reactor and the surroundings. The feed containing 5 mol% ethyl benzene and 95 mol% steam enters the reactor tube at a rate of 300 kmol/h. The other data are given below.

The temperature of the mixed feed entering the reactor is 630°C. The average pressure in the reactor tubes is 0.12 MPa. The heat of the reaction is 35,000 kcal/kmol. Assume the surrounding temperature is 20°C.

Determine the volume of the reactor for the production of 20 tons of styrene per day in the conditions (1) adiabatic, (2) isothermal, and (3) while heat is lost to the surroundings, which is at a temperature of 20°C. Assume that the heat transfer coefficient U of the insulation around the reactor is 100 kcal/h/m^2 °C based on the inside surface area. This value of U is based on the difference in temperature between the reaction mixture and the surroundings.

Solution

Plug flow model assumed:

Component material balance equation,

$$FC_{E0}dX/dV = R_E. \tag{a}$$

Heat balance equation:

For the adiabatic condition,

$$FCdT/dV = -R_E\Delta H. \tag{b}$$

For isothermal condition,

$$T = \text{unchanged}. \tag{c}$$

Non-isothermal with heat loss to surroundings,

$$FCdT/dV = R_E\Delta H - Ua(T - T_a), \tag{d}$$

where:

F molal flow rate = 300 kmol/h
C_{e0}, mole fraction of ethyl benzene in the feed, 0.05
X, moles of ethyl benzene converted per mole in feed
V, volume of the reactor tube occupied by the catalyst
R_E, rate of conversion of ethyl benzene
C, average specific heat, 0.55 kcal/kg °C
ΔH, heat of reaction, 35,000 kcal/kmol of ethyl benzene converted
U, heat transfer coefficient, 100 kcal/h/m^2 °C

a, surface area per unit volume, 4 m^2/m^3
T, reactor temperature, K
T_a, surrounding temperature, 293 K

For an adiabatic operation solution of Equations (a) and (b) are carried out numerically taking vary small fraction of reactor volume (0.001) and integrated until required rate of production of styrene (20 tons/day) is obtained. To save time, a small computer program is used to carry out the computations.

Reactor volume, m^3	Conversion, fraction	Reactor temperature, K	Tons of styrene/day
0.1	0.022	899	0.85
0.5	0.10	887	4.0
1.0	0.20	874	7.6
1.5	0.286	862	10.7
2.0	0.35	852	13.4
2.5	0.42	843	15.7
3.0	0.47	835	17.7
3.5	0.52	829	19.4
4.0	0.55	823	20.8

Hence, the reactor volume required is 4 m^3. However, if the reactor volume is more, the rate of production will increase, but after this no appreciable increase in production will be obtained as obvious from further study. The maximum conversion achievable under a perfect adiabatic condition is 69% and the rate of production is 25.87 tons/day in a reactor volume of 14 m^3.

For an isothermal operation at 630°C, using Equation (a), the numerical solution yields the following results.

Reactor volume, m^3	Conversion, fraction	Reactor temperature, K	Tons of styrene/ day
0.1	0.022	903	0.846
0.5	0.108	903	4.04
1.0	0.203	903	7.62
1.5	0.287	903	10.77
2.0	0.362	903	13.57
2.5	0.430	903	16.04
3.0	0.487	903	18.22
3.5	0.54	903	20.15

Hence, the volume of the reactor is 3.5 m^3. In such an isothermal reactor, the maximum conversion achievable is very high, up to 82.7%, though the rate of increase of production will not exceed 30.95 tons/day at a reactor volume of 9.1 m^3. An additional increase of reactor volume is not economically justified for the little increase in conversion after this volume.

For a non-adiabatic condition while heat loss is encountered, the following results are obtained, using Equations (a) and (d).

Reactor volume, m^3	Conversion, fraction	Reactor temperature, K	Tons of styrene/day
0.1	0.022	893	0.85
0.5	0.108	855	4.03
1.0	0.202	813	7.56
1.5	0.280	775	10.47
2.0	0.33	743	12.28
2.5	0.35	716	13.06
3.0	0.33	695	12.58
3.5	0.286	681	10.79
3.9	0.224	673	8.38

From the above result, it is observed that the desired rate of production is not achievable due to the fast cooling of the reactor for heat loss to the surroundings. The maximum conversion is 35%, the corresponding production rate is 13.06 tons/day, and the volume of the reactor is 2.5 m^3. Increasing the reactor volume will effect a reduction of conversion and is not justified.

It is understood from the analysis of heat effect over conversion for the three cases that an isothermal reactor may be used for maximising the conversion. However, an ideal isothermal condition may not be feasible for a long length of reactor, in which case a number of small reactors may be taken in series with intermediate heaters to maintain the highest reaction temperature. In fact, real reactors may experience a lower rate of production due to mass transfer resistances of the fluids and catalyst surface, temperature distribution in the reactor bed and catalyst surface, activity of the catalyst, etc.

Example 11.6

Hydrogenation of nitrobenzene is carried out in a 0.03-m (inside) diameter catalyst packed tubular reactor. A mixture of nitrobenzene and hydrogen enters at a rate of 0.065 kmol/h. The catalyst bed void fraction and pressure are 0.4 and 1 atm, respectively. The feed temperature is 427 K and the tube is kept in a constant temperature bath maintained at the same temperature. The heat transfer coefficient from the reactor and the bath is 86 kcal/h/m^2 °C. The exothermic heat of the reaction is 152,000 kcal/kmol. The nitrobenzene concentration in the feed is 5.0×10^{10} kmol/m^3. The net rate of disappearance of nitrobenzene is represented by the expression,

$$r_A = 5.79 \times 10^4 \exp(-2958/T)C^{0.578},$$

where:

r_A, kmoles nitrobenzene reacting/lit of void volume per hour
C, concentration of nitrobenzene, kmol/lit
T, temperature, K
Evaluate the conversion and temperature at various catalyst bed.

Solution

Material balance equation for the nitrobenzene (A) is given as,

$$FC_{A0}dX/dZ = \varepsilon a_x r_{A.} \quad \text{(a)}$$

Heat balance equation:
For adiabatic condition

$$FCdT/dZ = \varepsilon a_x R_E \Delta H - Ua_s(T - T_s), \quad \text{(b)}$$

where:

F, volumetric flow rate, $0.065 \times 22.4 \times 427/273$ m³/h
C_{A0}, feed concentration of A, 5.0×10^{10} kmol/m³
C, average specific heat of feed mixture and product, 7 kcal/kmol
Z, length of reactor, m
ε, void fraction, 0.4
a_x, cross-sectional area of tube, $3.14^*(0.03)^2/4$ m²
U, heat transfer coefficient, 86 kcal/h/m² °C
a_s, surface area per unit length of the tube, $3.14^*0.03$ m²
T, reactor temperature in kelvin
T_s, bath temperature, 427 K

A numerical solution of Equations (a) and (b) is presented below.

Reactor length × 100, m	Conversion, fraction	Temperature, K
0.10	0.013	432
1.0	0.035	439
4.0	0.18	478
6.0	0.32	506
10.0	0.66	564
12.0	0.83	573
15.0	0.97	547
16.0	0.98	536

The above results show that both conversion and temperature rise rapidly up to 0.12 m in height, but the temperature starts falling after this height.

Example 11.7

Styrene and butadiene are to be polymerised in a series of CSTR each of volume 26.5 m³. If the initial concentrations of styrene and butadiene are 0.795 and 3.55 kmol/m³, respectively, with a feed rate of 19.7 tons/h, estimate the total number of reactors required for polymerisation of 85% of styrene. The density of the reaction mixture is assumed to be constant at 0.87 g/cc.

Given that the rate equation is

$$R_A = -10^{-5}\, C_A C_B \text{ gm moles/lit sec,}$$

The molecular weight of styrene and butadiene are 104 (A) and 54 (B), respectively.

The reaction is,

$$A + 3.2\ B \rightarrow \text{product}.$$

Solution

Considering the first reactor and steady state operation,
Styrene balance:

$$FC_{A0} - FC_A - r_A V = 0. \quad \text{(i)}$$

Butadiene balance:

$$FC_{B0} - FC_B - r_B V = 0. \quad \text{(ii)}$$

From the stoichiometry,

$$r_B = 3.2 r_A, \quad \text{(iii)}$$

where $V = 26.5\ \text{m}^3$; $F = 19.7/(0.87 \times 3600)\ \text{m}^3/\text{sec}$; combining (i), (ii), and (iii),

$$C_B = C_{B0} + 3.2C_{A0} - 3.2C_A. \quad \text{(iv)}$$

Assume a value of $C_A = 0.01$ and determine C_B from Equation (iv), and then determine the rate,

$$r_A = 10^{-5} C_A C_B, \quad \text{(v)}$$

and re-evaluate C_A (i.e., C_{Acal}) until it equals the first guess of 0.01. If not, then assume the next guess of C_A as equal to C_{Acal}. Finally, we get the exit composition of styrene from the first reactor. We also get the value of the butadiene concentration simultaneously from Equation (iv). Taking these exit compositions as the feed composition to the second reactor and repeating the calculations similarly, the exit composition from the second reactor is determined. The calculations have to be repeatedly carried out until the conversion X_A equals 0.85 as desired.

This repeated numerical method can be programmed and the result is given in the following table. The number of required reactors is found to be 22.

C_A	C_B	X_A	No. of reactors
0.596	2.911	0.250	2
0.526	2.688	0.337	3
0.470	2.506	0.408	4
0.423	2.355	0.468	5
0.383	2.226	0.518	6
0.348	2.117	0.561	7
0.319	2.021	0.598	8
0.293	1.938	0.631	9
0.270	1.865	0.660	10

C_A	C_B	X_A	No. of reactors
0.249	1.800	0.685	11
0.230	1.742	0.709	12
0.213	1.691	0.730	13
0.198	1.644	0.749	14
0.184	1.602	0.767	15
0.171	1.564	0.783	16
0.160	1.529	0.797	17
0.150	1.498	0.818	18
0.140	1.469	0.822	19
0.132	1.442	0.833	20
0.124	1.418	0.843	21
0.116	1.395	0.852	22

The above problem could be done graphically, too. Plot r_A vs. C_A on graph paper at various values of C_A and the corresponding values of C_B from Equation (iv). Next, plot the operating line of the reactor given by Equation (i) as

$$r_A = C_{A0} - F/VC_A. \quad \text{(vi)}$$

This is an equation of a straight line while r_A is plotted against C_A.

The slope of the line is F/V, which is the same for all reactors. The points of intersection with the rate equation as plotted earlier will give the corresponding exit composition of styrene. Proceed until the exit composition coincides with the desired conversion as shown in Figure 11.3.

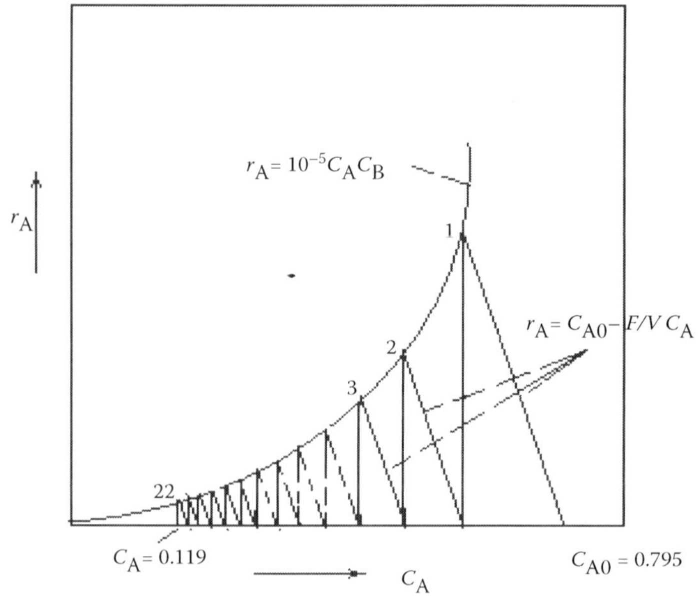

FIGURE 11.3 Graphical solution of Example 11.7.

The graphical solution is left as an exercise for the readers to try. However, the graphical solution is time consuming and erroneous as compared to the computer solution.

11.5 NAPHTHA REFORMER CALCULATIONS

Example 11.8

A naphtha reformer has to be constructed to treat desulfurised naphtha at a rate of 32 m^3/h in a catalyst packed cylindrical reactor. The pathways of reforming reactions are reported as

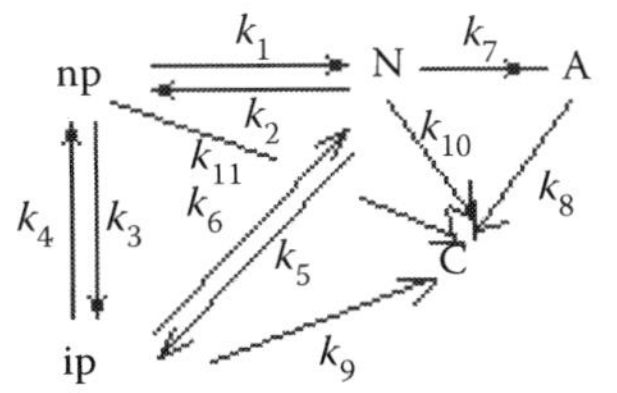

where the groups of hydrocarbons involved in the reactions are normal paraffins (np), iso-paraffins (ip), naphthenes (N), aromatics (A), and coke (C). The rate constants k is indicated in each path and rewritten separately as below for convenience,

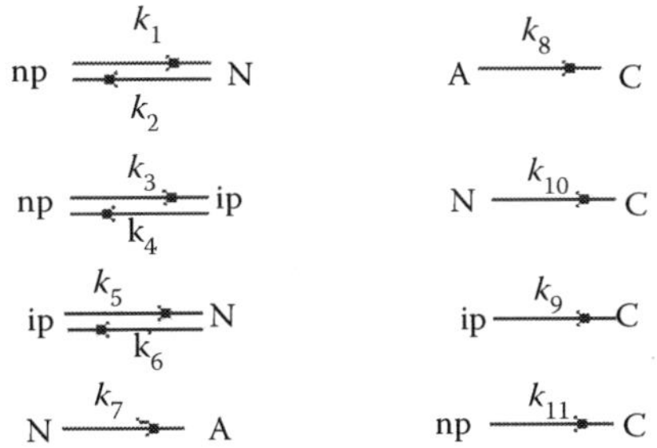

where k values are reported as

$K_1 = 1.07 \times 10^{10} \exp(-23{,}000/T)$,
$K_2 = 3.23 \times 10^{10} \exp(-23{,}000/T)$,
$K_3 = 3.24 \times 10^{9} \exp(-20{,}200/T)$,
$K_4 = 3.25 \times 10^{8} \exp(-20{,}200/T)$,
$K_5 = 1.073 \times 10^{10} \exp(22{,}727/T)$,
$K_6 = 3.227 \times 10^{11} \exp(-22727/T)$,
$K_7 = 7.24 \times 10^{8} \exp(-15{,}150/T)$,
$K_8 = 1.76 \times 10^{13} \exp(-28{,}000/T)$,
$K_9 = 1.76 \times 10^{13} \exp(-28{,}000/T)$,
$K_{10} = 1.76 \times 10^{13} \exp(-28{,}000/T)$,
$K_{11} = 1.76 \times 10^{13} \exp(-28{,}000/T)$.

The rate equations for the components are then in gm.moles per cubic meter per hour as

$$R_{np} = k_2C_N - k_1C_{np} + k_4C_{ip} - k_3C_{np} - k_{11}C_{np},$$

$$R_{Ip} = k_3C_{np} - k_4C_{Ip} - k_5C_{Ip} + k_6C_N - k_9C_{Ip},$$

$$R_N = k_1C_{np} - k_2C_N + k_5C_{ip} - k_6C_N - k_7C_N - k_{10}C_N,$$

$$R_A = k_7C_N - k_8C_A,$$

$$R_c = k_{11}C_{np} + k_9C_{Ip} + k_{10}C_N + k_8C_A.$$

Taking the operating temperature of 470°C, determine the volume of catalyst to be loaded in the reactor.

Solution

The component material balance for the *i*th hydrocarbon component in the packed bed is given as

$$\frac{dc_i}{dv} = \frac{r_i e}{F}$$

where

- c_i mole fraction of component "*i*"
- r_i rate of production of *i*-th component in moles/cu.m hr
- v volume of the catalyst bed, m^3
- e porosity as 0.90
- F feed rate in moles/hr

The solution is carried out numerically using the following steps.

Take a very small increment of 0.0001 in volume to integrate the rate equation. At the feed entry, v = 0 and the compositions are assigned as equal to the feed compositions.

Evaluate the rate constants as defined earlier at the given temperature of 470°C.

Evaluate the rates of formations r_{np}, r_{ip}, r_N, r_A, and r_c, respectively, using the rate constants *k* and the feed compositions.

Integrate the ordinary differential equation using Euler's method of numerical integration and determine c_i values, where *i*th species are np, ip, N, A, and C, respectively. The volume is then incremented by 0.0001 and repeat the above steps while the compositions evaluated by integration will replace the older values until the conversion of normal paraffins (np) becomes 80%.

A small program is then made following the above steps and results are obtained as

V	C_{np}	C_{ip}	C_N	C_A	C_C	X_{np}
0.5	0.56	0.025	0.13	0.088	0.198	0.20
1.0	0.44	0.038	0.087	0.076	0.357	0.37
1.5	0.35	0.042	0.057	0.064	0.485	0.50
2.0	0.28	0.042	0.038	0.054	0.587	0.60
2.5	0.22	0.039	0.025	0.045	0.669	0.68
3.0	0.17	0.035	0.017	0.037	0.735	0.76
3.47	0.14	0.031	0.012	0.030	0.785	0.80

$V = 3.47$ m^3 for 80% mole conversion of np in the feed at 470°C.

Hence, the desired volume of catalyst is 3.47 m^3. In fact, the above solution is representative of an ideal case assuming that there is no mass transfer resistance due to gas films, catalyst pores, and deactivation resulting from coke formation during the reaction. If these effects were accounted for, the required volume of catalyst would have been more than that calculated.

11.6 CALCULATIONS FOR A FLUIDISED CATALYTIC CRACKING REACTOR

Example 11.9

In a riser FCC converter, it is proposed to crack vacuum gas oil (GO) at the rate of 50 tons/h at a temperature of 500°C. The reaction pathways together with details of the rate equations and kinetic parameters are given below,

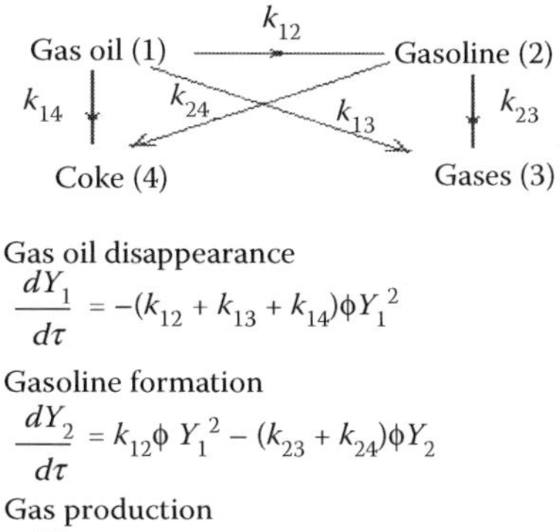

Gas oil disappearance

$$\frac{dY_1}{d\tau} = -(k_{12} + k_{13} + k_{14})\phi Y_1^2$$

Gasoline formation

$$\frac{dY_2}{d\tau} = k_{12}\phi\, Y_1^2 - (k_{23} + k_{24})\phi Y_2$$

Gas production

$$\frac{dY_3}{d\tau} = k_{13}\phi\, Y_1^2 + k_{23}\phi Y_2$$

Coke production

$$\frac{dY_4}{d\tau} = k_{14}\phi\, Y_1^2 + k_{24}\phi Y_2$$

where Y is the weight fraction and subscripts 1, 2, 3, and 4 correspond to the pseudo components as the GO, gasoline, gases, and coke, and τ indicates residence time V/F (volume of reactor/feed rate) and φ is the deactivation factor of the catalyst. The temperature-dependent kinetic parameters are

$$\Phi = \exp(-\alpha\tau) \text{ and } \alpha = 3.017 \times 10^8 \exp(-11{,}705/RT),$$

$$K_{12} = 7.978 \times 10^6 \exp(-68{,}250/RT),$$

$$K_{13} = 4.549 \times 10^6 \exp(-89{,}216/RT),$$

$$K_{14} = 3.765 \times 10^4 \exp(-64{,}575/RT),$$

$$K_{23} = 3.255 \times 10^3 \exp(-52{,}718/RT),$$

$$K_{24} = 7.957 \times 10^3 \exp(63{,}458/RT).$$

Assume that the riser reactor behaves like a plug flow reactor at a constant temperature of 500°C and that the reaction rate controls and diffusional resistances of the fluid film and pores in the catalyst are absent. Calculate the maximum conversion of GO and the height of the riser having an inside diameter of 3 m. Assume that the bulk density of the catalyst is 129 kg/m^3 with an average bed porosity of 60% of bulk volume.

Solution

The above problem has been solved numerically with the help of a small computer program following the steps below.

Assume that the entire riser reactor is divided into a large number of infinitely small reactors of the same cross section through which the reaction mixture flows at the same flow rate *F*. The space time is incremented from the feed entry by the amount of 0.0001 h through each of these incremental reactors and exit conversion from this elemental reactor is repeatedly evaluated using numerical integration, as done in the previous problem, until conversion achieves constant or maximum value. The results of this program are listed below.

Y_1	X_1	Φ	τ
0.90	0.10	0.982	0.004
0.80	0.20	0.966	0.01
0.70	0.30	0.943	0.017
0.60	0.40	0.911	0.027
0.50	0.50	0.867	0.042
0.40	0.60	0.801	0.065
0.30	0.70	0.690	0.109
0.20	0.80	0.469	0.22

Hence, the conversion is 80% mole of feed with about 50% deactivation due to coking. The space time is 0.22 h, i.e., $W/F = 0.22$ h, where *W* is the required weight of the catalyst hold up in the riser, $F = 50$ tons/h = 50,000 kg/h, and the area of cross section = $\pi/4 \times 3^2 = 7.065$ m^2. The weight of the catalyst hold up in the riser = 50,000 × 0.22 = 11,000 kg.

Hence the riser height, $h = 11{,}000/(129 \times 7.065 \times 0.60) = 20.0$ m.

Example 11.10

A riser reactor catalyst of an average diameter of 5.5×10^{-5} cm is fluidised with vaporised hydrocarbon feed at an average temperature of 500°C and 1.6 atm pressure. The feed and product hydrocarbons have the following properties,

Feed mixture (vapour):
mol wt: 300, viscosity at 500°C: 0.0068 cp
Products mixture (vapour):
mol wt: 90, viscosity at 500°C: 0.015 cp

Determine whether external fluid film diffusion will control the conversion process while the average superficial velocity of the hydrocarbons are (a) 80 cm/sec and (b) 400 cm/sec.

Solution

It is established that when Reynolds number of particles is < 30, external diffusion will prevail. Hence, let us evaluate the particle Reynolds' numbers for the desired cases.

Case a

For feed hydrocarbons:

Velocity = 80 cm/sec, dp = 5.5×10^{-5} cm, $T = 773$ k, $P = 1.6$ atm, $M = 300$; $\mu = 0.0068$ cp.

So, the density of feed is PM/RT (assuming ideal gas law) = $1.6 \times 300/(0.0832 \times 773) = 7.46 \times 10^{-3}$ gm/cc.

$Nre_p = 5.5 \times 10^{-3} \times 80 \times 7.46 \times 10^{-3}/(6.8 \times 10^{-5}) = 48.26 > 30$.

For product hydrocarbons:

Velocity = 80 cm/sec, dp = 5.5×10^{-5} cm, $T = 773$ k, $P = 1.6$ atm, $M = 90$, $\mu = 0.015$ cp.

So, the density of feed is PM/RT (assuming ideal law of gases) = $1.6 \times 90/(0.0832 \times 773) = 2.239 \times 10^{-3}$ gm/cc

$NRe_p = 5.5 \times 10^{-3} \times 80 \times 2.239 \times 10^{-3}/(1.5 \times 10^{-4}) = 7.389 < 30$.

This indicates that though reaction controls at the feed entry section and diffusion controls at the product exit section, the overall effect will be due to diffusion control.

Case b

For feed hydrocarbons:

Velocity = 400 cm/sec.

$NRe_p = 5.5 \times 10^{-3} \times 400 \times 7.46 \times 10^{-3}/(6.8 \times 10^{-5}) = 241 > 30$.

For product hydrocarbons:

$NRe_p = 5.5 \times 10^{-3} \times 400 \times 2.239 \times 10^{-3}/(1.5 \times 10^{-4}) = 32.84 > 30$.

Since Reynolds numbers both at the feed entry and product exit of the riser are each greater than 30, reaction controls the rate of conversion. In this case, the riser height can be determined using the method of calculation followed in Example 11.9.

Example 11.11

A bed of 36 tons of 100 mesh coked catalyst particles from the converter is to be fluidised with air at 400°C at a pressure of 250 psia (17 atm) in a cylindrical regenerator vessel of 10 ft (3.48 m) in diameter. The average density of the coked particles is 168 lb/cft (268.8 kg/m^3) and the viscosity of air at the operating conditions is 0.032 cp. Calculate (a) the minimum height of the fluidised bed, (b) the pressure drop in the bed, and (c) the critical superficial velocity of air. Given the average diameter of the catalyst particles is 0.0058 in (0.147 mm) and the minimum porosity of the bed is 0.55.

Solution

Density of air applying the ideal gas law is $\rho = 0.55$ lb/cft, taking the molecular weight of air as 29.

Viscosity of air, $\mu = 0.032 \times 0.000672 = 2.15 \times 10^{-5}$ lb/ft sec.

Volume of solids in the static bed = $36 \times 2{,}000/168 = 428.6$ m^3.

Height of static bed, $L_0 = 428.6/\pi/4 \times 10^2 = 5.45$ ft.

a. So, height (L) of the fluidised bed at minimum porosity, $\varepsilon = 0.55$ is 5.45/(1–0.55) = 12.1 ft {as ε = void vol/bulk vol = $(L-L_0)/L$}.
b. Pressure drop, $-\Delta p = L(1-\varepsilon)(\rho_p - \rho)$, where ρ_p is particle density.

   So, $-\Delta p = 12.1(1-0.55)(168-0.552) = 912$ lb/ft^2 = 6.33 psi = 0.43 atm.

c. Critical superficial velocity (V_{om}) is the velocity of air at which minimum fluidisation occurs corresponding to the minimum porosity of the bed, and is obtained by the following equation,

$$V_{om} = g(\rho_p - \rho)Dp^2\,\varepsilon^3/(150\rho(1-\varepsilon) = 0.141 \text{ ft/sec} = 0.49 \text{ cm/sec},$$

where $g = 9.81$ m/sec^2 (32.49 ft/sec^2), $Dp = 0.0058$ in $= 4.83 \times 10^{-4}$ ft, $\mu = 2.15 \times 10^{-5}$ lb/ft sec.

It is to be noted that the greater the fluid velocity, the greater the height of the bed and porosity. As long as the velocity is smaller than that corresponding to the maximum porosity of unity, the solids will be retained in the bed (batch fluidisation), but if velocity is higher the solids will be carried with the fluid from the bed (continuous fluidisation).

12 Elements of Pipeline Transfer Facilities

12.1 PIPES AND TUBES

Pipes and tubes are long cylindrical conduits (the *L/D* ratio of which is very large as compared to tanks or vessels) of a hollow material of uniform cross section through which liquid and gas can pass. Though the terms "pipes" and "tubes" are used interchangeably, pipes are of a larger diameter and are made of metal, e.g., steel, copper, brass, etc. Tubes are comparatively smaller in diameter and are usually non-metallic. Liquid and gases are transported through pipes connecting one tank to another, pump/compressor discharges to processing units, unit to unit, from one plant to another plant, etc. Thus, small or long pipelines are inevitable for transporting fluids. Liquid density does not change appreciably with pressure and is therefore known as incompressible fluid. Gas density directly varies with pressure to a great extent and hence gases are called compressible fluids. Liquids and liquified gases are transported by pumps, whereas gases are transported by compressors. Pipe diameters as small as 1/8 in and as high as 36 in are commonly employed in refineries and petrochemical plants. The wall thickness of a pipe is specified by the *schedule number*, which is defined as

$$\text{Schedule number} = 1000{*}P/S,$$

where *P* and *S* are the fluid pressure carried in the pipe and the working stress of the material of the pipe wall, respectively. In practice, schedule numbers varying from 10 to 160 are commonly employed. Pipe diameters inside (i.d.) and outside (o.d.) and the thickness corresponding to various schedule numbers are available in standard tables as given in the Appendix. The greater the schedule number, the greater the wall thickness. Tubes are usually specified by standard gauge numbers (British wire gauge number, BWG) where the wall thickness decreases with the increasing gauge numbers. A large number of tubes are used in heat exchangers where the tubes are specified by the Tubular Exchangers Manufactures' Association (TEMA) code. In fact, for hydrocarbon services, the American Petroleum Institute (API) codes are commonly employed. Therefore, when the selection of pipes for then for hydrocarbon services, the API codes should be consulted.

12.2 FITTINGS AND SUPPORTS

Fittings are shorter pieces of materials that are required for connecting pipes of the same or different diameters, connecting pipes leading in different directions without bending the pipes, providing multiple pipe connections, or closing a pipe end, etc. Examples of such fittings are bends, tee joints, elbow, etc., as shown in Figure 12.1.

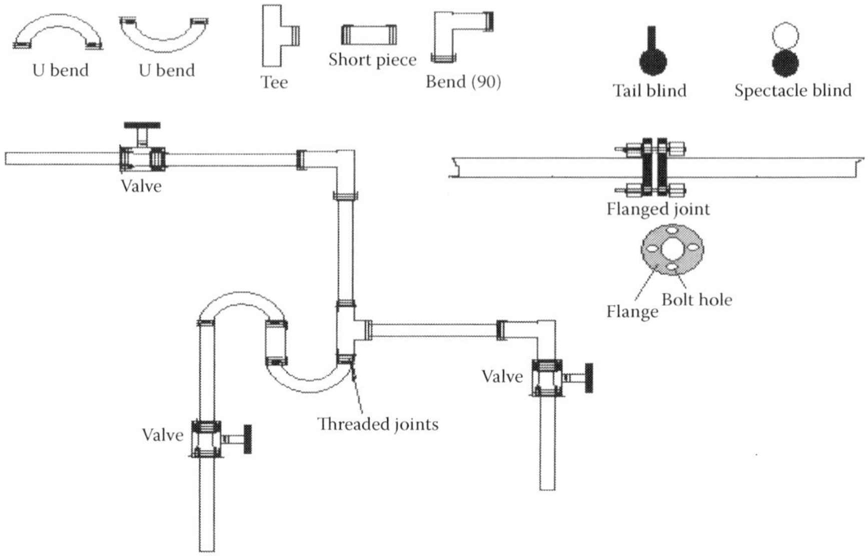

FIGURE 12.1 Common pipe fittings and joints.

Fittings are essential for connecting pipes with pumps, compressors, other machines, vessels, or equipment. These fittings may be joined by threading or by flanges. Pipes running overhead are supported by hangers and suitable columns. Pipes may be laid on the surface or underground (humed or buried). For long pipelines, it may be desirable to lay the pipes in loops to avoid sudden expansion, contraction, or vibration of the pipe. Valves are installed on the pipes connecting two pipe pieces and for manipulating flow rates through a pipeline. Valves may be small or large, depending on the pipe diameter, and may be manually or mechanically operated. A control valve is a special type of valve that is driven by the actuator. Control valves are discussed in more detail in Chapter 13. When valves are tightly shut, flow between the connected pipes is isolated. Since valves may leak because of erosion or corrosion of the valve plugs and/or seat due to prolonged use, it may be necessary to provide *blinds* (tail or spectacle) in between the pipe and valve joints or additional pipe joints specially provided for blinding. Pipes may carry gas or liquid, be cold or hot, or be bare or insulated. Pipe lines carrying liquids having a tendency to freeze are usually provided with a steam coil running over the bare surface of the pipe and are known as steam-traced lines. The entire pipe and the steam-traced surface are further insulated to avoid condensation of steam. Hot pipes are also sometimes provided with a temperature safety fuse or valve (TSV) to avoid damage to the pipeline due to a sudden temperature rise on the pipe surface. Gases are usually transported through pipes under pressure with the help of compressor. Pressure safety valves are provided on these pipes to avoid bursting of the pipes.

12.2.1 Corrosion Protection

The bare surface of pipes is exposed to oxide corrosion. Pipes are usually made of mild steel, which is vulnerable to oxide corrosion due to the galvanic action of iron and atmospheric oxygen in the presence of moisture. Pipes are usually protected from corrosion by painting or galvanising or by applying an electrical charge from an outside source to the bare surface opposing the oxide corrosion, most commonly by the *cathodic protection* method. Pipes passing through an environment containing fluids that may react with mild steel need special corrosion protection.

12.3 CRUDE OIL TRANSFER LINES

The density of liquid is almost unaffected by pressure, therefore liquid is called an incompressible fluid. Crude oil is liquid but it may partially or completely solidify at low temperatures (even at ambient temperature) due to the presence of waxy hydrocarbons. In fact, crude below sea level or at the formation to the surface undergoes a variation of temperature due to the geothermal effect. According to Newton's law of viscosity, the ratio of shear stress to shear rate is the viscosity and is unchanged during its flow while temperature is constant. Fluids that follow this rule are known as Newtonian fluids. Many classes of crude oil do not follow this and are known as non-Newtonian fluids. However, most of the varieties follow Newton's law at elevated temperatures. Some varieties may behave as thixotropic or dilatent, etc. For non-Newtonian crudes, the viscosity term is replaced by formation the consistency index. Different types of fluid viscous properties are presented in Figure 12.2.

12.3.1 Design Steps for Crude Pipes

As most crudes behave like Newtonian fluid at ambient temperature on the surface of Earth, the following steps can be extended for the design of pipelines.

Step 1: Assume the diameter and length of the pipe and the necessary fittings and valves, if any, from a standard table of commercially available pipe dimensions.

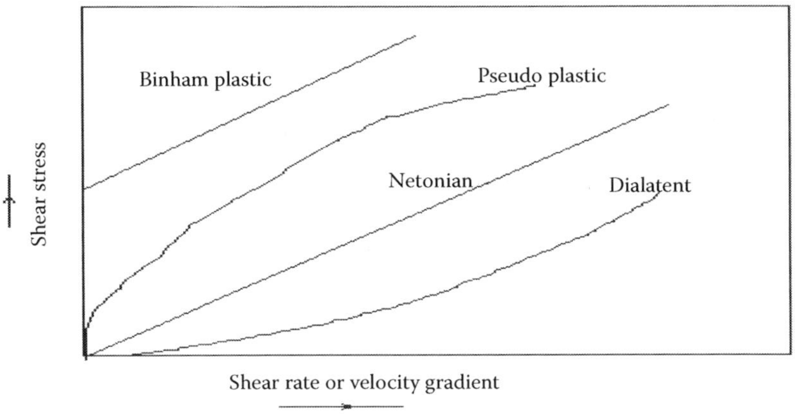

FIGURE 12.2 Newtonian and non-Newtonian fluids.

Step 2: Determine the Reynolds number (NRe), which is defined as the ratio of the product of density (ρ), average velocity (V), and pipe inside diameter (D) to the viscosity (μ), i.e.,

$$NRe = DV\rho/\mu. \tag{12.1}$$

In case the temperature varies along the pipe length, the arithmetic average temperature may be evaluated to determine the properties of the liquid at that temperature.

Step 3: Obtain the friction factor (f) from the NRe vs. f chart or evaluate f for the following ranges of NRe.

If $NRe \leq 2100$, the condition of the viscous streamline flow or parabolic velocity distribution is

$$f = 16/NRe, \tag{12.2}$$

if $3000 < NRe \leq 3 \times 10^6$, the turbulent condition is

$$f = 0.00140 + 0.125/NRe^{0.32}, \tag{12.3}$$

with ± 5% accuracy. However, a more accurate f can be obtained with the help of friction factor chart. No accurate determination of f has been established for NRe between 2,100 and 10,000, however an average value may be used from extrapolation from both the stream line and turbulent regions.

Step 4: Determine the pressure drop (Δp) using Fanning's rule, as

$$\Delta p = 2f\rho v^2 L/Dg, \tag{12.4}$$

where g is the acceleration due to gravity, ρ is the density, v is the average velocity of fluid, and L is the straight length or equivalent length in case of the presence of joints and valves.

Step 5: Verify the desired delivery pressure from the available upstream supply pressure at the entrance to the pipe. Otherwise, repeat from Step 1 with a new selection.

In case the crude is non-Newtonian, a modified Reynolds' number (NRe_n) must be used,

$$NRe_n = 2^{(3-n)}\{n/(3n+1)\}^n D^n \rho v^{(2-n)}/k, \tag{12.5}$$

and the friction factor is given as

$$f_n = 2^{n+1}k(3 + 1/n)^n/D^n \rho v^{(2-n)}, \tag{12.6}$$

where n and k are the power law indices for the flow behaviour and consistency index, respectively. After the inside diameter of the pipe is evaluated, it is essential to consult the API standards applicable for hydrocarbon fluids in order to decide on the material of construction and the thickness of the pipe. Other conditions, such as the reactive or corrosive atmosphere of both the outside and inside surfaces of the pipe, the operating temperature and pressure involved, etc., must be taken into account.

12.3.2 Economic Pipe Diameter

Since the pumping cost increases with an increase in pressure drop due to friction as the diameter is smaller and smaller, whereas the material cost of the pipe falls with the decreasing pipe diameter for the same length, economic consideration of the pipe diameter is essential. Thus,

Total cost per year = cost of pumping for the year + depreciation of pipe per year,

where the cost of pumping for the same throughput in the year will increase with the smaller diameter pipe, i.e., it will be inversely proportional to pipe diameter, whereas depreciation of the pipe per year is a portion of the total cost of the pipe divided by the useful life span of the pipe material, which decreases with a decrease in the pipe diameter for the same throughput.

$$C_{\mathrm{T,\,annum}} = C_{\mathrm{p}} K_{\mathrm{p}} D^{a} + C_{\mathrm{m}} K_{\mathrm{m}} / D^{b}, \tag{12.7}$$

where:

$C_{\mathrm{T,\,annum}}$: total annual cost per unit length of pipe
C_{p}: cost of pumping per unit mass throughput
K_{p}: constant relating properties of fluid and friction factor, etc.
D: inside pipe diameter
C_{m}: cost of pipe material (purchase price) per unit mass
K_{m}: constant relating material properties and thickness
a and b: the exponents of variation correlated with the pipe diameter

The minimum total annual cost can be mathematically evaluated by differentiating the $C_{\mathrm{T,\,annum}}$ with respect to D and equating to zero to get the corresponding diameter, which is the optimum diameter (D_{opt}). The economic pipe diameter of any pipe is available from a standard nomograph.

12.4 PRODUCT TRANSFER LINES

Liquified gases, such as liquified natural gas (LNG), liquified petroleum gas (LPG), and liquid fuels, e.g., motor spirit, kerosene, diesel, furnace oil, etc., are also Newtonian liquids and pipe design steps will be as described in the previous section. Usually, white oils are transported through a same pipeline separating the products by maintaining pressure and a predetermined delivery time schedule. High pressure pumps will be required to maintain the flow as per schedule. Sticky, congealing, and viscous black oil products need separate steam-traced pipelines for transfer and are usually not economical for long distance transport through pipelines. A pig, a tightly fitting leather or polymer ball moved by the high pressure of liquid in the pipe, is required to separate black oil products in pipeline transport. Such a piping requires high pressure design as per the API and ANSI standard.

12.5 GAS TRANSFER LINES

The density of gas is a strong function of pressure and temperature. It increases with pressure and decreases with temperature. As density of gas varies appreciably with the variation of pressure, this is known as compressible fluid, i.e., the volume of gas changes with pressure at a constant temperature. According to the ideal gas law, density (ρ) is related as

$$\rho = PM/RT, \tag{12.8}$$

where P is the pressure, M is the molecular weight, R is the universal gas constant, and T is the temperature in absolute scale. Gas is transported through pipes of selected diameters based on the flow rate, density, pressure, temperature, and viscosity of the gas. Sufficient pressure must be available at the upstream side of the pipe to overcome the necessary pressure drop due to fluid friction and the gas is delivered at the other end at the desired pressure. It is to be noted that the viscosity of gas increases with temperature, and as the friction manifests itself in the form of heat, which increases the temperature of the gas, this, in turn, magnifies the friction. In a long pipe with a low pressure, the velocity of gas may be so high that it may equal the velocity of sound in that gas. In this case, a further increase in the length of the pipe will reduce the gas flow rate and the pipe may be choked, i.e., little or no flow of gas will occur at the delivery end. The ratio of the velocity of gas (u) to the velocity of sound (u_s) of the same gas is known as the Mach number,

$$N_{Ma} = u/u_s. \tag{12.9}$$

If the Mach number increases, the mechanical energy of transport is lost in the form of heat and sound as well. Therefore, it is desirable that the pipe should be designed so that the Mach number is less than unity. The pipe may handle hot or cold gas in a well insulated or a bare pipeline. A well-insulated pipe resembles an adiabatic condition whereas a bare pipe is non-adiabatic or isothermal. In the isothermal flow condition, the temperature being constant, density is a function of pressure only and the pressure may be evaluated by integrating it over the line. It is desirable that the average temperature and the pressure of the gas should be evaluated to determine its properties, such as viscosity, density, etc., avoiding complications due to variable properties throughout the length of the pipe. The design procedure will then assume the steps described in Section 12.4. Liquified methane, better known as LNG, is usually transported by big tankers with the provision of an *in situ* refrigeration unit. However, compressed natural gas may be transported through a long pipeline at high pressure, but in gaseous form. Since the density is low, gaseous transfer may not be economical for a long pipeline.

12.6 PUMPS AND COMPRESSORS

Pumps and compressors are machines that deliver liquids and gases at the desired quantity and pressure through pipes connecting the storage to the delivery end, either short or long distances connected through pipelines. Gaseous fluid or gas is transported by machineries, known as compressors, as gases are compressible fluids. Thus, fluid moving machines are classified as pumps (for liquids) and compressors (for gases). These machines add necessary energy to make the fluid move in the pipeline by one of the following methods:

1. *Volumetric displacement* by reciprocating pumps or compressors
2. Applying *centrifugal force* by centrifugal pumps or compressors
3. *Mechanical* impulse by axial flow pumps and compressors
4. *Transfer of momentum* by using another motive fluid, by air lift or acid egg pumps
5. Applying electrical power and magnetic flux over an electrically conducting fluid with an *electromagnetic* pump

12.6.1 Centrifugal Pumps

The term "centrifugal" implies the centrifugal force generated by the blades of the impeller rotating at very high speeds used to deliver liquid to the pipeline. A typical centrifugal pump consists of a volute or a casing in which are housed a number of s-shaped impeller blades that look like a fan. Liquid enters the eye of the pump by gravity or due to the inlet or suction line pressure of the liquid. As the blade rotates, liquid from the eye of the pump moves to the tip of the blade and attains the kinetic energy imparted by the tip and finally strikes the delivery end of the volute. At the delivery end, kinetic energy is converted to pressure and thus the liquid discharges through the delivery end at high pressure energy or head. Such a centrifugal pump is shown in Figure 12.3. If the delivery end is shut off, i.e., there is no discharge from the pump, maximum pressure will be generated. While the delivery end is opened,

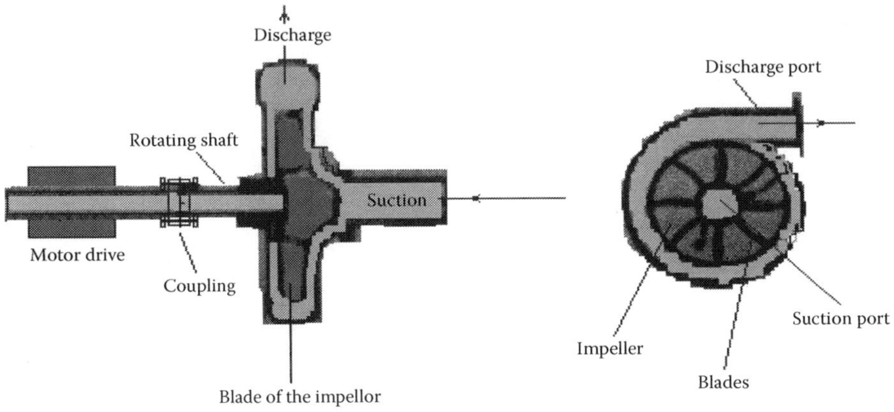

FIGURE 12.3 Centrifugal pump (single stage).

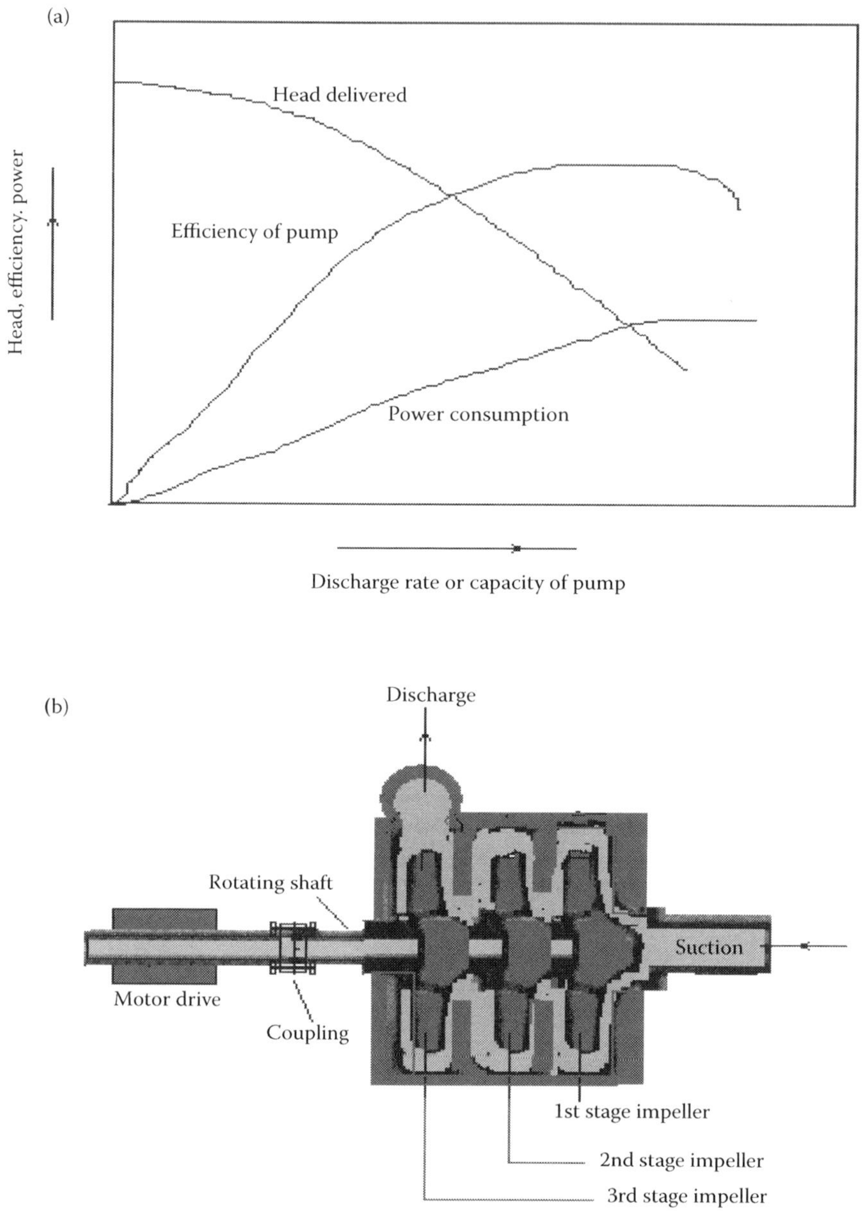

FIGURE 12.4 (a) Centrifugal pump characteristics. (b) Multistage centrifugal pump.

liquid will be discharged but the pressure will be less than the shut off pressure. Thus, the discharge pressure (or discharge head expressed in terms of height of liquid) will fall with the increasing flow rate and finally reduces to a very low discharge pressure. Such behaviour of a centrifugal pump is commonly encountered and is presented in Figure 12.4a.

In order to maintain a uniform discharge pressure with a high flow rate, a multistage pump is desirable. In a multistage pump, a number of impellers are connected in a series mounted on the shaft of the pump such that discharge from one set of impellers feeds the next set, which compensates for the fall in discharge pressure from the previous impellers. For a multistage pump, the discharge rate is the same as that of a single-stage pump, but the discharge pressure is the sum of the discharge pressures of all the stages. This is explained in Figure 12.4b.

For a single-stage centrifugal pump, the following relations hold good:

Discharge flow rate (Q) varies with the impeller diameter (D) and speed of rotation (N) as

$$Q = k_1 ND, \tag{12.10}$$

where k_1 is a constant depending on the type of liquid and the pump.

Thus, for two pumps having the same impeller diameter but of different speeds (N_1 and N_2), the ratio of their flow rates, Q_1 and Q_2, is given as

$$Q_1/Q_2 = (N_1/N_2). \tag{12.11}$$

The head developed (H) is proportional to the square of the speed (N) and the square of the diameter (D), thus

$$H = k_2 N^2 D^2, \tag{12.12}$$

where k_2 is a constant typical for the pump and liquid.

Power (W) consumption is the cubic power of the speed (N) and the diameter (D), thus

$$W = k_3 N^3 D^3, \tag{12.13}$$

where k_3 is a constant typical for the pump and the liquid. The shaft power is consumed to overcome various frictions, like bearing friction, disk friction, and fluid friction. Some amount of recirculation takes place from the impeller to suction and, as a result, some power is not useful for delivery. For all these reasons, the net amount of power delivered to the fluid is reduced, i.e.,

FHP = BHP − bearing loss − disk friction − fluid friction − loss due to leakage or recirclation.

The efficiency of pump (η) is defined as the ratio of power delivered to the fluid to the shaft power,

$$\eta = \text{FHP/BHP}. \tag{12.14}$$

12.6.1.1 Priming

In a centrifugal pump, the volute must be filled with liquid to displace air/gas in order to maintain a continuous supply of liquid at the lifting end of the blade to replenish the discharged amount. If the volute becomes dry or filled with vapour or gas, the impeller will not be able to deliver the flow rate as the density of the gas is too small to be lifted by centrifugal force. If the volute is partially filled, the pump will not be able to discharge because the liquid will be returned back to the volute, causing knocking, which may even damage the impeller blades. Hence, it is essential that the volute of the pump must be filled before starting. Self-priming pumps are also available in which provision is made within the volute and the impeller such that gas is expelled at the starting of the pump, which then discharges the liquid when the self-priming is complete. The formation of vapour or low suction pressure causes cavitations, which result in no discharge, as is the situation with an unprimed pump. *Cavitation* may occur if the pressure at the suction port is lower than that required suction head for the pump. Theoretically, the available suction head should be greater than the desired (fixed by the pump manufacturer) suction head. The available suction head is calculated from the pressure head after adding the kinetic head at the eye and deducting the vapour pressure head. This calculated head is usually expressed as the net positive suction head (NPSH). Thus, if the average velocity and static pressure at the eye of a pump are V_{eye} and P_{eye}, respectively, and the vapour pressure is P_v, then NPSH is given as

$$\mathrm{NPSH} = V_{eye}^2/2g + (P_{eye} - P_v)/\rho. \tag{12.15}$$

Since the pressure at the eye of the pump is not always available, it is convenient to evaluate NPSH from the surface of the liquid as shown in Figure 12.5.

According to the mechanical energy balance neglecting the heat effects (Bernoulli's equation), levels *a* and *b* as shown in Figure 12.5, considering level *a* as the reference, are

$$P_a/\rho + V_a^2/2g = P_{eye}/\rho + V_{eye}^2/2g + Z_a + h_{fs} = P_b/\rho + V_b^2/2g + Z_b. \tag{12.16}$$

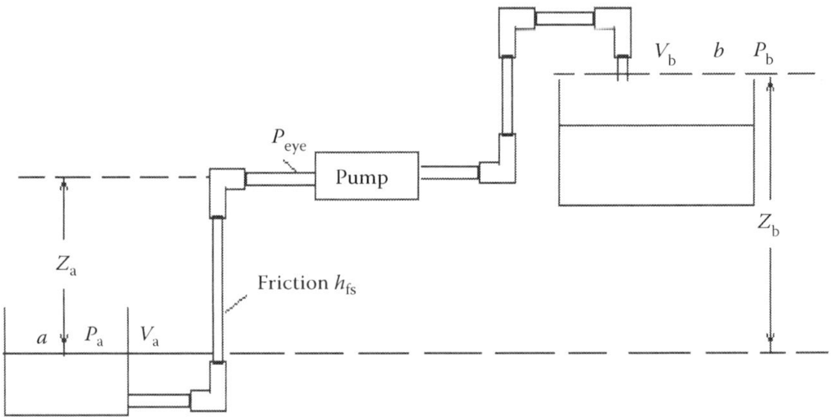

FIGURE 12.5 A typical liquid pumping system with a negative potential head at the suction.

(Note that h_{fs} is the fluid friction experienced by the moving fluid in the pipe and it is the difference of the total energy of the upstream and the downstream of the flowing fluid.)

So,

$$P_{eye}/\rho + V_{eye}^2/2g = P_a/\rho + V_a^2/2g - Z_a - h_{fs}. \quad (12.17)$$

Hence, from Equation 12.13,

$$\text{NPSH} = V_{eye}^2/2g + (P_{eye} - P_v)/\rho = (P_a - P_v)/\rho + V_a^2/\,2g - Z_a - h_{fs}, \quad (12.18)$$

where V_a is the velocity of the liquid level at a, which is practically negligible for a large tank. NPSH is evaluated as

$$\text{NPSH} = (P_a - P_v)/\rho - Z_a - h_{fs}. \quad (12.19)$$

This is true while the eye of the pump suction is above the liquid level, i.e., negative suction as Z_a, the potential head is deducted in Equation 12.19.

If the eye of the pump is below the liquid level, as shown in Figure 12.6, Z_a will be added and hence it becomes a positive potential head at the suction.

In this case, NPSH is given as

$$\text{NPSH} = (P_a - P_v)/\rho + Z_a - h_{fs}. \quad (12.20)$$

In both cases, h_{fs} is the head loss due to friction in the suction line.

The available value of NPSH in feet must be greater than the required NPSH by at least three feet.

12.6.1.2 Specific Speed

The head (H) delivered by a centrifugal pump depends on the speed (n) and the capacity (Q). A dimension analysis is then applicable to find a general relationship as follows.

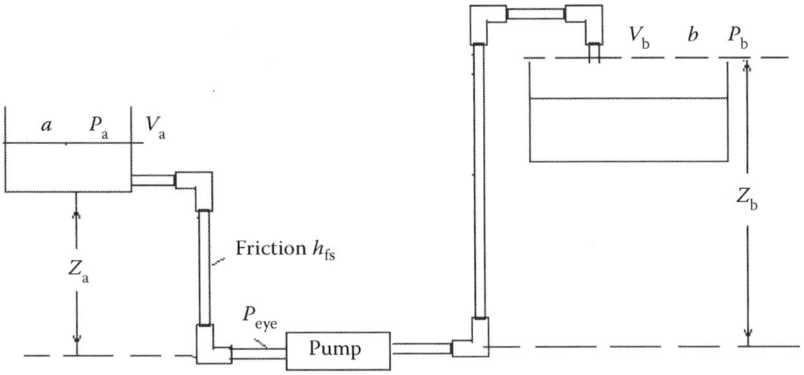

FIGURE 12.6 A typical liquid pumping system with a positive potential head at the suction.

$$H = f(n,Q). \tag{12.21}$$

Taking

$$H = kn^a Q^b, \tag{12.22}$$

where the dimensions of H, n, and Q are expressed in fundamental units, length (L), mass (M), pound mass (lb_m) and time (θ) as,

$$H = \text{pressure/density} = lb_f/ft^2/lb_m/ft^3 = lb_m ft/sec^2/ft^2/lb_m/ft^3 = ft^2/sec^2 = L^2/\theta^2,$$

$$n = \text{rpm} = 1/\text{sec} = 1/\theta,$$

$$Q = ft^3/\text{sec} = L^3/\theta.$$

Substituting L and θ terms in relation 12.22, we obtain

$$L^2/\theta^2 = k(1/\theta)^a(L^3/\theta)^b = kL^{3b}/\theta^{(a+b)}. \tag{12.23}$$

Hence, equating the indices,

$$3b = 2 \text{ and } a + b = 2, \text{ or } a = 2 - b = 2 - 2/3 = 4/3.$$

Substituting $a = 4/3$ and $b = 2/3$ in relation 12.22,

$$H = kn^{4/3}Q^{2/3}, \tag{12.24}$$

or

$$H^{3/4} = k^{3/4}nQ^{1/2} = k^1 nQ^{1/2},$$

or

$$k^1 = nQ^{1/2}/H^{3/4}. \tag{12.25}$$

Thus, $nQ^{1/2}/H^{3/4}$ is a dimensionless entity.

For $Q = 1$ and $H = 1$, $k^1 = n$.

Thus, this index is equivalent to the speed of the pump while the capacity is 1 gal/min and the head is 1 ft of fluid to be discharged. Therefore, this index is expressed as the specific speed (index) or N_s, as,

$$N_s = n\sqrt{Q}/H^{3/4}, \tag{12.26}$$

where, if Q is in gallons per minute, H should be in feet, and n should be in revolutions per minute. Other units may be used with the appropriate conversion factors to make the number, N_s, dimensionless.

For various pumps, the specific speed is available from published data for a variety of impeller designs. For example, for a radial vane-type impeller, the specific speed will be from 500 to 1000, and the maximum limit for the specific speed is also obtained against the head for a given suction lift. Thus, for a particular design, *Ns* is obtained from relation 12.26 and the corresponding *H* expected to be delivered by this design is obtained from published data to find the operating efficiency of the pump.

For example, from a standard pump manufacturers' data, the following information is available for a pump type: single suction, with shaft passing through the eye of the pump.

Maximum specific speed, N_s	Suction lift, 25 ft	Suction lift, 20 ft	Suction lift, 15 ft	Suction lift, 10 ft
	Total head, ft	Total head, ft	Total head, ft	Total head, ft
1500	170	260	352	450
2000	100	155	220	280
2500	70	110	150	200
3000	52	85	110	140
3500	40	65	90	110
4000	32	50	70	85
5000	25	35	47	60

12.6.2 Positive Displacement Pumps

These pumps add energy by direct pressure to the fluid or by displacement of the fluid (positive displacement). Two groups of pumps fall into these categories—reciprocating and *rotary*. In the reciprocating pump, force is exerted by a piston compressing the fluid contained in a cylindrical chamber. During suction stroke, the piston depressurises the chamber and allows fluid to enter the chamber, which is then pressurised, in the discharge stroke, to the desired pressure head followed by delivery of the fluid.

Thus, suction stroke and discharge stroke complete a cycle, where half the cycle is for suction and the other half is for delivery of the fluid. Pumps that deliver fluid in half the cycle are known as single acting or stroke or single cylinder pumps. If two cylinders are connected in such a way that one cylinder discharges while the other completes the suction stroke and thus the delivery of fluid occurs in a single cycle, this type of pump is called a two stroke or double acting or two cylinder pump. For a large pressure head with a small capacity, reciprocating pumps are suitable. For very high pressure with very small capacity, usually the dozing pumps are equipped with a plunger in place of a piston, as shown in Figure 12.7. These pumps are not suitable for sticky fluids and slurries, which may damage the cylinder and/or the piston. However, viscous but clear, even molten, liquid may be pumped without damage. Corrosive acid or similar chemicals should not be pumped by piston pumps. However, a diaphragm is used to replace the piston and cylinder (Figure 12.8).

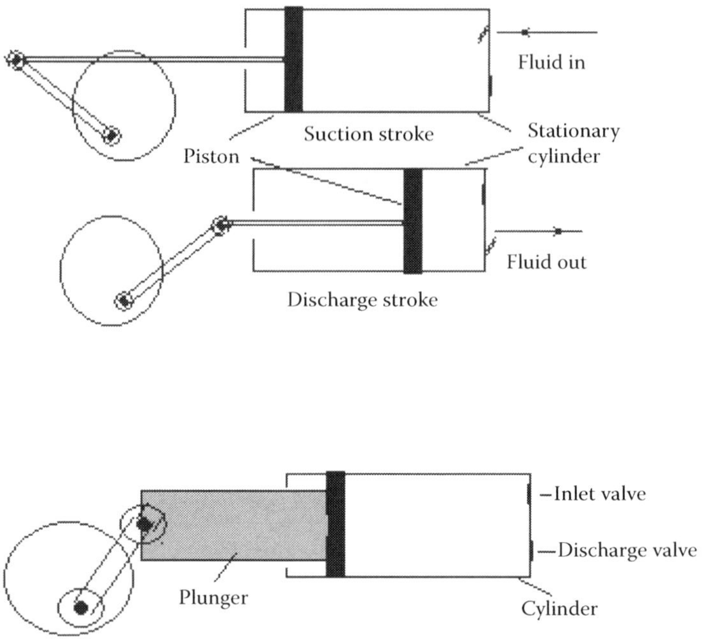

FIGURE 12.7 Reciprocating pump activities.

12.6.3 Rotary Pumps

The positive displacement of fluid is carried out by rotating gears, screws, cams, etc. In a gear pump, the space between the gears is so closely meshed such that the fluid is trapped and carried out to the next mesh and finally discharged at high pressure. In a screw pump, twin screws can also act similar to the gear pump. In a single screw system, the liquid is carried by the helical space of the screw and just lifted. Such a single screw pump is used to discharge a large amount of liquid at low pressure. Some of these rotary pumps are gear pumps, screw pumps,

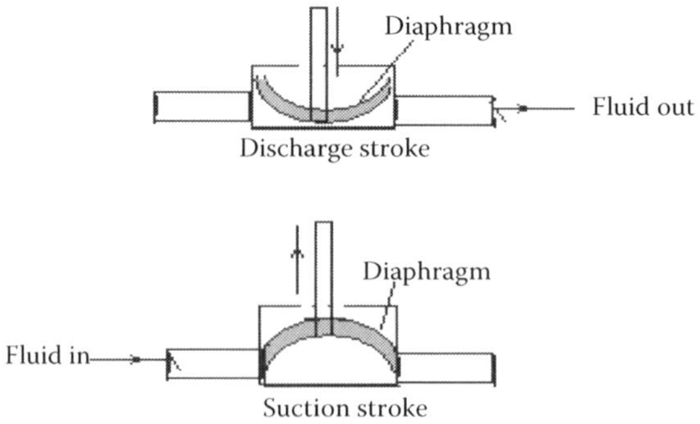

FIGURE 12.8 Reciprocating diaphragm pump activities.

lobe pumps, cam pumps, etc. Gear pumps that consist of two gears in close mesh directly pushes the fluid trapped in between them. All these pumps are suitable for viscous and sticky liquids. Diaphragm pumps can be used to deliver slurries and corrosive liquids.

In a screw pump the helical surface of the shaft pushes the fluid. Screw pumps are widely used in wastewater treatment plants to lift a large quantity of water.

12.6.4 Compressors

Gases are compressed from their suction pressure to the discharged pressure either by centrifugal or reciprocating compressors. These are also available as single stage and multistage compressors. During compression, heat is developed (it is to be noted that heat is generated due to adiabatic compression and also the mechanical and fluid friction within the moving parts of the compressor), which increases the temperature of the gas, causing density reduction from that in the suction. If this heat is not removed (adiabatic), compressed gas will attain a high temperature causing low delivery rate as far as the mass flow rate is concerned. Therefore, the work required to deliver the same mass flow rate will be more for such an adiabatic condition as compared to an isothermal condition when heat is removed to bring the temperature of gas to the suction temperature. In fact, this is easier said than done because of the resistance to heat transfer from compressed gas to cooling fluid. Practically, the temperature of the compressed gas will be between the adiabatic (while no heat is removed) and isothermal (while all heat is removed) temperature, which is also known as the *polytropic* (both adiabatic and isothermal conditions) temperature. For a multistage compressor, a cooling arrangement is desirable in each stage, known as the interstage cooler (stage cooler), to reduce the power consumption for a given mass flow rate of the gas. A single-stage reciprocating compressor is shown in Figure 12.9. An accumulator

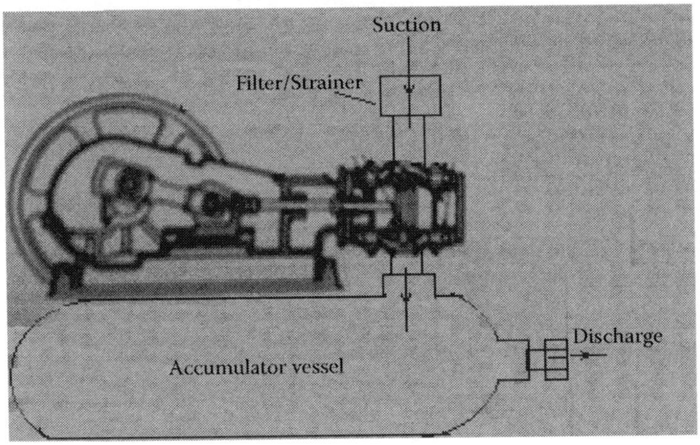

FIGURE 12.9 A single-stage reciprocating compressor.

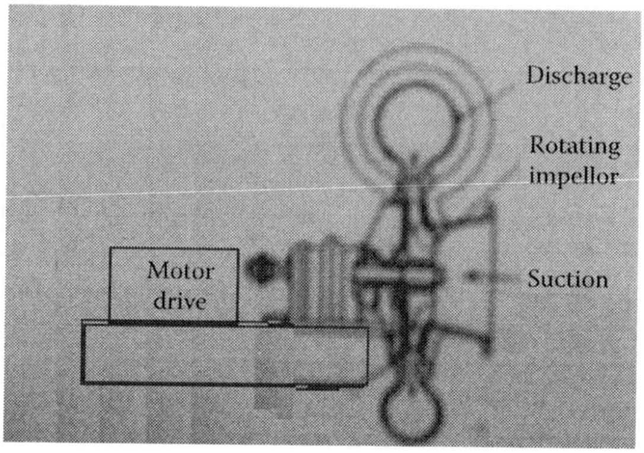

FIGURE 12.10 A single-stage centrifugal compressor.

drum is required to store compressed gas to avoid pressure fluctuation. A single-stage centrifugal compressor is presented in Figure 12.10.

12.7 POWER CALCULATIONS FOR PUMPING AND COMPRESSION

The power required for pumping and compressing can be estimated based on the following steps.

Power for pumping a liquid or a liquified gas:

$$\text{Power in horse power or BHP} = \text{mass flow rate } (m) \times \text{head of fluid } (\Delta H)/\text{efficiency}(\eta) = m\Delta H/\eta$$

where the BHP is the total power exerted to turn the shaft of the pump. The amount of energy actually transferred to the fluid is a fraction (equal to the efficiency, η) of the BHP and the rest is consumed mainly for overcoming the bearing friction.

In the case of a compressor, fluid ΔH cannot be the same as that for liquid as the density varies with pressure. In this case, the FHP is calculated as the mass flow rate and the integrated value of vdp, neglecting the kinetic head, the potential head from the total energy balance equation (Figure 12.11).

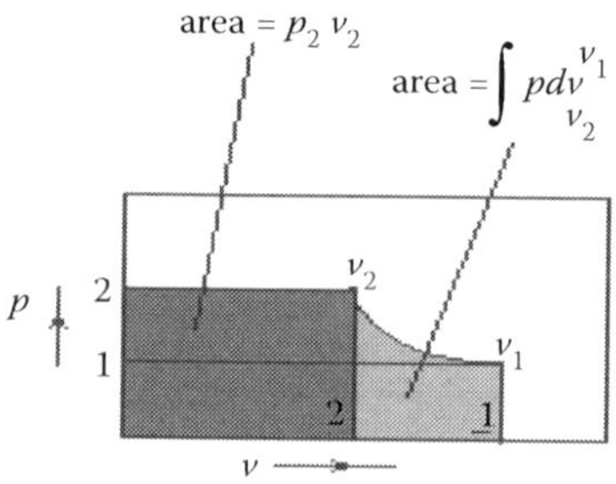

FIGURE 12.11 p-v diagram for the compression cycle.

Work done by the system = p_1v_1 (const.press expansion)	Work done on the system = $\int_{v_1}^{v_2} pdv$ (work to compress from v_1 to v_2)	Work done on the system (work to compress at const pressure) = $-p_2 v_2$ (return stroke)

$$\text{Net work done on the gas for compression} = p_1v_1 + \int_{v_1}^{v_2} pdv - p_2v_2$$

$$= -(p_2v_2 - p_1v_1) + \int_{v_1}^{v_2} pdv$$

$$= -\int_1^2 d(pv) + \int_{v_1}^{v_2} pdv = -\int_{p_1}^{p_2} v\, dp.$$

This value will be different for the adiabatic, isothermal, and polytrophic compressions.

$$W + E_1 + P_1V_1 + u_1^2/2g + Z_1 = E_2 + P_2V_2 + u_2^2/2g + Z_2$$

which is valid for liquid only. For gas, the same should be written in the differential form as gas is a compressible fluid, i.e., density is a strong function of pressure as

$$W + dE + \frac{dP}{\rho} + d(\frac{u^2}{2g}) + dZ = 0$$

Work is then obtained by integration under steady flow conditions as

$$W = -\int_{p_1}^{p_2} vdp$$

This relation is valid for the adiabatic or isothermal or polytrophic conditions (Figure 12.12).

Adiabatic work:

$$pv^y = c \text{ or } v = \left(\frac{c}{p}\right)^{1/y},$$

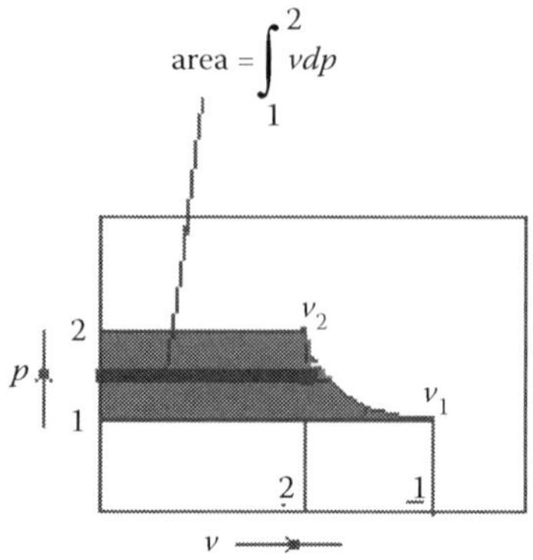

FIGURE 12.12 Net area indicating work of compression required.

$$W = \int_{p_1}^{p_2} vcdp = \int_{p_1}^{p_2} \left(\frac{c}{p}\right)^{1/y} dp = c^{1/y} \int_{p_1}^{p_2} p^{-1/y}\, dp = \frac{yp_1v_1}{1-1/y}\left[\left(\frac{p_2}{p_1}\right)^{1-1/y} - 1\right].$$

W can also be evaluated directly from the difference in enthalpies ($H_2 - H_1$).

Isothermal work:
PV = constant and $V = RT/P$.

$$W = \int_{p_1}^{p_2} vdp = \int_{p_1}^{p_2} \frac{RT}{P} cdP = RT \ln\left(\frac{p_2}{p_1}\right).$$

The above work is the work required to compress per mole of the gas, where γ = ratio of specific heats = c_p/c_v. This is applicable for a single-stage compressor.

In the case of a multistage compressor of *N* stages with stage cooling provision, the minimum work of compression per unit mass (*W*) is given as

$$W = \frac{Nyp_1v_1}{1-1/y} \cdot \left[\left(\frac{p_2}{p_1}\right)^{\frac{(y-1)}{Ny}} - 1 \right]$$

polytropic work:
In practice in the multistage compressor, the compression is neither truly adiabatic nor isothermal and a polytrophic coefficient *k* is used in place of γ. The value of *k* will be between 1 and γ.

Exercise 12.1

A pump draws crude oil having an API of 36 and a viscosity of 6 cp from a storage tank at 30°C through a 2 in schedule 40 steel pipe at a rate of 1.68 kg/sec and is delivered to a location in the process plant at a height of 15 m from the level of the storage tank. Determine the power required by the pump at an average efficiency of 60%.

Solution

Using Bernoulli's mechanical balance equation:

$$w_p\eta = z_2 - z_1 + \frac{u^2}{2g} + h_f$$

where $\eta = 0.60$, $z_1 =$ (ground level) $= 0$, $z_2 = 50$ m, $u =$ velocity in the exit pipe, and $h_f =$ friction in the line.

Density of liquid = 850 kg/m^3.

So, volumetric flow = 1.68/850 = 1.97×10^{-3} m^3/sec.

Inside diameter of the 40 schedule 2 in pipe = 2.067 in = 0.0525 m.

Area of cross section = $3.14/4 \times (0.0525)^2 = 9.64 \times 10^{-5}$ m^2.

So, the average velocity, u = volumetric flow/area of cross section = $1.97 \times 10^{-3}/9.64 \times 10^{-5}$ = 20.4 m/sec (66.9 ft/sec),

$$\mu = 6 \text{ cp} = 0.06 \text{ poise} = 0.06 \text{ gm/cm sec} = 0.006 \text{ kg/m sec}.$$

So

$$NRe = DU\rho/\mu = 0.0525 \times 20.4 \times 850/0.06/0.006 = 151{,}725.$$

From the friction factor chart for Newtonian liquids, $f = 0.003$.

Using Fanning's equation for pressure drop,

$$h_f = 2fu^2L/Dg = 2 \times 0.003 \times (20.4)^2 \times 50/(0.0525 \times 9.81), = 242.4 \text{ m(794 ft)}$$

where

$$U^2/2g = 20.4^2/2 \times 9.81 = 21.2 \text{ m (69.4 ft)},$$

$$Z_2 - Z_1 = 15 \text{ m (49.2 ft)}.$$

So

$$W_p = (242.4 + 21.2 + 15)/0.60 = 463.3 \text{ m (1,020 ft)}.$$

Hence, the power required is

Mass flow rate × W_p = 1.68 × 464.3 = 780 kg m/sec = 7,800 J/s = 7.8 KW = 10.45 hp,

(as 1 hp = 550 ft lb/s = 746 W = 74.6 kgf m/s, 1 kgf m = 9.82 J.)

Exercise 12.2

A hydrocarbon solvent at 38°C is to be pumped at a rate of 150 L/min from a tank at atmospheric pressure. The discharge pressure of the pump at the end of the line is 4.4 kg/cm² abs. The discharge pipe is 3 m and the pump suction is 1.2 m, respectively, above the liquid level in the tank. Calculate (a) the developed head of the pump, (b) the BHP, and (c) the NPSH. Assume that the inner diameter of the discharge pipe is 1–1/2 in schedule 40 pipe and the friction at the suction and discharge lines is 0.034 and 0.37 kg/cm², respectively. Take the efficiency of the pump as 60%. The density and vapour pressure of the solvent are 0.864 kg/lit and 0.25 kg/cm², respectively.

Solution

$$W_p\eta = V_{dis}^2/2gc + Z_{dis} + P_{dis}/\rho + h_{fric} - P_a/\rho,$$

where

Inner diameter of the pipe obtained from std table = 1.38 in

Area of cross section A: 3.14/4 × (1.38/12 × 0.3048)² = 9.64 × 10⁻⁴ m²

V_{dis} = 150 × 10,000/10⁻³(60 × 9.64) = 2.593 m/sec

$V_{dis}^2/2gc$: kinetic head at the discharge = (2.593)²/2 × 9.81 = 0.34 m

Z_{dis}: potential head at the discharge = 3.0 m

P_{dis}/ρ: pressure head at the discharge = 4.4 × 10⁴/864 kg/m²/kg/m³ = 50.9 m

h_{fric}: total friction head in the line leading from the tank to the end of the discharge line = (034 + 0.37) = 0.404 × 10⁴/864 kg/m²/kg/m³ = 4.67 m

P_a/ρ: pressure head above the liquid level in the tank = 1.0133 × 10⁴/864 = 11.72 m

$W_p\eta$ = 0.34 + 50.9 + 4.67 − 11.72 = 47.19 m = 47.19 kgf m/kg

Head to be developed by pump = 47.19 m

Power required = 47.19/η = 47.19/0.60 = 78.65 kgf m/kg

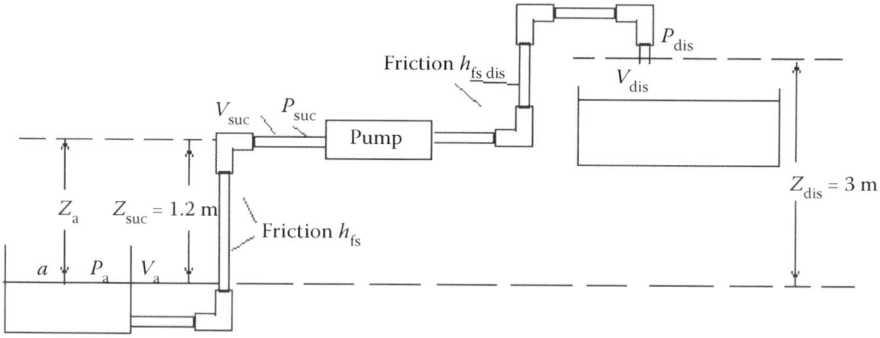

Mass flow rate = 0.150 × 864/60 = 2.16 kg/sec
Hence, the power required = 2.16 × 78.65 kgf m/sec = 169.8 kgf m/sec = 169.8 × 9.81 J/sec (as 9.81 J = 1 kgf m)

$$= 1{,}665.738 \text{ J/sec} = 1.665 \text{ KW} = 1{,}665.738/746 \text{ hp} = 2.2 \text{ hp}$$

$$\text{NPSH} = V_{suc}^2/2gc + P_{suc}/\rho - P_{vap}/\rho$$

Since P_{suc} is not known, alternatively,

$$P_{suc}/\rho = P_a/\rho - Z_{suc} - V_{suc}^2/2gc - h_{fsuc}.$$

So

$$\text{NPSH} = V_{suc}^2/2gc + P_a/\rho - Z_{suc} - V_{suc}^2/2gc - h_{fsuc} - P_{vap}/\rho$$

$$= P_a/\rho - Z_{suc} - h_{f\,suc} - P_{vap}/\rho$$

$$= (11.7 - 2.89) - 1.2 - 0.39 = 7.22 \text{ m}.$$

Ans.

Exercise 12.3

Heavy hydrocarbon oil having an API gravity of 23 is to be pumped at a rate of 38 barrel/h through a 520-m long pipeline of 3 in schedule 40. If there are two full open gate valves and six 90° elbows in the line before entry into the tank at the delivery end, determine the pressure drop in the line. Assume the pumping temperature along the pipeline is 15°C and the viscosity of the oil at this temperature is 80 cp.

Solution

From the API gravity relation,

Sp.gravity = 0.916
Flow rate = 38 barrel/h = 38 × 158 = 6004 L/h = 6.004 m^3/h
Pipe i.d. from std table = 3.07 in = 0.08 m
Area of cross section = 3.14/4 × $(0.08)^2$ = 5.024 × 10^{-3} m^2
Velocity (avg) = 6.004/0.005024 = 1195.06 m/h = 0.332 m/sec
Total friction head

$$h_f = (4\,fL/D + K_f)V^2/2g,$$

where L = 520 m.

For elbow K_f = 0.9 and full open valve K_f = 0.2
K_f for all the elbows and valves = 6 × 0.9 + 2 × 0.2 = 5.8
F, the Fanning's friction factor, is obtained by the following steps:

$$NRe = DV\rho/\mu = (0.08 \times 100) \times (0.332 \times 100) \times 0.916/80 \times 10^{-2} = 304 < 2100$$

Hence,

$$f = 16/NRe = 16/304 = 0.0526$$

Hence,

$$h_f = (4 \times 0.0526 \times 520/0.08 + 5.8) \times 0.332^2/2 \times 9.81 = 7.715 \text{ m}$$

$$\text{Pressure drop} = 7.715 \text{ m} \times 916 \text{ kg/m}^3 = 7066.94 \text{ kg/m}^2 = 0.707 \text{ kg/cm}^2$$

Ans.

Exercise 12.4

A three-stage reciprocating compressor is to compress 5 Nm^3/min of methane from 1 to 62 atm. The gas temperature is 27°C at the suction. Assume that the average C_p and γ values are 9.3 cal/gmol °C and 1.31, respectively. What is the BHP of the compressor? Take efficiency as 60%.

Solution

According to thermodynamic analysis, the minimum work required for a multistage compressor with intercooler facilities is given as

$$W = \gamma\, nm\, RT_1/(\gamma - 1)[r^{(\gamma-1)/\gamma} - 1],$$

where $\gamma = 1.31$
n = number of stages = 3
m = moles flow rate = 5000/22.4 × 1/1 × 303/273 = 247 gmol/min
R = universal gas constant = 1.98 cal/gmol K
r = compression ratio = $(61/1)^{1/3} = 3.936$
$W = 1.31 \times 3 \times 247 \times 1.98 \times 303 \times [3.936^{0.31/1.31} - 1]/(1.31 - 1) = 719460.4$ cal/min = 67.50 hp
Hence, the power required at 60% is 67.5/0.6 = 112.5 hp.
Ans.

13 Instrumentation and Control in a Refinery

13.1 CONTROL HARDWARE

Modern chemical plants are automatically controlled with the objective of achieving the highest productivity with minimum human intervention. Both the process control logic and strategies of control have undergone a sea change owing to the transformation of the analogue environment to a digital one. The automatic process control in a plant consists of three groups of elements—hardware, software, and the transmission lines. Hardware includes sensors/transducers, transmitters/signal conditioners, controllers, and final control elements (control valves, switches, solenoids, motors, etc.). The software is the control algorithms or programs residing (or downloaded from the host computer) in the controllers, according to which output signals of controllers are generated.

13.1.1 Hardware

Hardware instruments are classified as four elements: primary, secondary, controlling, and final control. The sensor or transducer generates the primary signal and the transmitter (secondary element) sends the signal (carrying the primary signal) to the controller (sometimes trimmed through a signal conditioner to make the signal acceptable to the controller). The primary signal is usually obtained or converted in the form of electrical voltage or current (analogue signals). The controller receives the signal (the process variable, PV) in terms of electric voltage (milli to few volts) or current (4–20 milli ampere) as the signal accepted by the controller, compares it with the desired set point (SP), and generates the corrective analogue signal or power, which actuates the final control element. This process of measurements and actuating actions is repeatedly carried out in the controlled system (control loop) as a guard to maintain the desired SP (or to follow the SP) under any circumstances of load fluctuation or SP variation. Current computer controlled systems, such as a supervisory control and data acquisition (SCADA) system, uses the analogue transmitters with "stand alone" microprocessor controllers in the field, which are capable of communicating to the supervising computer/computers located in the control room. Communication is carried out by the digital signal only. Hybrid analogue remote terminals (HART) are also being used for handling both analogue and digital signals for communication. With the ever-reducing size and cost of IC chips in the digital industry, almost all the analogue electrical gauges, transmitters, controllers, and even control valves are being rapidly replaced. Today's controllers are capable of receiving and generating both analogue and digital signals. These usually communicate through an RS485 or standard IEEE communicating port through which computers can communicate to large numbers (as high as 50 or more loops) of controllers at a distance of a few feet to a few kilometres. Microprocessors or programmed logic controllers (PLC) are being widely used in the industry because of their low cost (Tables 13.1 through 13.3).

13.1.2 Cables

Apart from the controlling instruments, transmission wires or cables play an important role in deciding the speed and cost of automation. A variety of cables are required for a controlled system. For example, (1) a cable connecting the sensor to the controller, (2) a cable connecting the controller to the control valve, (3) a cable connecting the controller to the host computer, and (4) a cable connecting the computer network. The transmission of signals from the sensor to the controller and from the controller to the final control element is carried out by conducting copper wires (for electrical or digital signal). In order to enhance digital communication between the loop controllers and the host computers, optical fibres are the most suitable medium. In many

TABLE 13.1
Some of the Common Transducers

Process Variable	Sensor/Transducer	Phenomenon
Temperature	Thermocouple	Thermoelectric
Temperature	Resistance temperature detector (RTD)	Resistance change of metallic conductor
Temperature	Bimetal	Differential thermal expansion
Temperature	Thermistor	Resistance change of semiconductor
Pressure	Strain gauge	Resistance change for deformation
Pressure	Piezoelectric	Static electricity generation of quartz crystal
Pressure	Linear variable differential transformer (LVDT)	Variation of output voltage of the secondary transformer due to changing inductance between the primary and secondary transformers
Flow rate	Orifice plate with differential pressure transmitter	Pressure drop due to the flow rate across the orifice. The DPC transmitter generates electric current or voltage in relation to the pressure drop
Flow rate	Venturi n/meter with DPC	Pressure drop due to the flow rate across the throat of a venture tube. The DPC transmitter generates electric current or voltage in relation to the pressure drop
Flow rate	Magnetic flow meter or turbine flow meter	Induced voltage due to rotating impeller with magnetic material
Flow rate	Doppler meter	Variation of frequency of reflected ultrasonic sound
Flow rate	Corriolis meter	Vibration current due to corriolis flow in bent tubes
Level	DPC	Differential static pressure varies with level
Level	Displacer gauge	Buoyant force on a submerged body
Level	Float gauge	Variation of position of floating object
Level	Ultrasound gauge	Time variation with reflected ultrasound wave

TABLE 13.2
Some Common Controllers

Controller	Input Signal	Output Signal	Logic
Pneumatic	3–15 psi air	3–15 psi air	PID
Electrical	4–20 mA	Current or voltage	PID
Microprocessor	Current or digital	Current or digital	PID
Microprocessor	Current or digital	Current or digital	Fuzzy
Microprocessor	Current or digital	Current or digital	Ladder

plants, optical fibres are replacing traditional metallic conductors, but the cost is prohibitively high. In addition to these, where pneumatic control valves are used, copper tubes are used to connect the current/pressure (I/P) converter to the control valves.

13.2 CONTROL LOOPS

Traditionally, controllers had limited intelligence and control actions were on/off, three term PID, and SP control logics in a feedback control loop.

A feedback flow control loop is presented in Figure 13.1, where the pressure drop across the orifice plate in the pipe is transmitted with the help of a differential

TABLE 13.3
Some Common Final Control Instruments

Final Control Instrument	Actuator	Input Signal	Action	Characteristics
Pneumatic control valve (PCV) used for gas or liquid lines	Pneumatic	Air pressure, 3–15 psi	Stem travel to open or close valve	Linear, equal percentage, increasing or decreasing sensitivity
Motor operated valve (MOV) used for gas or liquid lines	Electric	Current or voltage	Stem travel to open or close valve	Linear, equal percentage, increasing or decreasing sensitivity
Solenoid-operated valve (SOV) used for gas or liquid lines as safety relief or power cut off	Electric	Current or voltage	Full open or full close the valve	On-off
Hydraulic valve used for handling solids or liquid	Hydraulic	Liquid pressure	Stem travel to open or close valve	Linear, equal percentage, increasing or decreasing sensitivity
Electric switch or contact for power supply only for safety cut off	Electric	Current or voltage	Connect or disconnect power supply	On-off tripping device

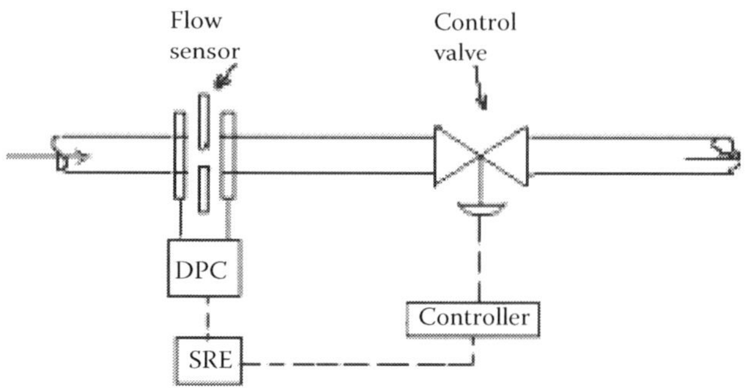

FIGURE 13.1 A feedback control loop of a flow control system.

pressure cell (DPC) to the controller, which manipulates the flow by actuating the control valve. Operators were accustomed to *manual*, *control*, and *cascade* modes only. A cascade temperature control of a pipe heater is shown in Figure 13.2. Here, the temperature of the coil outlet is measured by the master controller, which generates the output signal as the SP for the fuel flow controller, which is the slave of the master controller. In the digital control system, the strategy of such feedback control is still in practice with increased speed of response. The feed forward control is another strategy of control that is applicable for a process where the controlled variable is either not available or not measurable in a continuous manner. For instance, the composition of a reactor effluent or the composition of a distillate drawn from a distillation column are not available online, rather these are periodically sampled and analysed in the laboratory. In fact, the feed forward control strategy is an age-old logic, but it is comfortably implemented in the digital control system only.

A typical naphtha pretreatment unit is shown in Figure 13.3, where the temperature of the feed heater is controlled by the temperature controller (21 TIC01), the flow rate of the gas is maintained by the flow controller (21 FIC 01), and the level of the liquid in the vapour-liquid separator drum is maintained by a liquid level

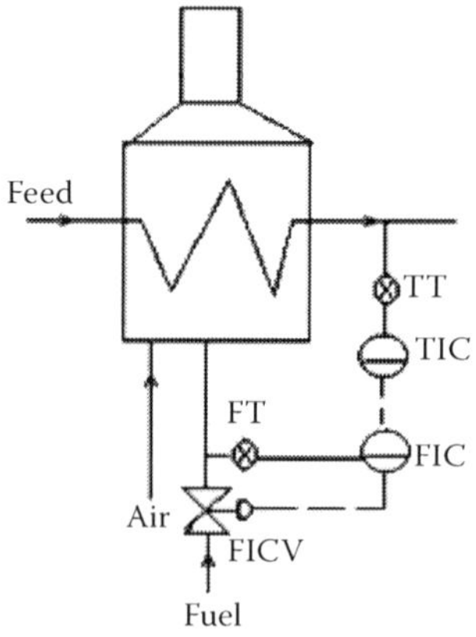

FIGURE 13.2 A cascade control loop of a pipe-still heater.

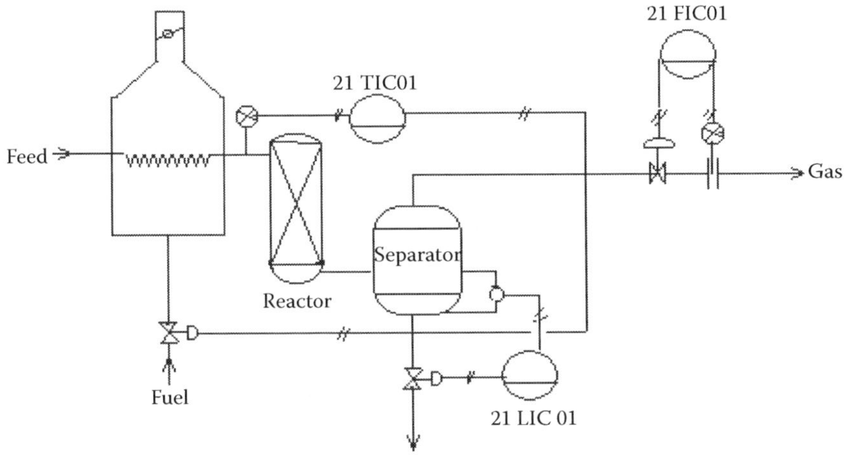

FIGURE 13.3 A process piping and instrumentation diagram of a naphtha pretreatment unit control.

controller (21 LIC 01). The liquid drawn from the separator drum is further distilled in a separate column (not shown in the figure) and the bottom product from this column is the desulfurised naphtha, which is sampled for routine laboratory testing. Of course, the operating temperatures of the furnace, gas flow, and the liquid level are adjusted to maintain the quality of the product drawn.

13.3 THE PROCESS PIPING AND INSTRUMENTATION DIAGRAM

It is noted in Figure 13.3 that the sensors, controllers, control valves, control signal lines, and the process flow pipelines have been depicted with certain symbols as it is necessary to distinguish the process equipment, connecting processes, and the corresponding controlling instruments in the figure. In addition, code names like 21 TIC 01, 21 FIC 01, and 21 LIC 01 have been used where TIC, FIC, and LIC are, respectively, types of controllers as temperature, flow rate, and level indicating controllers. The process unit, the naphtha pretreatment unit, is designated as "21" and the instrument number is given an address of "01." These symbols and code numbers help in identifying the processing unit and instruments as well. The process piping and instrumentation diagram with these symbols of instruments help in understanding the connectivity of processing steps and the control strategy. Though these symbols of instruments and process may differ from vendor to vendor, some commonly accepted standard symbols of instruments are presented in Figure 13.4.

13.4 CONTROL SOFTWARE

The control programs residing in the memory of the controller is the control software. For instance, a traditional control equation is a proportional-integral-derivative (PID) relation as mentioned below.

$$c = A + K_c \varepsilon + \frac{K_c}{\tau_i} \int \varepsilon \, dt + K_c \tau_D \frac{de}{dt} \tag{13.1}$$

where,

C: the output signal of the controller
ε: the difference between the SP and the PV
K_c: the proportional gain of the controller
τ_i: the integration time constant
τ_D: the derivative time constant
A: the bias signal from the controller while $\varepsilon = 0$

In the controllers, these parameters are known by different names. For instance, proportional gain, K_c, is expressed by the term *proportional band* (PB), which is the

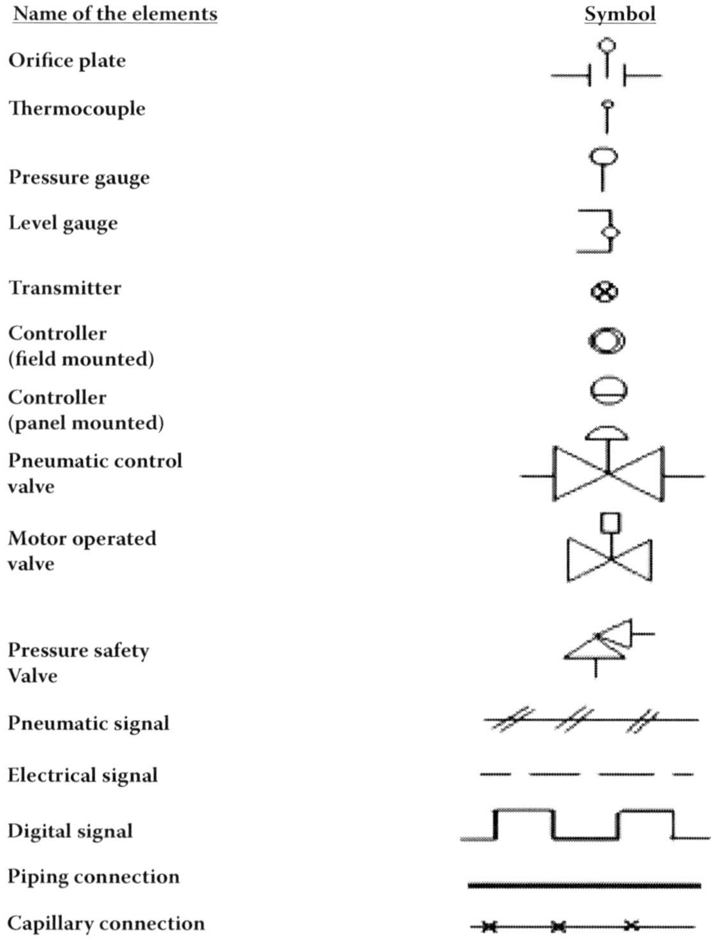

FIGURE 13.4 Standard symbols of instruments.

ratio of the error to the span of the controller output scale, which causes the control valve either to fully open or fully close. In fact, this becomes the inverse of K_c, i.e., the greater the value of PB, the smaller the value of K_c. In order to increase the speed of response, it is desirable to select a lower value of PB. Of course, too low a PB value will cause wide fluctuations in the response, and control will be difficult. The integration time constant, τ_i, is expressed by the term as the reset time which is, in fact, the inverse of τ_i. The derivative time constant, τ_D, is also known as the rate time, which is the same as τ_D. If these parameters, PB, rest time, and rate time are not properly selected, the controller will not act properly. Selection of these parameters is tricky and they are selected by certain tuning methods, like the Ziegler–Nichol method or the Cohen–Coon method. Discussions of these methods will be found in books on control dynamics. Modern controllers are capable of self-tuning these parameters. Hence, if the controllers' performance are not satisfactory due to delays, too many fluctuations in the controlled variable, abnormal overshoot or undershoot, etc., tuning must be done and if the problem does not improve, instrument maintenance personnel must be called in to rectify the problem.

Nowadays, digital programs are used to change the control equations of the input and output signal of a controller, as a result of which a suitable control algorithm is implanted in the controller. Even PID control actions are carried out successfully with an auto-tuning facility (both offline or online) in certain control variables, but PID control is not successful for many processes due to non-linearity and other complex dynamic problems, rather they are being replaced by a process-friendly model predicted algorithm, which has the capability of online process identification and tuning. Artificial intelligence is also implemented for online fault diagnosis and action. In certain processes where control variables are not measurable, they are predicted or inferred and controlled by an inferential control model. The expertise of the plant operators' knowledge can also be implemented for such a complex plant control. Uncertain external events, such as fluctuation in market price and demand for products, exigencies, etc., are also taken care of in the controlled software. SPs that were earlier decided by the operators, engineers, or the managers, are now replaced by control software (a management information system) where the standard decisions are chalked out by the intelligent software itself and target SPs are delivered to the controllers without the need for human intervention.

Finally, the success of the process control relies on the functions in the analogue controllers (electrical circuits) or the programs (software) in the digital controllers, according to which an output signal is generated by the controller. Various types of programming languages are involved in building control software. For example, high level languages like Visual Basic and Visual C++, are commonly employed for the programs loaded in the host computers, assembly languages are used in the loop controllers, and ladder logic programs are loaded in the PLC controllers. Programming skill of the control panel operators is not required as they are involved in changing certain entries in the front panels of the consoles or the controllers. Good programming skill is essential for addition or modification of the programs loaded in the host computers. If it is necessary to build or modify the control software by the automation personnel in the users' company, it is essential to collect the *protocols* of the controllers and its accessories from the vendors. The protocols of the communicating controllers are the string of data sent and received by them to the computer, which must be compatible with the BAUD rate, parity, data structure, etc. This can save spiralling expenditure of software maintenance in the near future. Much of the software loaded in the host computers' library is maintained by the software vendors on a contract basis. Operators and managers are trained to handle the parameters as needed for plant operation and management. Hence, little programming skill is required for the users of this software. Required assembly programs or ladder logic programs are usually downloaded from the archives of the host computers to the loop controllers' memory by authorised operators or engineers of the plant.

13.5 DISTRIBUTED CONTROL SYSTEM

In the beginning, digital control was introduced as a direct digital control system where the input signals were directly fed to the computer and the output signal was used to drive the control valve. An analogue to digital converter (ADC) was used to convert the input signal from the transducer and a digital to analogue converter (DAC) was used to convert the digital output of the computer to an analogue signal to drive the control valve. In such a system, computer workstations with more than two or three CPUs were used to drive the control system. In this system there was a danger of upsetting the controlled system due to any fault in the computer, which acts as the controller. This problem has been overcome in the distributed control system (DCS). Today, the majority of modern plants use a DCS. In a DCS, a workstation with supervising computers are connected to the loop controllers (microprocessors or PLC), which are distributed throughout the process, each performing locally to control the respective control variables. High-speed communication is through a two-way field bus to transmit process data (in the coding and decoding of appropriate protocol) from the local controllers to the computer/computers in a centralised control room and vice versa. Data includes sensed PVs, SPs, controller parameters or program parameters, alarms, sensors' type, address and the status of online connectivity of sensors (like sensor open or online, etc.), output signal from controller delivered, control valve position (full open or closed, etc.), etc. PV signals are the read-only (RO) type, which cannot be changed from the console but can be recorded and visualised. Output power, modes of action (manual, control, cascade, tuning, alarms, etc.), parameters of the controllers, SP, etc., are read-write (RW) type signals. These can be changed from the consoles in the control room. Controller parameters can be changed in the configuration mode or tuning mode from the consoles. Panel operators are permitted to change certain intermediate SPs (RW) to adjust the process conditions and can make a print out and save the information. Only authorised personnel or managers are allowed to change the vital SPs, such as target throughputs in a plant, control programs or parameters, accessing of records of vital operating data saved, configured values, addresses of controllers, etc., which are not accessible to the plant operators. An additional data bus is provided to access the operating conditions of the plant for management decisions. An operating system like WINDOWS, or any other operating system may be chosen for digital communication and control. Any fault in the host computer will not upset the controlled system as it is the controllers' chip memory that will maintain the control action in their respective loop at the previous set operating conditions. Redundant paths are always provided to avoid failure in communication and control actions.

13.6 THE CONTROL ROOM

The control room is where the controller can be operated to change the SPs, record the PVs, tune the parameters of the controller, and have correcting signals delivered to the control valves. Sensors and control valves are mounted over the appropriate field or location on the equipment and pipelines of any process unit to be controlled. Controllers are usually placed in the control room far from the field locations. Sensors are connected through long cables of wires or optical fibre lines, ranging from a few feet to more than a kilometre in length, terminating at the controllers located in the control room. Output signal lines from the controllers are then drawn from the control room to the actuators of the final control elements. All these controllers, in fact, fetch measurement signals or PVs continuously, evaluate error (ε) by comparing with the desired SPs and deliver an output signal as a function of the error (ε). In the DCS, these controllers are interfaced with high-speed desktop computers with large memory space. The necessary software is loaded onto these computers to access the individual controller's parameters, the signals received, and the signals delivered. The controllers communicate with the computers through digital input/output signal lines connected through RS485 or IEEE standard ports. In the DCS, the controllers are supervised by the computers (i.e., the computers can access any individual controller by their respective address and recover the measured PVs, change the SP values, tune the controller as and when required, record the data, etc.). In case of computer failure due to some unforeseen event, the controllers existing data will be unaffected and the control action will continue without fail. The control room is provided with a clean and temperature-controlled environment in order to protect the computers, controllers, and recording/printing devices and also to provide a comfortable working space for the panel operators and engineers for decision making. A modern control room in a refinery.

13.7 CRUDE THROUGHPUT CONTROL

Control of crude throughput means that the flow rate or throughput is maintained at the desired throughput or flow rate. As shown in Figure 13.1, an orifice plate is placed in the pipeline through which crude is pumped. Upstream or high pressure (HP) tapping and downstream or low pressure (LP) tapping across the plate are connected through a DPC transmitter, which delivers a current signal in relation to the pressure difference (hence relating the flow rate). This current signal enters the controller, which is compared with the SP, the desired flow rate and computes an output signal according to the PID equation as given in Equation 13.1. However, the adjustment of the flow rate must be done carefully by smooth SP adjustment since a change in the SP does not guarantee an immediate response without fluctuations above and below the SP. It will take some time for these fluctuations to subside before a stable response is achieved. For example, if the present throughput is 100 m^3/h and it has to be raised to 400 m^3/h, the panel operator will adjust the SP in gradual steps, e.g., 50 m^3 in 1 h, until the process measurement reaches 150 m^3/h, then 50 m^3 in another hour to 200 m^3/h, and so on until 400 m^3/h is achieved. This type of gradual SP increase is called a *ramping algorithm*, where the slope of the ramp, in this case 50 m^3 in 1 h, is selected. This is done to avoid a wide fluctuation in throughput if an attempt was made to increase it from 100 to 400 in one go in a short space of time. Similarly, while throughput has to be reduced from 400 to 100 m^3/h, a gradual reduction, selecting a ramping algorithm as described above, is carried out to avoid a violent fluctuation in the flow rate. Typical SP adjustments from 100 to 400 m^3/h of throughput are presented in Figures 13.5 and 13.6 for a sudden change and a gradual change in SP, respectively.

13.8 DESALTER CONTROL

As already discussed about an electrical desalter, as shown in Figure 13.7, pressure and interface level control are widely practised. However, pH control is rather wieldy as compared to pressure or level control due to non-linearity of the pH.

Pressure control is desirable to avoid vaporisation within the desalter because vapour formation may cause explosion due to the heat dissipation from the electrode plates as soon as the liquid phase disappears. A pressure controller measures the pressure within the desalter and manipulates the discharge flow of desalted crude oil. Usually, a minimum pressure, depending on the type of crude, must be set (typically about 8 bar) to avoid vaporisation. In case the pressure rises above the maximum allowable operating pressure, a safety valve pops (opens) for safe discharge of fluids (gas or liquid), usually to a flash zone of the distillation column nearby or to a flare line. In both the events when low pressure (LP) or high pressure (HP) occurs, the transformer is switched off with the help of a power tripping device.

The other important control is the oil–water interface level below the electrode plates. In many refineries, this level is usually controlled manually by monitoring

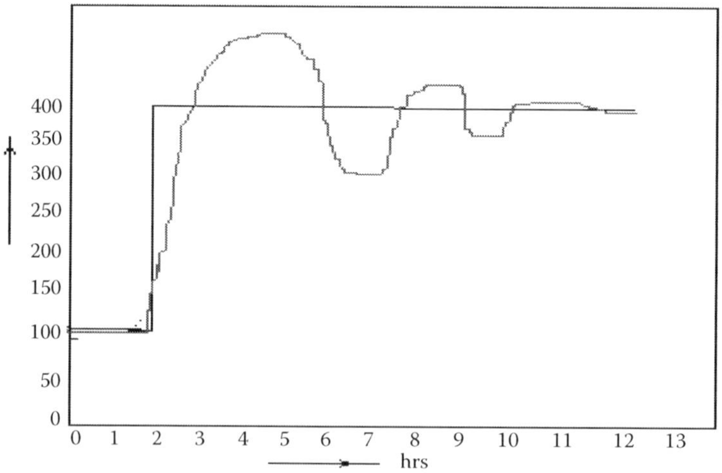

FIGURE 13.5 Throughput fluctuation due to a sudden rise in set point. The thick line indicates set point and wrinkled line indicates measured value of throughput.

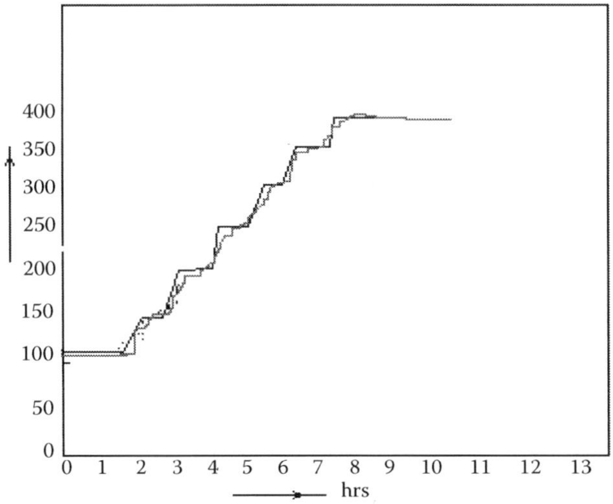

FIGURE 13.6 A gradual increase in set point using a ramping algorithm.

the presence of water through multiple sampling pipes within a distance from the minimum to maximum allowable limits. In case the water level touches the electrodes, there will be a short circuit and it may cause explosion. In order to avoid such situations, the transformer is switched off by a power tripping device as soon as the water level reaches a certain percentage of the level, which is well above the danger level that may cause a short circuit. Too low a level of water may cause draining of the crude from the discharge line for water. However, in case of automatic control of the interface level, a level controller and a control valve at the discharge must be provided, as shown in Figure 13.7. Alarms are usually fitted with cutoff valves (usually

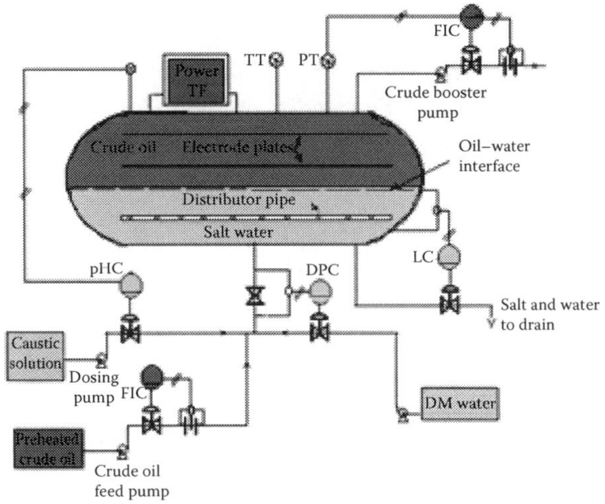

FIGURE 13.7 A process piping and instrumentation diagram of a desalter having controls of drum pressure, oil–water interface level, pH, and mixing valve pressure drop.

solenoids) at the low and high values of pressure and water level. The other vital control is the water injection rate, which is crucial for the desalting operation. This rate is controlled by a controller that measures the pressure drop across the mixing valve.

13.9 ATMOSPHERIC DISTILLATION COLUMN CONTROL

In the crude distillation unit, the distillation column has two sections, the rectification section and the stripping section, respectively, above and below the flash zone. Gases and naphtha are drawn as vapour from the top of the column, kerosene and diesel (with additional distillate heavier than diesel like jute batching oil (JBO) in some columns) are drawn as the liquid products from the rectification section, and residue (reduced crude oil, RCO) is drawn from the stripping section bottom. However, kerosene, diesel, and JBO are further stripped in the side strippers. Control of the top pressure and temperature, flash zone temperature and pressure, flow rate of over flash (which is a split stream taken from the plate above the flash zone), overhead reflux and circulating refluxes, liquid levels in the plates of the column and strippers, steam rates at the bottom of the column and the strippers, are vital to maintain the quality of the products.

13.9.1 Reflux Drum Pressure Control

The top pressure of the column is dependent on the pressure in the reflux drum. Control of the pressure drum is carried out by the *split range control* strategy. The drum pressure is increased in two ways—by the vapours leaving the top plate and from the pressure of the vaporiser drum or from a high pressure source like liquified petroleum gas (LPG) storage vessels. On the other hand, drum pressure is reduced by expelling it to return to the vaporiser or fuel gas consumption points, e.g., furnace or flare. Such a connection is presented in Figure 13.8. There are two control valves, one allows gas entry from the vaporiser and the other allows gas exit to the vaporiser or furnace or flare. If these valves are air pressure operated in the range of 3–15 psi instrument air, the following actions will be carried out to manipulate the pressure in the drum. The valve at the gas entrance is air operated and is normally open, i.e., air to close, thus it fully opens at 3 psi and starts closing as the pressure rises from 3 to 9 psi. It is completely shut when the pressure rises to or above 9 psi. This valve is normally closed, i.e., air to open from 3 to 9 psi air pressure, starts opening as the pressure rises from 9 psi, and completely opens when the pressure rises to 15 psi. The controller generates a signal (direct acting mode) as the pressure of the drum increases, but below the SP, in the range of 3–9 psi, gas enters the drum to fill the drum while the gas exit valve is shut. While the drum pressure goes above the SP, the controller will still increase the signal pressure above 9 psi, actuating the exit valve to start opening to reduce the pressure in the drum and fully opens the valve if the pressure signal rises to 15 psi.

13.9.2 Reflux Drum Level Control

Control of the liquid level in the reflux drum is essential because a sufficient level must be maintained to make it pumpable as the overhead product and as reflux as

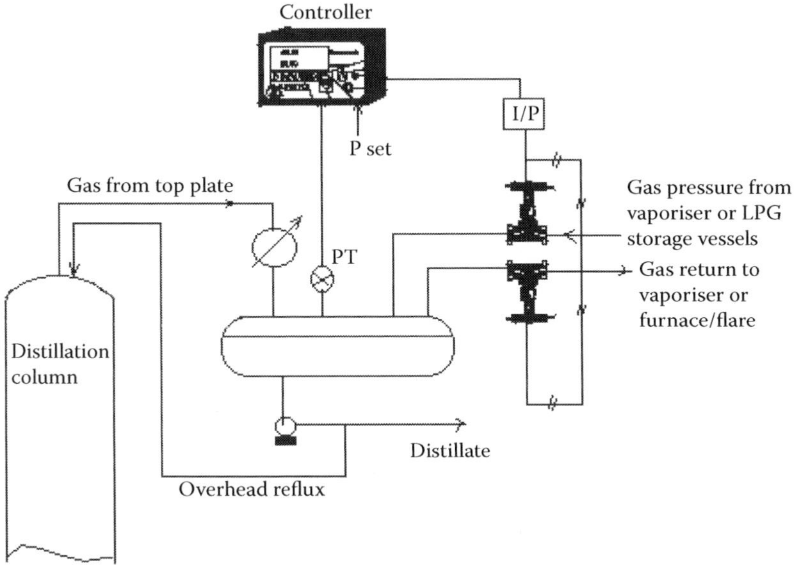

FIGURE 13.8 A split range pressure control of a reflux drum.

well. At the same time, it must be ensured that the liquid level should not fill the entire drum and reach the gas lines. Usually, a controller is needed that acts after getting the signal from the level transducer, and generates the output signal, which actuates the control valve sitting at the overhead product discharge line. This control valve is usually an air-to-close valve, i.e., normally open type, and the controller must be set in the reverse acting mode. Thus, as the level goes above the SP, the controller decreases the signal and opens the valve to release more liquid to maintain the level and vice versa. Such a level control system is shown in Figure 13.9. Water of the condensed steam is settled below the hydrocarbon liquid level and is drained

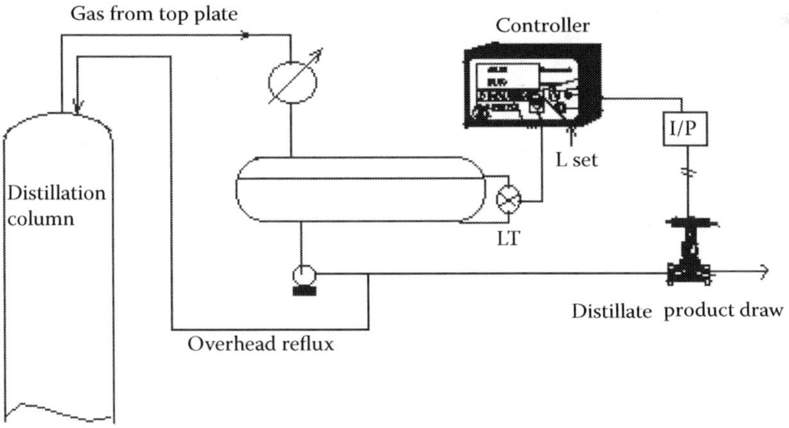

FIGURE 13.9 A level control loop for liquid in the reflux drum.

off through a conical pipe or boot (not shown in the figure). High-level and low-level alarms are also required in case of problems with the condenser, discharge pump, or column troubles causing widely fluctuating vaporisation and condensation, or even due to the effect of a malfunctioning controller.

13.9.3 Top Plate Temperature

This is controlled by a *cascade* control strategy, as shown in Figure 13.10. The temperature of the vapour leaving the top plate is measured and transduced to a controller, which generates the necessary signal to actuate another controller to manipulate the reflux flow rate to the top. A complete assembly of the pressure, level, and temperature control loops for a reflux drum is presented in Figure 13.11.

13.9.4 Draw Plate Temperature

Liquid is held up over a plate to a height provided by the weir. This is provided to maintain a certain liquid level on the plate through which upcoming vapours pass through distributors, such as bubble caps, valves, nozzles, etc. This helps to intimately mix the vapours and liquid for maximum mass and heat transfer, aiding purification of the more volatile components. Liquid overflows the weir through a downcomer slit, down to the next plate below. Certain plates are selected for drawing a portion of the liquid as the side product, such as kerosene, diesel, JBO, etc. Usually, drawn products are further stripped in a side stripper where the liquid level is maintained by a level controller, as shown in Figure 13.12. The controller increases the drawing rate as the level increases over the desired SP and vice versa. If the level is maintained low by setting a low level, the draw will increase. The effect will reduce condensation and increase vaporisation and the plate temperature will rise. The end point of the draw will increase. In case no level controller is used, a flow controller

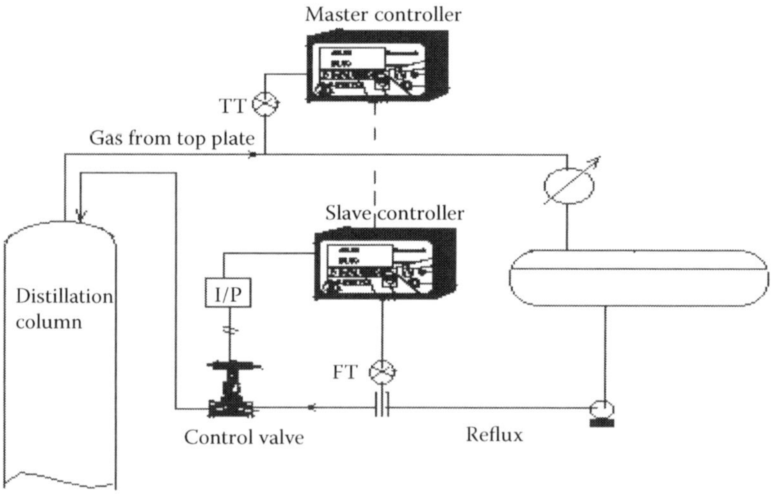

FIGURE 13.10 Top temperature control loop of the distillation column.

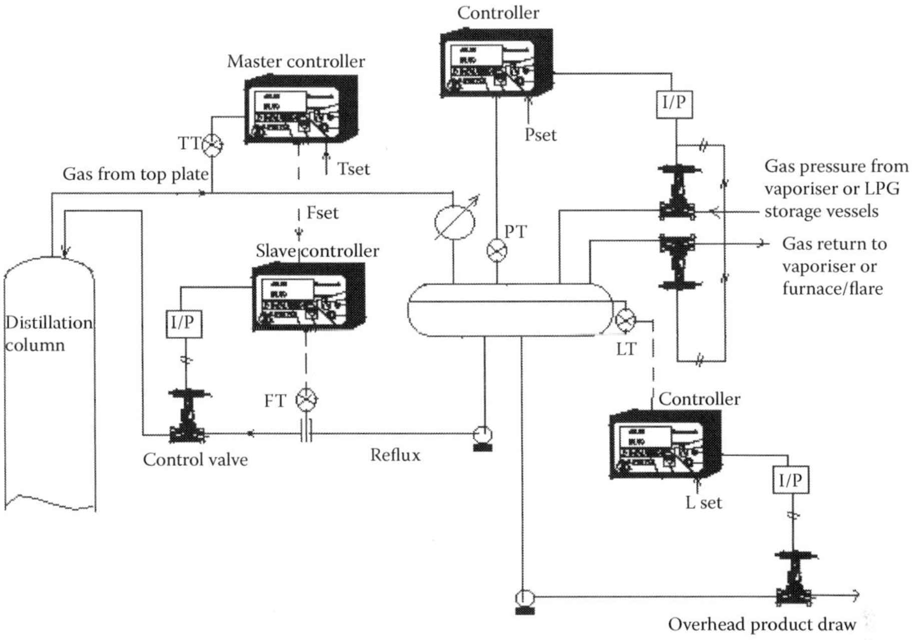

FIGURE 13.11 Complete control loops for a reflux drum.

is provided at the pump discharge, increase in the SP of the flow will increase the draw and the temperature of the plate will increase. Thus, the increase and decrease of the plate temperature will be in the hands of the operator, who must take care to maintain a certain liquid level on the plate. A similar control is essential while circulating reflux (or pumparound) is drawn from the same plate and externally cooled by crude preheaters. The greater the draw, the more the preheat of crude and the less

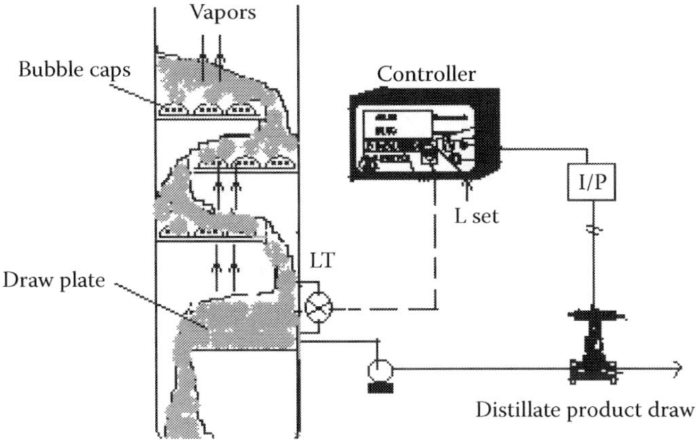

FIGURE 13.12 Draw plate level control loop.

fuel consumption in the crude heater. But the operator must not make the plate dry, which will upset the entire column operation.

13.9.5 Overflash Rate

The flow rate of overflash, which accounts for 2%–3% of the crude throughput, is controlled by a flow controller that allows flow from the plate above the feed plate down to the flash zone, according to the SP selected by the operator.

13.9.6 Flash Zone Pressure and Temperature

The pressure and temperature of the crude oil exiting the crude heater determines the flash zone temperature and pressure, i.e., the manipulation of the discharge pressure and temperature at the upstream of the flashing nozzle determines the flash zone conditions.

13.9.7 Bottom Temperature

Like any other draw plate, the bottom temperature is dependent on the RCO draw, which is manipulated by a level controller to maintain the level at the bottom of the column. However, steam rate through the bottom also affects the bottom temperature. A steam flow controller is used to manipulate the flow rate of steam, which needs to be changed for adjustment of the amount of light distillate components in the RCO. Too high a steam will cause heavier fractions to be carried over with the upper plate draws and too low will cause carryover of light distillates with RCO. A high steam to crude ratio may lead to a high water condensation rate, which may cause water in the top and side products. A typical piping and instrumentation diagram of an atmospheric column with side stripper is shown in Figure 13.13.

13.9.8 Furnace Control

The coil outlet temperature of the furnace is controlled by a cascade control system where the master controller gets its input signal from the thermocouple from the coil outlet fluid and a slave controller manipulates the fuel to the furnace. This is shown in Figure 13.14. A complete piping and instrumentation diagram of a crude distillation unit of a refinery is presented in Figure 13.15.

13.10 VACUUM DISTILLATION CONTROL

Vacuum distillation of RCO in a multiplated distillation column is carried out at a top pressure of around 60–80 mm Hg abs, which is created by more than one steam eductor. The vacuum is controlled by the steam flow rate in the eductors. Usually, a cascade control is advised for accurate control owing to fluctuation in the supply pressure of steam. A conceptual control system is shown in Figure 13.16. *Draw plate temperature and level* are controlled similarly as described for the atmospheric

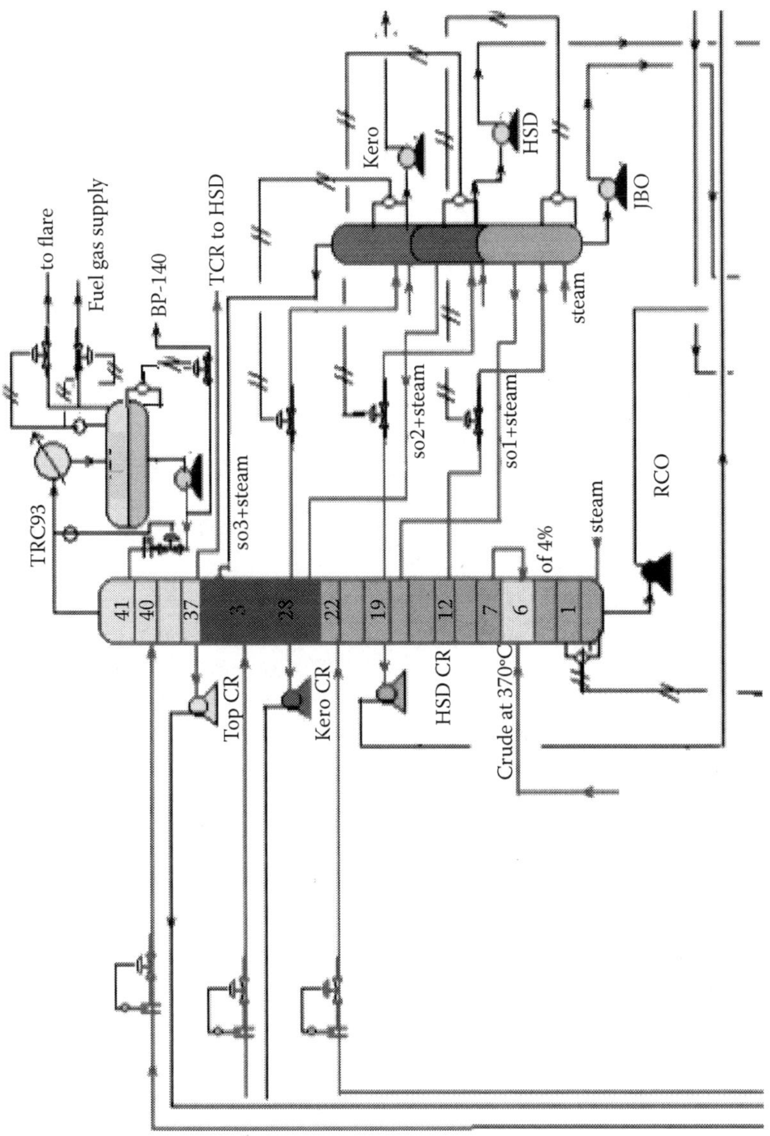

FIGURE 13.13 A typical atmospheric distillation column with side stripper control loops.

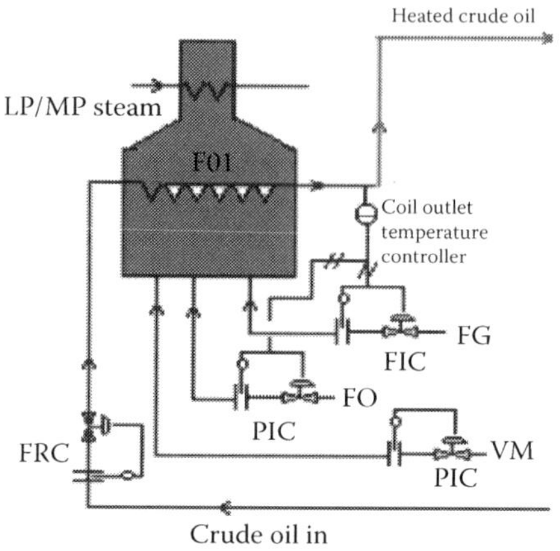

FIGURE 13.14 A typical crude oil furnace control loop.

distillation column. A simplified control system of a vacuum distillation unit is presented in Figure 13.17.

13.11 REFORMER UNIT CONTROL

There are a variety of reforming methods, however, a discussion of the control system of each type is out of the scope of this book. A fixed bed three-stage platformer will be discussed here. The details of the process have been discussed earlier. The desulfurised feed naphtha flow rate is controlled by a flow controller. After preheating with the hot product from the last reactor, feed is further heated in a furnace before entering the first reactor. Because the endothermic reaction temperature of effluent from each reactor falls, reheating is done by the intermediate furnaces. The temperature of the furnaces is controlled by cascade controllers. The level in the flash drum separator is maintained by a level controller before the liquid from the drum enters the debutaniser column. This debutaniser column is, in fact, a multiplated distillation column. The top pressure and temperature are controlled by separate pressure and temperature controllers. The bottom temperature is controlled by a separator controller at the reboiler drum. This is explained in Figure 13.18.

As shown in Figure 13.18, local flow controllers installed at the feed entry, recycle gas and products flow lines. The coil outlet temperatures of the furnaces are controlled by cascade TIC and FIC controllers. The vapour liquid separator (flash drum) level is controlled by a cascade control with LIC and FIC controllers. The debutaniser bottom temperature is controlled by a TIC controller. The pressures of the flash drum and the debutaniser column are controlled by PIC controllers connected with the off gas/flare line.

13.12 FLUID CATALYTIC CRACKING UNIT CONTROL

The fluid catalytic cracking (FCC) unit, as described elsewhere in this book, contains a reactor and a regenerator. A modern reactor has a long riser tube, where the

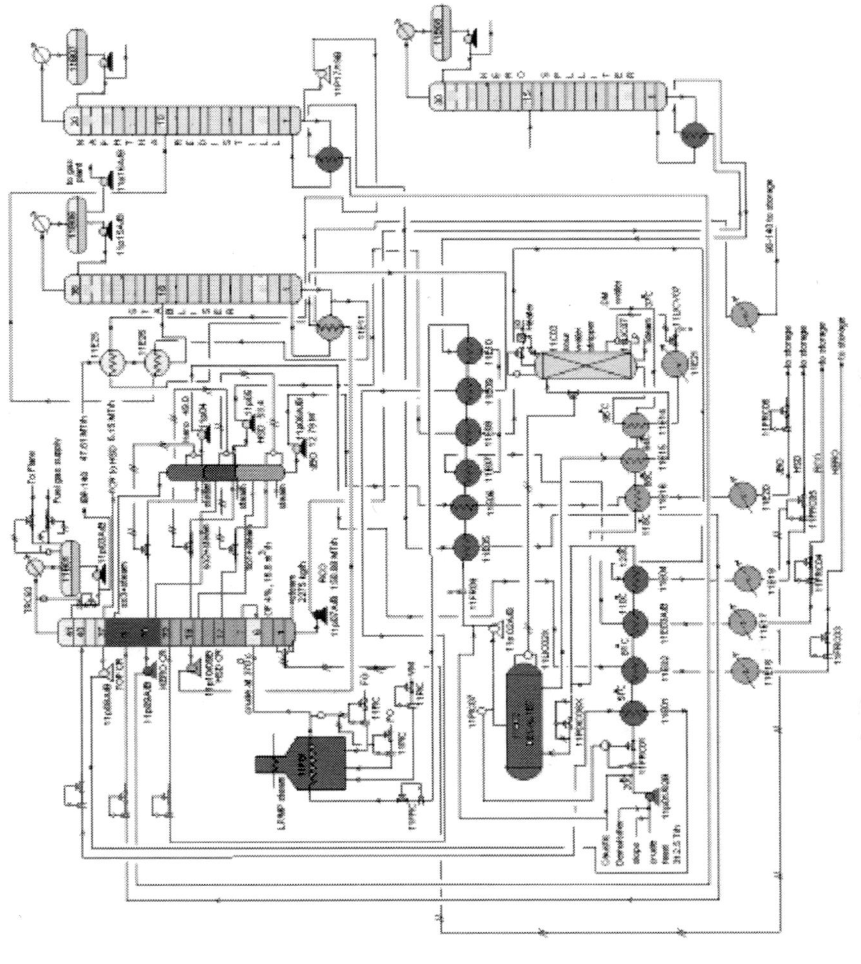

FIGURE 13.15 A process piping and instrumentation diagram of a typical crude distillation unit of a refinery.

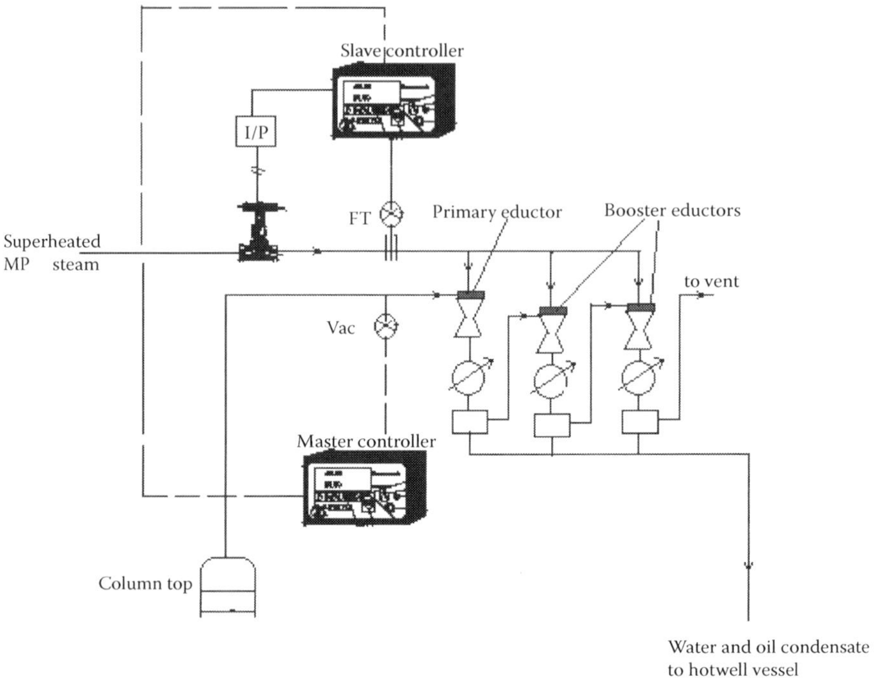

FIGURE 13.16 A schematic vacuum control loop of a vacuum distillation column.

cracking reaction is carried out in a continuous fluidised bed of catalyst and the feed hydrocarbon oil is atomised with medium pressure (MP) steam. The residence time or catalyst–hydrocarbon contact is about 2 sec at a temperature of around 600°C. As soon as the reaction is over, the catalyst is separated from the cracked product in a large diameter vessel where the riser terminates. The spent catalyst falls into the annular space between the riser end and the disengagement vessel. This accumulated catalyst is stripped from the remaining hydrocarbons and overflows to a regenerator vessel, where coke on the catalyst surface is burnt in air. The hot regenerated catalyst is then lifted to the reactor along with preheated feed and steam. The reactor-regenerator unit is controlled for various operating variables. Vital control strategies are listed below.

13.12.1 Reactor Outlet Temperature Control

The exit temperature of the cracked product is measured and a temperature controller (TIC) manipulates the flow of the hot regenerated catalyst from the regenerator to the reactor with the help of a slide valve (control valve), the stem of which is hydraulically actuated. A pressure drop controller (PDC) across the same control valve is also used to measure and manipulate the catalyst flow. Thus, there are two controllers in action with a single control valve. As the temperature controller increases its signal for raising the catalyst flow rate to the reactor, the pressure drop across the control valve increases. Since too high a velocity of catalyst through the valve may

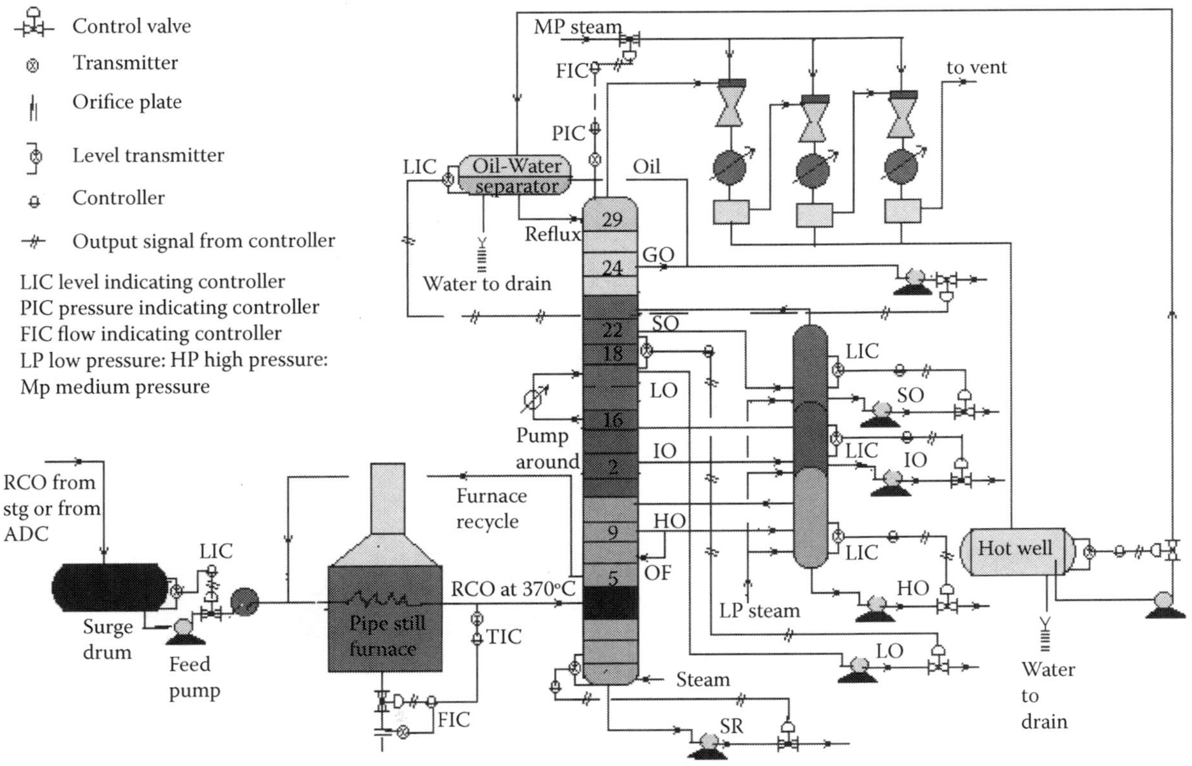

FIGURE 13.17 A simplified process piping and instrumentation diagram of a vacuum distillation unit.

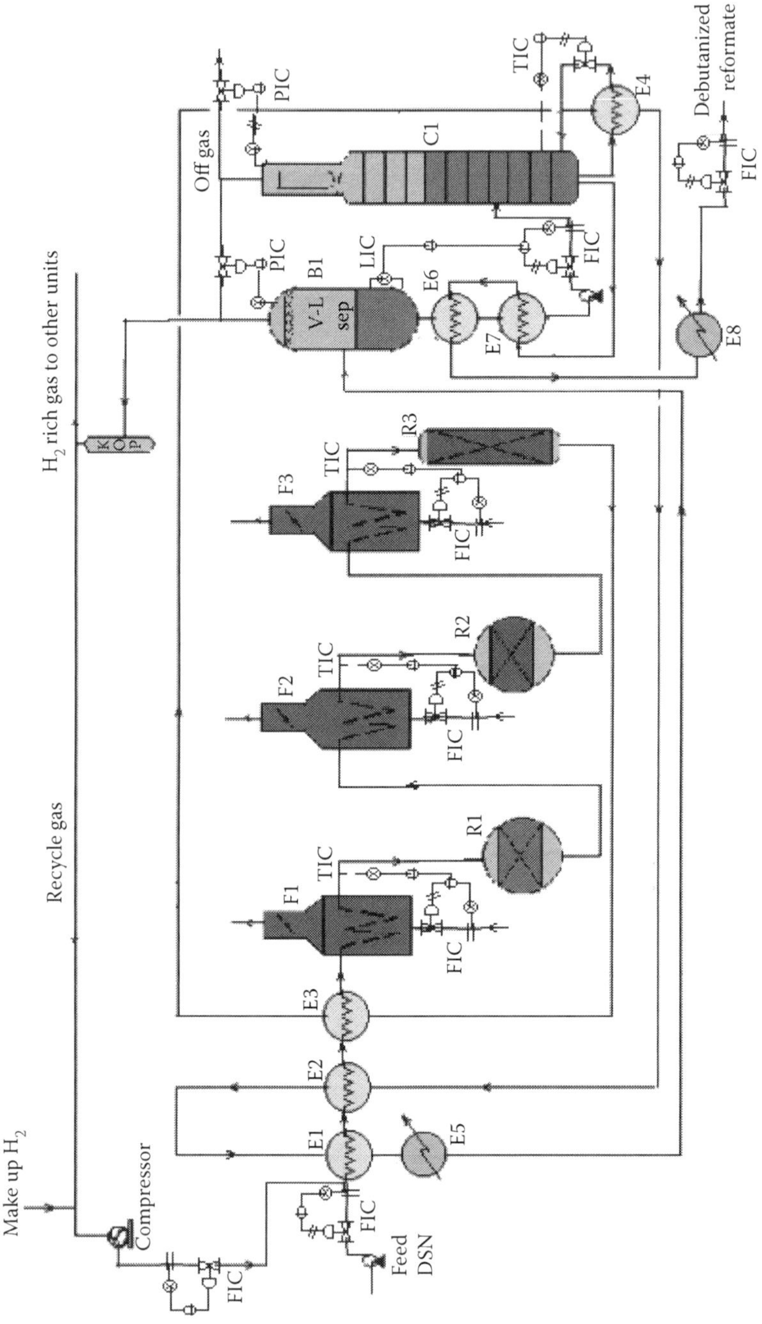

FIGURE 13.18 Process control in a plat reformer plant for naphtha.

cause erosion of the control valve and pipeline and will also disturb the desirable fluidisation behaviour, it is necessary to maintain the allowable limit of pressure drop across the control valve. In this event, a limit switch acts to disable the TIC signal and enable the PDC signal instead, i.e., while the upper limit of pressure drop across the valve is reached due to the action of the temperature controller, the PDC controller will actuate the valve to reduce the flow rate, *overriding* the action of the temperature controller. An additional temperature controller is also provided to reduce the reaction temperature with the help of the flow of quench oil (usually light cycle oil) to avoid unwanted overcracking of the products.

13.12.2 Level Control of the Catalyst Bed in the Stripper Section of the Reactor

A level controller is used to manipulate the exit flow of the spent catalyst through a slide control valve. In this case also, a PDC is used to manipulate the flow rate through the valve, overriding the level controller. The action is similar to the strategy described for the reactor outlet temperature control.

13.12.3 Pressure Balance between the Reactor and the Regenerator

The pressure in the reactor section is governed by the pressure of the distillation column, where the cracked products are separated and no additional pressure control loop is added to the reactor top. However, pressure controllers are provided in the flue gas lines of the first and second stage regenerators. The pressure over the regenerators is so controlled that the flow of spent and regenerated catalyst is maintained. The control strategy can be understood from the simplified process piping and instrumentation diagram of a typical modern RFCC unit, as shown in Figure 13.19.

13.13 FAIL-SAFE DEVICES

During failure of the supply of power, steam, and instrument air and also due to accidents, certain safe operating conditions are maintained with the help of inbuilt safety arrangements. Usually, the failed situations are classified into three categories:

1. During normal running conditions
2. During planned shutdown
3. During accidents or emergency shutdown

13.13.1 Normal Running Conditions

Normal running conditions assume that the power and utilities supply are uninterrupted while the equipment and machineries need to be protected so that the maximum or minimum limits of operating conditions are not violated. Controlling instruments are used to maintain the normal limits of operation only, but additional instruments, known as *securities*, such as safety valves or switches or varieties of tripping devices, are used to maintain maximum or minimum limits for the safety of the equipment and machineries. Some of these common safety arrangements are listed below.

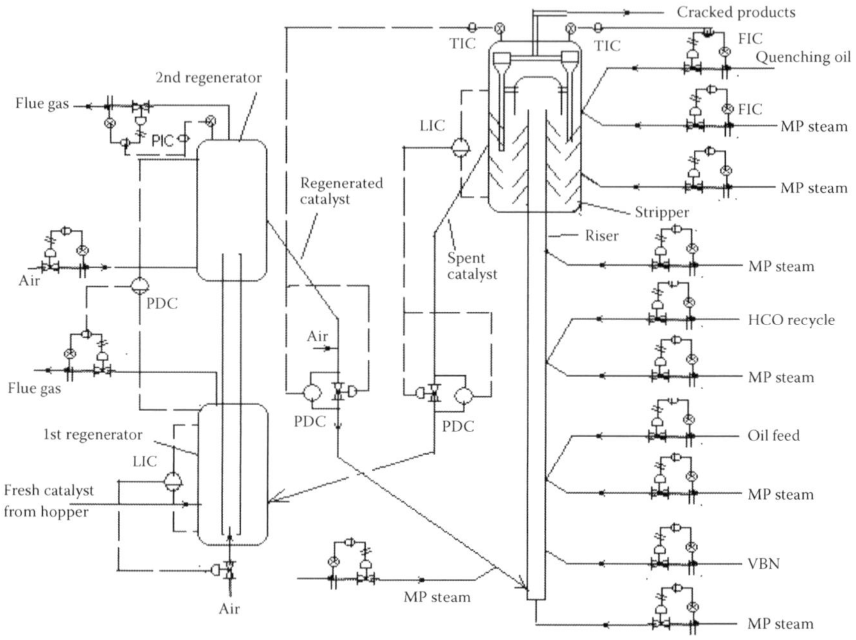

FIGURE 13.19 A simplified residue fluidised catalytic cracking plant control system.

1. If the motor of a pump is overloaded, this is automatically switched off by a tripping device, e.g., a pair of contact plates in the power supply line. These plates separate from its contact while air between the plates is expanded due to heat. A typical solenoid switch is shown in Figure 13.20.
2. If a pump is vapour locked, a mercury switch cuts off the power supply to the motor drive. A typical arrangement of a mercury tripping switch is shown in Figure 13.21.

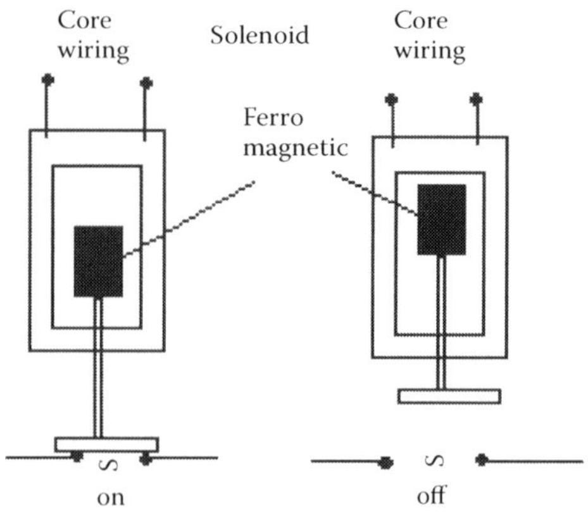

FIGURE 13.20 A solenoid cutoff switch.

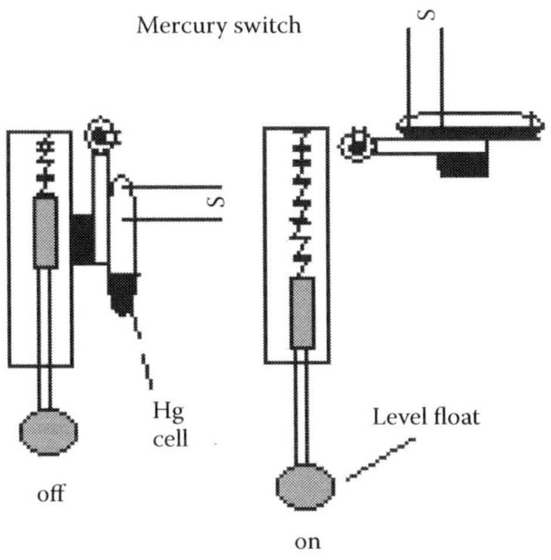

FIGURE 13.21 Mercury-filled cells as cutoff switches.

3. In case a vessel is filled with gas at HP and the pressure rises to approach the maximum limit, a pressure safety valve will pop to release the gas from excess pressure.
4. In case of a rise in temperature approaching the maximum limit, a safety fuse is used to protect any equipment from overheating.
5. In case a furnace flame is extinguished for a certain reason, the fuel flow to the burners must be shut with a solenoid-operated valve in the flow line. Alternatively, the solenoid valve disconnects the instrument air supply to the flow control valve (which is normally closed type (N/C) valve so that it closes in absence of instrument air) to shut. It is to be noted that there is a common fuel circulation circuit for all furnaces and hence it is not wise to shut fuel pumps while only one furnace is extinguished.
6. In case of a high level of water, excess pressure exceeding the limits, feed pump or booster pump failure in a desalter unit, the power supply to the transformer will be switched off with the help of a cutoff switch. There are each such operations carried out for the safety with the help of instruments known as securities, however discussion of all of these operations are out of the scope of this book.

13.13.2 During Planned Shutdown

Planned shutdown is carried out for large-scale repair, cleaning, painting, or modification jobs. This shutdown is judiciously carried out in sequence. First, the processing units are shut down in a sequence from the downstream units to the main unit. During this operation, power generation is also reduced gradually as the demand comes down and, finally, the power plant is shutdown. The safe operations involve a gradual reduction in furnace temperature to avoid thermal shock to the equipment.

The SPs of the controllers are mostly handled by the ramping algorithm for a reduction in the temperature of the furnaces and the flow rates of the feed streams. Manual operations and monitoring are increased during these shutdown steps. A brief account of these steps is given in Chapter 14. Control valves are so selected that these will be fully open or fully closed when the instrument air supply is stopped. Usually, the selection of these valves must be judged by the designer so that any mishap can be avoided in control signal failure or instrument air failure. For instance, in a flash drum, the vapour line should be provided with an air-to-close valve, i.e., a normally opened (N/O) control valve so as to permit exit of the vapour from the drum at the time of failure of the instrument air or control signal. Similarly, the fuel supply line in a furnace or the steam (or hot fluid) supply to the reboiler of a debutaniser should be provided with an air-to-open control valve, i.e., a normally closed valve, to assist in stopping the fuel or steam flow, respectively, at the time of instrument air supply or controller signal failure. Typical pneumatic air-to-open and air-to-close valves are shown in Figure 13.22.

13.13.3 During Accidents or Emergency Shutdown

Sudden failure of machineries or equipment causes an emergency shutdown. Accidents due to fire in a refinery are commonplace. An emergency shutdown of the affected unit and sometimes other adjoining units becomes inevitable to contain the devastation. In this situation, the power supply is stopped either by inbuilt security instruments or by the power supply grid. An automatic sprinkler or water spray may be started by the temperature sensitive fuses or evacuation of fluid through normally open control valves, as the case may be. All the cutoff solenoid or switches should work in tandem to protect the unaffected plant and machineries while fire fighting is carried out manually.

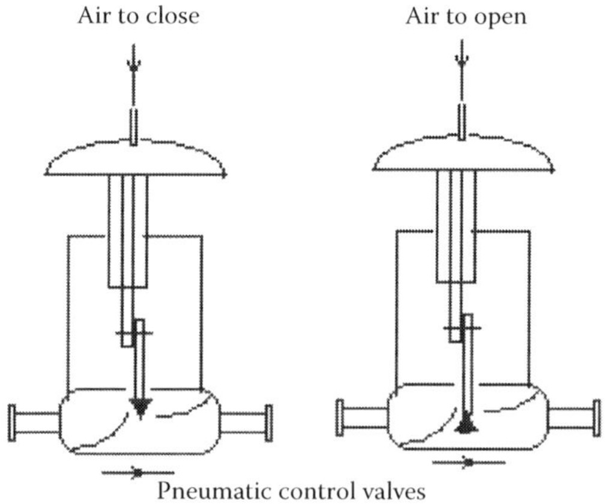

FIGURE 13.22 Typical pneumatic A/C and A/O control valves.

13.13.4 Power Plant Failure

Large refineries and petrochemical plants have their own power plants (captive power plants) to provide an uninterrupted supply of electricity and steam. If a power plant is suddenly shutdown for any inadvertent exigency in the plant other than a planned shutdown, necessary inbuilt safety must be envisaged in the plant. It is easily understood that if the running machineries with electric drives suddenly stop, immense damage will be caused to the plant, even accidents may occur. For this, it is essential to have certain safety devices to allow necessary actions to eliminate any catastrophic situation. In case of partial failure of a power plant, e.g., a power grid problem, a frequency fluctuation, etc., or the generation of electricity while the steam generation is unaffected, all the vital machineries should have steam-driven spares that should automatically start, to protect the equipment and machineries. Such a sudden shutdown of a power plant will affect the supply of steam, instrument air, and cooling water, in addition to the power. As a result, the cooler or condenser will be out of service and the gases or hot fluid need to be evacuated to avoid pressure development (temperature increase) by the normally open control valves or pressure safety valves or temperature safety valves.

Thus, process instrumentation and control of any plant encompasses a thorough knowledge of instrumentation hardware and software with the necessary inbuilt securities strategically located to save the plant and machineries from the various kinds of failures as discussed in this section.

13.14 STANDARD SIGNALS IN PROCESS CONTROL

Electrical voltages or current (dc) signals are generated by the transducers and transmitted, recorded, and processed by the remote controllers. The standard current signal lies in the range of 4–20 mA dc. Sometimes, 0–20 mA is also used. Common voltage signals are 0–10 V and 1–5 V, all in dc. Most modern controllers are capable of accessing and delivering milliamperes, millivolts, and voltage signals. All the instruments connected must be compatible with the respective signals to operate them. Pneumatic control valves are actuated by pneumatic air pressure signal in

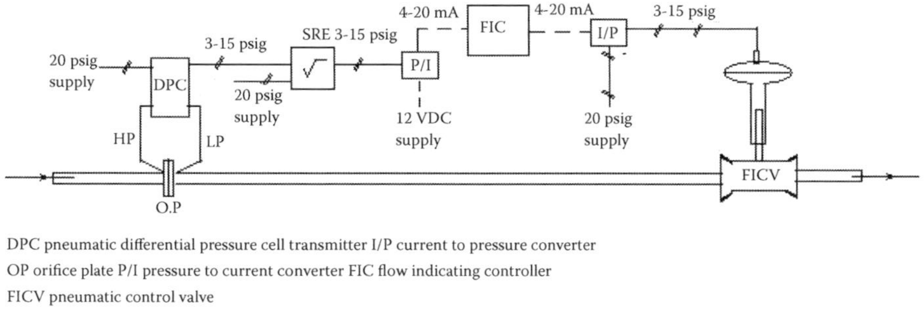

FIGURE 13.23 Standard signals in a flow control system involving both pneumatic and electric instruments.

the range of 3–15 psig from the current to pressure or voltage to pressure converter. Controllers are powered by 24–30 V dc or 230 V ac with a dc converter. Some of the common standard signals are shown in a typical flow control loop in Figure 13.23. Here, the DPC transmits an air pressure signal in relation to the pressure drop across the orifice plate. This signal is linearised by a square root extractor (SRE) that generates a pneumatic signal in relation to the signal received from the DPC transmitter. This signal is then converted to an electric current signal by a pressure to current converter (P/I), which enters the flow indicating controller (FIC). The output signal from the FRC is again converted back to the air pressure signal by a current to pressure converter (I/P) in order to actuate the flow indicating control valve (FICV).

In refineries and petrochemical plants, explosion or fire may occur due to high temperatures and electric sparks. Therefore, it is essential to protect the electric lines by flame-proofing gas pressure or with proper insulation and safety fuges, etc. Such type of barrier between the explosive area containing hydrocarbon vapours or liquid or even explosive dusts and the high voltage electric cables or machines is known as the Zenner barrier.

14 Miscellaneous

14.1 STARTUP

Any plant operation can be divided into three types of operating steps, namely, startup, production, and shutdown. Startup and shutdown operations are unsteady state and unproductive operations. Usually, production operation is carried out for 330 days a year and 35 days are left for repair and maintenance of the plant and equipment; during this period no production is carried out. After shutdown, the startup operation is carried out until continuous production is achieved. Thus, downtime between shutdown and startup operations should not exceed 35 days. Of course, these time limits vary from plant to plant. Rarely, the partial shutdown and startup operation of a plant unit may be carried out without affecting other units such that the production from the rest of the plant may continue throughout the year. At the beginning, the power plant must be started up to supply power and steam to the plant. When the power and utilities are available in full swing, next the processing plants are started up in a sequence. For example, in a refinery, the crude distillation plant is started first. When the products from this plant meet the desired specifications (on spec), the secondary plants, e.g., the naphtha pretreatment and reforming plants, are started up. In fact, for old plants where sufficient storage of intermediate products is available, all the secondary plants are started up simultaneously along with the crude distillation unit. Startup of every process plant is unique, depending on the design and operating conditions of the equipment in that plant. Though it is not possible to discuss startup procedures for every plant in a refinery or a petrochemical plant, some of the common steps will be discussed here.

14.1.1 Power Plant Startup

As described earlier, in a power generation plant, boiler feed water (BFW) is heated up and converted to steam to drive the turbine-alternator to generate electricity. The boiler startup is a crucial step carried out by a team of experts. Initially, pumps, blowers, compressors and other rotary machines are started after acquiring the necessary power from power plants of other companies or organisations. The instrument air supply compressors are started, the instruments are checked before putting them online. Feed water is allowed to flow in the furnace coils of the economiser, boiler, and superheater, and circulation is established. Next, the water level in the boiler drum is monitored and controlled. Cooling water flow is established through the condenser. The draft fans are started next to ensure the desired draught in the flue gas circuit from the superheater, boiler, economiser, and then to the stack, in this order. Fuel gas lines are opened through the pilot burners of the furnace and lit up with automatic sparking. The temperature of water in the boiler rises gradually to its saturation temperature after some time. As the steam attains at its required pressure and temperature, steam valves to the turbine are opened and condensates are collected through the condenser circuit. The alternator poles are supplied with a DC power through a DC converter from the external power supply or through a battery of storage cells. Steam pressure and level of water in the boiler drum are controlled with the help of automatic controllers until a steady production of desired steam quality is generated. After establishing the desired speed of rotation of the turbine, power generation is observed for steady supply. When sufficient steam pressure is available, the atomising of liquid fuel is started in the furnace burners and fuel gas to the pilot burners is withdrawn so that the furnace is run with a steady supply of furnace oil or internal fuel oil (IFO) from storage. As the power is generated in full swing, the external power supply along with the dc supply from battery to alternator poles is withdrawn. Note that, during this production period, the alternator is synchronised with a dc generator set to supply necessary dc power for the poles while the turbine rotates and drives both the alternator and the dc generator simultaneously while power generation is peaked up by the alternator.

14.1.2 Startup of a Crude Distillation Unit

After the supply of power and utilities are ensured, the crude distillation unit is started. Initially, steam lines are opened to maintain the necessary temperature of the steam-traced lines for the pipes, pumps, and equipment, wherever steam-tracing lines are provided. Care must be taken to check that the condensate draws through the steam chest containing the condensate drain valves. It should be noted that if steam leaks without condensing through the condensate release valves (float valves), heat equivalent to the latent heat of the steam leaked will be lost. Establishing this steaming rate will take some time as the lines will be cold initially. Certain pumps are operated at different temperatures, i.e., a crude feed pump is a cold pump while the booster pump drawing desalted crude from the desalter is a hot pump. Certain pumps are operated by steam turbines, therefore it is essential to check the steam pressure and its flow through the condensate drain points while the steam inlet valves are kept closed, otherwise these pumps will start running suddenly and will upset the plant. Hence, it is essential to check the valves (manually operated) as some of these must be fully open and others must be shut as per the processing requirements. The startup lines must be checked for any leakage, damage, undesirable openings, etc. Cooling water through the condensers and coolers must be opened through their respective valves (mainly manually operated). Steam purging is carried out through the process lines and startup lines, if any, before any hydrocarbon liquid is introduced. Crude is allowed to fill the desalter by gravity through the crude discharge line of the feed pump and a sufficient level is built up. Water draining from the desalter drum and other points is carried out. The crude pump is then started after ensuring suction from the storage tank to the heat exchanger trains, desalter drum, furnace, and the distillation column. Since crude is cold and the hot fluid streams in the heat exchanger trains are unavailable, no vaporisation will start in the column. However, the column is filled with propane or butane or liquified petroleum gas (LPG) from a storage bullet to maintain the pressure within the column in order to avoid any ingress of air from the atmosphere. Thus, liquid crude will accumulate at the bottom of the column from where it will be pumped out under level control, back to the suction of the feed pump and cold circulation will be established. The furnace is then lit with the pilot burners and a gradual temperature rise is observed. As the coil outlet temperature (COT) of the furnace rises, care must be taken to avoid vapour locking of the bottom draw (reduced crude oil (RCO)) pump as the crude at the bottom of the column starts vaporising. Usually, as the COT of the furnace reaches 200°C, the rate of heating should be increased quickly to raise the temperature above this, which will ultimately help vaporise the crude oil to the maximum extent and make the bottom product free from vaporisable components. Hot bottom products return through the crude preheaters and the water cooler before they move to the feed pump suction. This explains why a cold feed pump can tolerate RCO during startup, which was hot while drawn from the column, but sufficiently cooled before reaching the feed pump, thereby avoiding vapour locking of the feed pump. But this will be stopped as soon as the products from the column are withdrawn and routed to their respective slop tanks. During this circulation, as the temperature from the preheaters reaches 95°C, the desalter transformer is switched on after establishing water and caustic injections. The bottom draw is gradually routed to the RCO tank with complete cut off from the suction. If the furnace reaches a high temperature, but the crude flow is below the minimum flow rate, the RCO pump will recirculate RCO through the furnace coil to avoid coking, which may otherwise lead to devastation if the coils are choked with coke. As the furnace temperature rises, vaporisation of crude occurs in the column and the temperature and pressure of the top condensate level over the reflux drum are monitored and controlled. While a sufficient level is built up in the reflux drum, the top reflux pump is started, followed by starting of the circulating reflux pumps, taking the utmost care to avoid vapour locking due to drying of the draw plates. Next, the top product is partially drawn along with the draws of kerosene, diesel, and jute batching oil through the strippers in sequence. These products are sampled for analysis and routed through their respective intermediate slop tanks until these are certified as on spec products. Later, slops are reprocessed with crude when there is an accumulation of sufficient slops or when there is a shortage of crude stock in future. Of course, the startup procedure is successfully carried out only by a team of experts, novices are advised to follow the instruction manuals. Proper planning is essential to prepare flow paths, such as leak testing, steam purging, blinding/deblinding operations, crude filling and circulation, furnace light up, desalter startup, slop and product routine, etc.

14.1.3 Starting a Naphtha Pretreatment Plant

After ensuring power, steam flow, cooling water flow, and warming up of the steam-traced lines, the feed pump is started. Utmost care is taken during leak testing of the process lines and equipment by compressed air pressure, followed by repeated vacuum and purging with nitrogen. Cold feed from the storage is pumped directly to the stabiliser column through a separate startup line, bypassing the train of heat exchangers, the furnace and the vapour liquid separator (flash drum), and the reactor. As a sufficient liquid level is accumulated in the bottom of the column, bottom liquid is recirculated. The reboiled naphtha is then pumped to the feed pump after being sufficiently cooled by the aftercooler. The stabiliser column is sufficiently pressurised with fuel gas and reflux condenser is started. There will be no vapour as the furnace is not lit and the liquid accumulated at the bottom of the stabiliser is returned to the feed pump and cold circulation (low pressure circulation section) is established. Hydrogen from storage is used up to pressurise the V–L separator drum by forcing hydrogen through the feed preheaters, the furnace, and the reactor, and the hydrogen is circulated back to the recycle compressor suction, establishing high pressure circulation. Next, the furnace is lit up and hot hydrogen is circulated through the reactor. The catalyst is also sulfided during this circulation for a specified time at a desired temperature. As the reactor reaches reaction temperature, feed is introduced along with hydrogen through the preheaters, the furnace, and the reactor and, finally, to the V–L separator. When a sufficient level is built up in the separator drum, it is routed to the stabiliser column and the circulation through the startup line is stopped. Finally, the bottom product is routed to the slop tank until the bottom product is tested for the extent of desulfurisation. In fact, if a sufficient quantity of hydrogen is not available during startup, it is advisable to startup the reforming unit prior to the pretreatment unit provided that a sufficient stock of desulfurised naphtha is available.

14.1.4 Starting a Naphtha Reforming Plant

Startup of a naphtha reforming plant is similar to the startup procedure of a pretreatment unit. Feed is introduced to the debutaniser and circulated through a startup line, bypassing the reactor section (high pressure section) consisting of the preheaters, furnaces, reactors, and the V–L separator. Hydrogen is circulated in the reactor section and the furnaces are lit up. Necessary catalyst additives are also injected during the hot circulation for a specified period. As the temperature of the reactors rises to 470°C, the reaction temperature, feed is switched to the high pressure section along with recirculating hydrogen. As the level in the bottom of the separator drum is built up, the liquid from the separator drum is routed to the debutaniser column and the flow through the startup line is stopped.

14.1.5 Starting a Fluid Catalytic Cracking Plant

A fluidised bed catalytic cracking plant (FCC) may be divided into three units, i.e., the reactor (converter), the regenerator, and the distillation unit. These three units must be positively separated by blinding from each other. After ensuring the supply of power and utilities, the air flow is established in the regenerator while the flue gas lines to the waste heat boiler must be kept closed. Air is blown out to the atmosphere. After draining the water and sufficient air pressure development and leak testing are carried out, the performance of the control valves are checked. The catalyst is loaded and fluidised in the regenerator vessel. When a sufficient level is built up, the reactor side is opened to allow the catalyst to flow to the riser and back to the regenerator. During fluidisation with air and catalyst circulation, air in the reactor is blown out to the atmosphere. A sufficient level of catalyst is built up in the stripper section of the converter and a steady catalyst circulation is maintained. The slide valves and controllers are checked for the overridding control action for temperature performance for temperature and differential pressure drop over the control valves. Steam is then introduced in the reactor riser, the stripper, and at the upstream of the slide valve of the spent catalyst, stand pipe and downstream of the slide valve of the regenerated catalyst flow lines. Fluidisation of the catalyst with steam is established. The waste heat boiler is commissioned with a water flow rate under level control. The distillation column is pressurised with flue gas and cold circulation is established with feed hydrocarbon so that as soon as the products are generated from the reactor, the distillation unit will process them. The startup procedures for the distillation unit and the crude distillation unit are the same. When steam purges out air from the reactor through the vent to the atmosphere, preheated feed hydrocarbon is introduced to the riser and the steaming rate is further adjusted to maintain fluidisation and catalyst circulation. Cracked products (after deblinding the feed to the distillation unit) are routed to the distillation unit under cold circulation. As the reactor temperature is established, the cold circulation is stopped and products from the distillation column are routed to their respective slop tanks until these meet the specifications. Flue gases from the regenerator are then routed to the waste heat boiler. Details of the startup procedure for the FCC unit are lengthy and a discussion of it is out of the scope of this book. Readers are advised to follow the instruction manual and emphasise the specific points of safety, such as positive isolation of hydrocarbons from air, sequences of the blinding/deblinding schedules, gradual rise of temperature in steps (ramping of temperature rise), taking care of the refractory linings, differential pressure drops over the slide valves, maintaining freeness in the distributors for air and steam, respectively, in the regenerator and the reactor, etc.

14.2 SHUTDOWN

As already mentioned in the foregoing section, a plant is down for about 30–35 days a year for necessary repair and maintenance jobs. Various jobs are performed, such as

1. Cleaning of equipment, e.g., heat exchangers, furnaces, columns, vessels, etc.
2. Repair of pipes, equipment joints, supports, etc.
3. Painting of external surfaces of pipelines and equipment for protection from oxide corrosion
4. Painting of internal surfaces of certain vessels, relaying protective lining inside certain equipment, e.g., column, furnace etc.
5. Repair and testing of pumps, compressors, and other machines
6. Checking of controlling instruments, sensors, control valves, instrument air lines, signal cables, etc.
7. Catalyst regeneration and loading for fixed bed reactors
8. Fixing column plates and valves, distributors, etc.
9. Steam pipes, steam valves, condensate drain valves, steam-traced lines, etc.
10. Checking electrical meters, e.g., ammeters, voltmeters, power meters, integrators, switches, fuses, tripping devices, motor control centres (MCC), etc.
11. Checking transformers, electrical supply cables, flame proof arrangements, etc.
12. Modifications and alteration of existing plant and equipment

Shutdown of a running plant is a procedure opposite to that followed during startup.

14.2.1 Shutdown of a Crude Distillation Unit

The first step is to reduce the crude throughput in the unit by gradually reducing the set point. As the flow rate comes down, the fuel consumption automatically comes down in the furnace. The pumps for circulating reflux, overhead, and draws are stopped when the levels over the respective draw plates have fallen. As the throughput comes to a minimum flow, the products are routed to their respective slop tanks and gases to flare. The temperature controller is put on manual and the fuel valves are closed. After the burners are switched off, steam is used to purge off the furnace chamber. The RCO is circulated to the furnace coils along with crude, while the crude flow is further cut and a steady circulation of the RCO is established in the furnace. As the temperature of the furnace comes down, the RCO is routed to the RCO tank until the column bottom is nearly empty. The positive pressure is maintained by keeping the vapours in the column and steam is simultaneously introduced into the column. It is to be remembered that none of the equipment or process lines are open to the atmosphere nor is any ingress of air allowed by maintaining a positive pressure either by the hydrocarbon vapours or by steam. After sufficient cooling of the lines and equipment, steaming is replaced by nitrogen and is allowed to cool to room temperature. The feed pump connections, slop lines, product lines, fuel and flare lines, etc., are blinded. The equipment and lines are then opened and the necessary cleaning and repair jobs are carried out.

14.2.2 Shutdown of a Naphtha Pretreatment Unit

The naphtha feed rate is gradually reduced and when it reaches minimum, the furnace temperature is reduced and the product naphtha is routed to a raw naphtha tank. The flash drum bottom valve is isolated and the bottom of the stabiliser is recirculated to the feed pump via the cooler and low pressure side circulation is established. While hydrogen circulation is continued through the feed preheaters, the furnace, the reactor, and the flash drum, hot circulation (high pressure side) is continued until all the hydrocarbons are swept from the reactor bed and purged through the flare line. When the level in the stabiliser bottom is nearly empty, the column is bottled up, keeping some positive pressure. After sufficient sweeping of the catalyst bed is achieved by confirming the sample of recycled gas, the furnace is switched off. The hydrogen circulation is continued to cool the reactor bed and later the compressors are stopped. A positive pressure is maintained in the circulation line. Next, the catalyst is regenerated in situ in the bed, which is discussed in the following section.

14.2.3 Regeneration of the Catalyst

The steam ejector is started to draw is next connected to draw the contents and vacuum is drawn followed by the introduction of steam to break the vacuum. The process is repeated until all the hydrocarbons in the high pressure section are removed. The furnace is lit up and air along with steam is introduced in the reactor section. At a temperature above 200°C, combustion is initiated with the coke and air over the catalyst. The rise in temperature due to combustion is monitored and controlled by manipulating the air rate. The flue gas is exhausted to the atmosphere. The temperature of the reactor bed is sensed by a number of thermocouples to track the uniformity of the temperature over the catalyst surface. A sudden increase in temperature over a desired limit will indicate hot spots, which may damage the catalyst and immediate air flow reduction must be maintained. The temperature profile over the bed height is recorded during the regeneration process and compared with that specified by the catalyst vendor. Steaming also helps to dislodge the coke and hydrocarbons, if any, over the catalyst surface. It is noted that the catalyst is not affected by steam.

14.2.4 Shutdown of a Naphtha Reforming Unit

Decommissioning of a naphtha reforming unit is similar to that followed in the pretreatment unit. The throughput to the unit is first reduced gradually to the minimum rate and then the furnace temperature is gradually reduced. Product reformate from the debutaniser is routed to the naphtha tank and later circulated back to the feed pump through a narrower startup line. Thus, low pressure side cold circulation is established. A low liquid level at the bottom of the flash drum is maintained and the discharge to the debutaniser is closed. Hydrogen is circulated through the preheaters, furnaces, reactors, and flash drum and returned to the recycle compressor. Thus, high pressure section circulation with hydrogen is established. Hot circulation is continued for some time to remove any hydrocarbon remaining in the reactor beds and lines. The hydrogen along with the hydrocarbon gases and vapours are sent to the flare, bypassing the fuel gas and hydrogen consumption units. When the purged gas sample shows minimum hydrocarbon vapours, the furnaces' temperature is reduced and finally fuel injection is stopped. Hydrogen is then replaced by nitrogen and circulated through the high pressure side and released to flare, while the debutaniser level is reduced to minimum and the fuel gas line is replaced by nitrogen to keep some positive pressure within the column. The cold circulation is stopped and the bottom of the debutaniser column is gradually dried out by draining to the naphtha tank. When the reactor temperatures reach room temperature, the compressors are stopped and regeneration of catalyst is started.

14.2.5 Regeneration of Reforming Catalyst

A reforming unit uses platinum as the catalyst with or without a promoter like rhenium (for bimetallic catalyst) in a fixed bed reactor. During reforming reactions, coke formation cannot be avoided, although at a reduced rate in the presence of hydrogen pressure. Coke deposited on the surface of the catalyst causes temporary deactivation of the catalyst and, as a result, the catalytic action ceases to produce the desired quality of reformate. If coke is burnt by oxidation to free the catalyst surface, catalytic activity is regained to the maximum extent and can be reused. The unit is decommissioned as described in the earlier section, followed by drawing vacuum with a steam ejector to evacuate any residual hydrocarbons in the reactor section, consisting of the preheater, furnaces, reactors, and the flash drum. The reactor section is isolated from the debutaniser column before vacuum is drawn. Nitrogen (not steam, as it will poison the catalyst) is introduced through the recycle compressor discharge line to break the vacuum. This process of drawing vacuum and filling with nitrogen is repeated until the oxygen content is negligibly small. The furnaces are then lit up to raise the temperature of nitrogen to around 200°C and air is introduced with caution so that combustion in the reactor bed is started slowly. The temperature of the entire bed is monitored by a number of thermocouples inserted at various locations within the catalyst bed. As the temperature at any spot increases suddenly above a certain specified limit, overheating or a hot spot is located and immediately the air flow rate is reduced. A plot of temperature against time during the regeneration period is recorded continuously to ascertain the uniformity of regeneration. The flue gas with nitrogen is continuously ejected to the atmosphere. As the temperature of the catalyst bed does not rise above the furnace temperature, the end of combustion is ensured. *(At this stage, oxychlorination of the catalyst is also carried out for the necessary isomerising function of the catalyst. A halogen source, usually carbon tetra chloride, along with air is introduced at the desired reaction temperature).* Next, the furnace temperatures are reduced and finally the fuel injection is cut off while nitrogen circulation is continued until the temperature reduces to room temperature. The compressors are then stopped and isolated from the unit and the equipment is opened for cleaning and repair jobs. Reloading of a fresh catalyst is also carried out in the reactors.

14.3 MAINTENANCE OF PLANT AND EQUIPMENTS

The maintenance jobs of a plant involves a team consisting of various engineering and technical experts from fields, such as mechanical, electrical, civil, metallurgical, instrumentation, computer, electronics, etc. The repair and maintenance jobs of machines against their *mechanical failures* may be carried out in three ways.

1. *During the annual shutdown period*: Repair and maintenance of equipment and machines, which cannot be isolated from the process without affecting production, are carried out. However, if the plant is integrated in such a way that each unit can be run independently for a certain period without affecting the other units in the plant, partial shutdown can be done at any time.
2. *Periodically carried out throughout the year*: Usually, the running hours of the pumps, compressors, and other machines are continuously recorded in any plant. After certain running hours, these machines must be checked and the necessary rectifications made. In fact, while these machines are removed from service, spare machines must be taken in line to maintain production. In other words, machines that have at least another duplicate machine (spare machine) in operation can only be repaired periodically. This type of maintenance is essential for any plant to prevent an accident during production. Hence, this is also known as *preventive maintenance*.
3. *At the time of failure or accident*: In case of sudden damage to any machine during the running period, it is inevitable to run the spare machine and the failed machine must be repaired. In fact, this type of repair and maintenance

is unwanted in any plant, but unfortunately these failures are either due to lack of supervision and operating skill or improper repair during annual or periodic maintenance. These types of failures may even lead to an accident small or big and cause huge loss to the organization.

Maintenance of instruments is similarly carried out for all sensory elements, cables, filters, controllers, control valves, and control panels. Periodic tuning of the controllers is also carried out. For a DCS control system, the control software must be checked for performance and reproducibility. A necessary update is also essential for any change in the plant due to the addition or modification of the plant and equipment. An inspection must be carried out on the security instruments, e.g., safety valves, safety switches, fuses, trippers, alarms, etc.

Civil maintenance looks after the repair of underground or overground bridges, culverts for pipe racks, pits, roads and approaches for the movement of light and heavy vehicles, for the removal and positioning of machines and equipment, painting and lining jobs, addition or modification of buildings, etc.

The *inspection* of materials of construction, testing piping and structures, inspection of column shells, vessel walls, etc., is carried out by a group of experts from metallurgical engineering. Common, sophisticated methods, such as ultrasonic and x-ray, are used for testing the plate thickness of vessels. Chemical composition analysis may also be required to determine the quality of the steel and other building materials. Correlations of composition and thickness may predict the permissible limits of the operating load, temperature, corrosion, and life of the materials. Inspections are carried out annually during *shutdown*, *periodically*, and also at the *time of failure*.

14.4 FIRE AND EXPLOSION

Fire is the result of an exothermic reaction between combustible matter and oxygen (from air) at a certain temperature. Fire, big or small, may be caused by oxidation of hydrocarbons (combustibles) ignited by a small spark or heat. The cause of fire may be small, but it can lead to a small or large fire and may be associated with explosion. Hydrocarbons are classified as *most dangerous*, *dangerous*, and *non-dangerous* goods as per explosive rules. Liquid hydrocarbons are also classified according to their increasing flash points as class A, B, and C. Class A products are highly inflammable and have flash points below 23°C, e.g., crude oil, naphtha, motor spirit, viscosity-broken-gasoline, pyrolysis-gasoline, benzene, toluene, xylene, etc. Class B products are also inflammable, but less volatile than class A products. Class B products have flash points greater than 23°C but less than 65°C, e.g., kerosene, aviation turbine fuel, mineral turpentine oil, high speed and light diesel oils, furfural, etc. Class C products are almost non-volatile at atmospheric pressure, but can be vaporised under vacuum. These products have flash points greater than 65°C but below 95°C. Examples of class C products are jute batching oils, RCO, vacuum distillates, short residues, asphalts, furnace oil, etc. It is found that a minimum temperature required for ignition of the combustible matters must be attained for the production of fire even in the absence of any spark or flame. This temperature is known as the *auto ignition temperature*, which is a property of inflammable substances, whether solid, liquid, or gas. For example, a typical gasoline (class A product) has an auto ignition temperature of 246°C, whereas the auto ignition temperatures of a typical kerosene (class B product) and bitumen (class C product) are, respectively, 254°C and 485°C. Besides auto ignition temperatures, the relative concentration of the combustibles and air necessary for the combustion or explosion are known as the limits of explosion or inflammability. The concentration of air is known as the lower limit of explosion and if it is much lower than that of the combustibles, incomplete combustion occurs. If air is present in large amounts with respect to the combustibles, and complete combustion occurs, this concentration of air is known as the upper limit. Hydrocarbons become explosive when the air concentration falls above or below the limits of inflammability. During incomplete combustion, gas expands in volume followed by a quick reduction in volume owing to the cooling effect of the expansion of the unconverted combustible gas or liquid. Such a sudden expansion followed by compression can cause knock or explosion. Whereas during complete combustion, the heat generated is huge and can cause a sudden expansion of the combusted gas mixture, leading to explosion in a closed container. If sufficient space is available, a continuous fire will take place without explosion. Thus, fire and explosion are dependent on the flash point (which means ease of vaporisation), auto ignition temperature (reaction temperature for combustion), the presence of oxygen (usually from air), and the space occupied during such a reaction. The degree of explosion depends on the heat of the reaction (heating value) of the combustibles and the volume of the container. The extent of fire (duration) is also dependent on the amount of combustibles present while sufficient oxygen is available. The mechanism and pathways of combustion reactions are rather complicated by the nature of the combustible matters involved, coupled with the heat and mass transfer phenomena of the oxygen and product of combustion. Thus, fire can be doused by any or combined means as

1. Cooling by water, to quench the heat of reaction.
2. Blanketing air by water spray or steam or inert gas or foam or sand (for a small fire) to starve the reaction of oxygen.
3. Spaying of oxygen-absorbing chemicals.
4. Reducing the supply of combustibles by quickly cutting off the remaining amount by pumping it out of the equipment or vessel, if possible.

The causes of fire may be due to sparks or flames generated from mechanical friction, electrical short circuit, static charge accumulation, natural lightning, or human error, e.g., smoking, lighting lamps, etc. Back fires from natural draught furnaces are very common during furnace light up operation with torches and due to fluctuation of draught.

14.4.1 Pyrophoric Iron

Certain metallic compounds, usually of iron, can generate heat due to an oxidation reaction sufficient to cause a fire. Such autothermal iron compounds are called pyrophoric iron. For example, ferrous compounds of sulfur are easily oxidised if exposed to air and generates heat due to reaction with oxygen. Usually, iron rusts or oxides of iron are commonly formed due to oxidation by dissolved oxygen in water or due to atmospheric air. This oxide then forms iron sulfide in the presence of hydrogen sulfide and is readily oxidised (the reaction is exothermic) in the presence of air, giving rise to heat and fire. This usually happens when equipment is opened for repair. The reactions are listed below,

$$Fe_2O_3 + H_2S \rightarrow FeS + H_2O + S$$

$$FeS + O_2 \rightarrow Fe_2O_3 + SO_2 + \uparrow \text{heat}$$

14.5 FACTORIES ACT

Hydrocarbon processing plants, whether refineries or petrochemical plants, must adhere to certain factory rules and explosive rules. Some of the usual practices relevant for safe operations are discussed here.

This act provides guidelines for the benefits of the employees as far as working conditions, safety, and contractual obligations are concerned. This act was initially implemented in 1948 and has been amended from time to time. Discussions of the provisions laid down in this Act are out of the scope of this book. Engineer supervisors and managers are the authorised representatives of the employer or management. Hence, they should know the obligations towards the workers or the employees under their supervision. They should provide each employee with adequate training to operate the specific plant and machineries, provide appropriate tools, garments, protective gear, etc., properly apprising the employees about the dangerous properties of the fluids, materials, and machines to be handled and the necessary precautions to be taken, about emergency plans and actions, etc. In the event of an accident, the employer or its representative engineers and managers must report to the factory inspector of the local area with details of the accident, such as the nature of the accident, e.g., fire or explosion, major or minor, casualties if any, type of casualties, wounded or dead, the number of such casualties, the possible causes of the accident, the amount of compensation for the affected employees, etc. The factory inspector must be apprised of the accident immediately it occurs, followed by a detailed report as mentioned above in a stipulated time frame. The factory inspector will then visit the area and scrutinise the veracity of the information received by him. He also will verify whether any deviation as laid down in the Factories Act has taken place for which the owner is responsible to answer or liable for punishment. A few common provisions of the Factories Act are listed below.

1. Definition of a factory: A place or premises where more than 10 people work with power driven machines for the last 12 months.
2. Definition of a worker: A person employed directly or through a contractor or agent with or without any remuneration involved in any manufacturing process or in any other cleaning job connected with the manufacturing activities. Worker does not include any member of the armed forces.

3. Obligations of workers: A worker is punishable for wilful interference, misuse of any appliance meant for securing the safety of self and others, any act of wilful negligence or damage that endangers himself or others and affects the manufacturing activities of the employer.
4. Definitions of classes of worker according to age: Workers considered as adult, adolescent, and child are, respectively, 18 years (and above), 15 years but below 18 years, and below 15 years of age. No person should be employed below the age of 18 years.
5. Calendar year, day, and week are defined as 12 months from the first of January, 24 h, and 7 days, respectively.
6. The owner of the factory must register its name, premises, and details of the manufacturing activities well in advance before any construction and also before manufacturing activities are to commence. The owner must obtain a license to carry out manufacturing activities from the chief inspector of factories.
7. Power of factory inspectors: The authorised persons of the chief inspector, i.e., the factory inspectors, are allowed to enter any factory premises to inspect, examine, or ask questions to be answered by the employers and employees or the owner. Any attempt by the owner or his/her employees to obstruct the inspectors in the discharge of their duties is a punishable offence. The inspector or inspectors may inquire about the ages of the workers, health, safety, working machines, working environment, etc. For instance, whether proper cleanliness is maintained or not, which may otherwise affect the health and safety of the worker; if disposal of wastes and treatment facilities is carried out or not; if proper ventilation and temperature are provided or not; protection from dust, fumes or fire, provision of toilets and urinals, rest rooms, etc.
8. Safe facilities: Proper fencing and protective covers for the machineries must be provided to protect the working personnel from any accidental contact with any movable part or hot spot or electric shock and the like from the machine or equipment. Proper measures must be taken to protect the workers from high pressure systems. Well maintained and appropriate floors, steps, stairs, ladders, and bridges must be provided for movement of the workers and easy access to the emergency exits. Manhole covers for the underground storage vessels, tanks, or pits must be provided to protect the workers from accidental falls. No persons shall be allowed to lift or carry a load that may cause him injury. Workers must be provided with protective gloves, eye goggles, safety shoes and gum boots, appropriate clothing, etc. Precautions must be taken as per factory rules against dangerous fumes and gases in explosive area. Battery-operated torches or lamps made of non-sparking materials, e.g., brass or aluminium, should be provided. Proper maintenance of the building and structures must also be carried out to prevent any accidents. Safety officers must be provided in the factory to supervise the safe facilities, fire fighting, safe practice training to the employees, etc. These officers must possess the necessary relevant qualifications as laid down by state or central government authorities.

9. Welfare for the workers: Provisions are laid down for the various welfare facilities, e.g., washing and drying cloths, sitting arrangement, first aid appliances, canteen, rest rooms, dining hall, crèches for women workers, etc. Welfare officers must be provided in each factory to look after the welfare facilities in the factory. Provisions for working hours, normal wages, overtime wages, leave and holiday entitlements, etc., are also available in the Factories Act. Compensation due to partial or full physical disabilities and death are handled by a separate act, known as the Workmen's Compensation Act 1923.

Engineers should know the foregoing discussions about Factories Act in more detail before they take the responsibility of management of the plant. However, separate safety rules and compensation must also be provided for the adjoining localities and factories.

14.6 SAFETY ANALYSIS

Safety means protection against any occurrence of accidents. Accidents do not just happen, they are caused. The causes are the potential hazards and the risks involved in any plant. Safety starts with the identification of hazards that may involve fire, explosion, toxicity, break or crack, fall or slip, etc. The entire pathways of hazards leading to accidents has to be identified. This is known as hazard analysis or HAZAN. In this method, hazards are quantified by numbers or weightage for specific risks. According to the DOW index, the ratings are evaluated and assigned to various hazards. The DOW index is evaluated based on various factors involving the type of materials, hazards associated with the materials, hazards associated with the processing or production scheme, quantity in storage, and the layout of the plant and equipment. These are classified as

1. Material hazards and special material hazards
2. Process hazards and special process hazards
3. Quantity hazard
4. Layout hazards

Materials hazards factor (A): Flammable, explosive, solid or particulate, inert, etc., factors are evaluated for each identified material, e.g., flammable materials will have factors based on heating value and molecular weight. Similarly for non-combustibles, another factor is assigned. When exothermic reaction other than combustion occurs, a separate index is evaluated based on the heat of the reaction and the heat of combustion separately, and the higher value is selected.

Special material hazards factor (B): These are the materials that are neither combustibles nor inflammables, but assist combustion and explosion, e.g., oxidants (oxygen, chlorates, nitrates peroxides), reactives with water (carbides, sodium, magnesium, certain hydrides), pyrophorics (ferrous sulfide, carbon disulfide), explosive gases or liquified gases, highly viscous furnace oil or asphalt, etc.

Process hazards factor (C): These are hazards involved in the storage, heating with or without steam, distilling, or reacting etc., involving inflammable materials under normal pressure and temperature. These are given certain weightage.

Special process hazards factor (D): Separate weightages are assigned for operating conditions other than normal, e.g., high pressure or vacuum operations, high temperature or cryogenic operation, corrosive operation, etc.

Quantity hazards factor (E): The greater the quantity in storage in tanks, in processing equipment, or in pipelines, the greater the hazards. This also involves value of loss. Factors are assigned on the amount and values.

Layout hazards factor (F): These are hazards due to overhead and underground storage or structure, discharge lines, etc. Separate numbers are allotted for the identified hazards of this category.

The Dow Fire and Explosion Index (DOW), developed by Dow Chemicals, is then calculated as,

$$\text{DOW} = A(1 + B/100)(1 + C/100)\{1 + (D + E + F)/100\}.$$

A typical listing of hazards are given as

Dow index	Hazards
0–20	Mild
20–40	Light
40–60	Moderate
60–75	Moderately heavy
75–90	Heavy
90–115	Extreme
115–150	Highly extreme
150–200	Potentially catastrophic
Above 200	Highly catastrophic

The DOW is applicable for identifying hazards like fire and explosion, and hazards from a chemical reaction during the design of a new plant or used for an existing plant. A similar numerical procedure has been developed for identifying wider scopes for hazards analysis by MOND index, developed by ICI. The *assessment of risks* involved in any plant and machineries, and the materials are also classified by *severity* and *probability of occurrence*. Severities are classified as *catastrophic, hazardous, major, minor*, and *safe*. A specific amount of loss in each category of severity is defined qualitatively by the degree of injury to operators and loss of the system. The probability of occurrence of hazards is classified as *most probable, least probable or remote*, and *extremely remote and improbable*. Probability values are assigned for such occurrences.

Hazards and operability, or HAZOP, is applicable for the operation of a batch or continuous plant. This is applied over a selected portion of any plant or the entire plant. This method applies in stages during the design, commissioning, and operation of a plant. For an existing plant, ranking of hazards is carried out using the DOW or MOND index to identify the potential hazards. A *safety audit* is then carried out to list the possible safety facilities. Past occurrences of unusual events are listed. Usually, a check list of the materials, unit operations, layout, and hazards are prepared. A flow sheet of the plant is then studied with respect to certain *guide words* or *events*. Possible reasons for the deviation or faults at these events are diagnosed and the consequences of such a deviation are eliminated by certain modification or action. Modification may involve the introduction of a controlling instrument, or the removal or extension of a pipe, provision of a vent or drain, etc., which were not originally present in the flow sheet or piping and instrumentation diagram of a plant. The following is a list of some of the guide words used in the HAZOP. These guide words are carefully prepared by experts relevant to the plant and equipment under study.

Guide word	Inherent meaning with respect to the parameters
NO or NOT	No flow, low pressure, low temperature, etc.
MORE	Increase of flow, increase in certain temperatures or pressure, etc.
LESS	Decrease in flow, decrease in temperature, pressure, etc.
AS WELL AS	Composition of the system which is/are redundant or more differential pressure, etc.
PART OF	Composition of the system which is/are missing or less
REVERSE	Back flow
OTHER THAN	Unwanted material, explosive pressure, unwanted reaction, etc.

Example of an application of HAZOP:

Let us take a flow sheet of a crude oil desalter, as shown in Figure 14.1. The following guide words are suggested by the team of experts.

Guide words	Cause	Consequences	Prevention
No flow	Feed pump trips	Transformer trips	Auto start of spare feed pump
No flow	Discharge pump trips	Transformer trips	Auto start of spare discharge pump
No flow	Water injection stops	Salt carryover with crude discharge	Auto start of spare water pump
No flow	Caustic injection stops	pH will fall, high current flow, transformer may trip	Auto start of spare caustic dosing pump or install low pH alarm
Low pressure	Low discharge pressure of feed pump	Drum pressure falls may cause vaporisation and transformer trips	Pressure controller to reduce discharge flow. Check the controller and discharge pump flow

(continued)

(Continued)

Guide words	Cause	Consequences	Prevention
High pressure	Low discharge rate	PSV will pop and transformer will trip	Check discharge pump flow. Check pressure controller
Low temperature	Low feed temperature	Poor desalting	Reason external (check crude preheaters). Low temperature alarm is suggested
High temperature	High feed temperature	Vaporisation, transformer will trip	Reason external, (check crude preheaters). High temperature alarm suggested
Low interface level	Low water injection rate	Oil carry over with brine discharge	Interface level controller or low level alarm is suggested
High interface level	High water injection, or feed contains more water	Transformer trips to avoid short circuit	High interface level alarm is suggested or feed rate is to be cut down or water injection to be reduced

HAZOP analysis of a crude pipe-still heater in a distillation unit:

Consider desalted and preheated crude is heated in a pipe-still furnace to raise the temperature of the crude to about 360°C–370°C before flashing into the distillation column. The following HAZOP analysis is carried out during the production period while liquid fuel is fired in a natural draught furnace, as shown in Figure 14.2.

Guide words	Cause	Consequences	Prevention
No flow of feed	Feed pump trips	Furnace pipes will be choked with coke	Auto start of RCO pump to circulate RCO to pipes to avoid coking and fuel firing to stop and steam purging to start
Low flow of feed	Strainer of feed pump choked, available head is low at the suction	Feed pump may trip suddenly if flow is below minimum discharge	Low flow alarm is suggested
High flow of feed	Discharge valve is fully open	Pump may trip due to high amperage	High level alarm or high power load alarm is suggested
No flow of fuel	Fuel pump tripped or solenoid valve off or the control valve stuck closed	Low coil outlet temperature	Check fuel circulation system, or check solenoid valve of fuel line, or check temperature controller at the coil outlet and control valve

Guide words	Cause	Consequences	Prevention
No flow of atomising steam	Steam control valve stuck closed	Burners will extinguish. Explosion may occur due to accumulated fuel oils	Low flow alarm in atomising steam flow and auto tripping of fuel firing
Low flow of atomising steam	Steam control valve malfunctioning or pressure in the steam header is low	Incomplete combustion and smoke generation	Steam to air ratio controller to be installed with low steam rate alarm
As well as low flow of primary air	Low draught	Incomplete combustion and loss of fuel and back fire accident probable	Air to fuel ratio controller is suggested
As well as high flow of primary air	High draught	Stack loss increases, furnace efficiency comes down	Air to fuel ratio controller is suggested

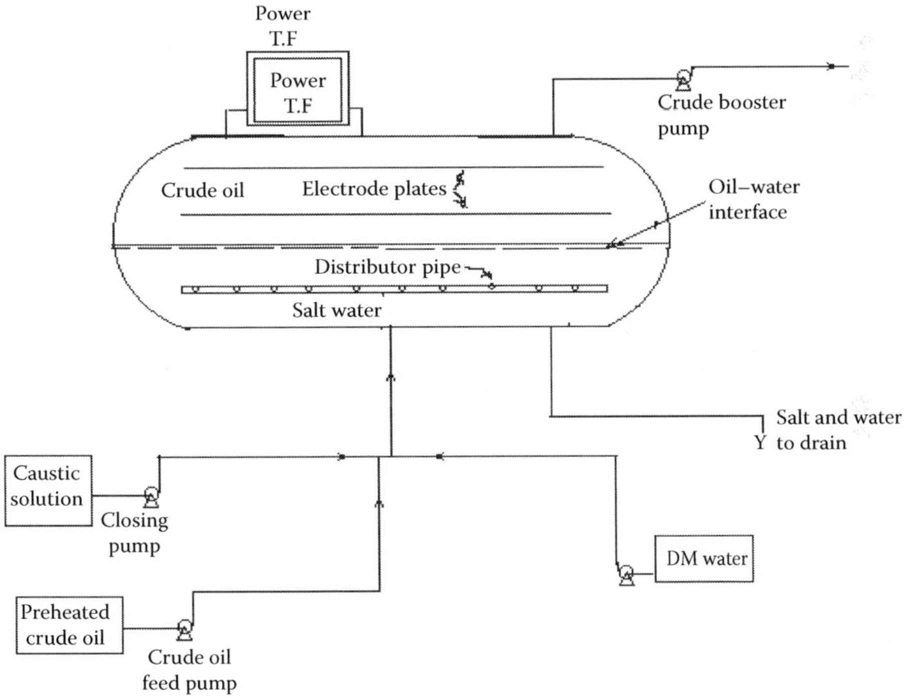

FIGURE 14.1 A crude oil desalter flow sheet under study.

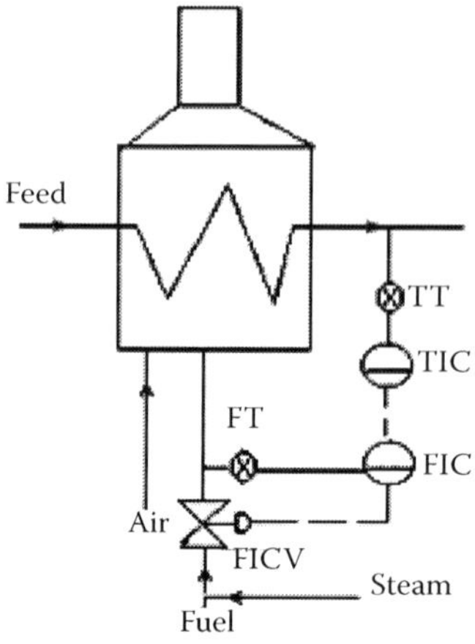

FIGURE 14.2 A natural draft pipe-still heater for crude oil heating.

15 Plant Management and Economics

15.1 COST OF EQUIPMENT

The cost of equipment may be obtained in two ways: (a) by estimation and (b) the purchased price. Cost estimation is required when the equipment is fabricated.

Estimated cost will include the material, labour, and the overhead components involved in manufacturing. Material cost will include the direct and indirect components. Direct cost will include the cost or price of the materials, e.g., plates, tubes, rods, beams, etc., which can be identified with the equipment. Indirect materials will include the cost or price of those materials that cannot be identified with the equipment, rather for large number of pieces of equipment manufactured, e.g., the costs of nuts, bolts, screws, welding, threading materials, etc. These are, in fact, required for fabrication of a number of pieces of equipment and cost is apportioned by dividing the total cost of these materials by the number of pieces of equipment manufactured. The labour cost, too, involves direct and indirect elements. For example, the labour involved in fabricating a particular piece of equipment is a direct labour cost, but the cost of labour involved in loading, transporting, welding, threading, lifting, rigging, etc., are indirect labour costs, as these are involved for a number of equipment manufactured. Additional costs, e.g., power, steam, chemicals, etc., which are also identified for a single piece of equipment are the other direct costs and those that cannot be identified as indirect costs or expenses. All these indirect costs are summarily taken as the overhead cost. The sum of direct costs is called *prime costs* and the sum of the indirect costs is called *overhead costs*.

Thus, the estimated cost of any equipment = [direct material + direct labour + direct expenses] + [indirect material + labour + expenses] = prime costs + overhead cost. This method is used while new equipment is manufactured.

15.1.1 Capacity Ratio Method

The cost of any equipment can also be obtained by a short-cut method using a capacity ratio based on the known value of the equipment at a given capacity. This is defined as

$$\textit{cost of the equipment} = P \times \left(\frac{c_2}{c_1}\right)^{0.6}$$

where p is the known a piece of equipment of capacity, c_1, and where the cost of the equipment is of the desired capacity, c_2. In case the known price is quoted in the past, the present price may be obtained by multiplying the ratio of the cost index of the present price to the past price. Thus, the estimated cost of the equipment is more accurately obtained by multiplying the cost index ratio.

15.1.2 Purchased Price

The price paid for purchasing equipment is the cost of the equipment. Sometimes, the cost of installation of the equipment is also included with the purchase price, which is the *installed price* of the equipment.

15.1.3 First Cost of the Equipment

This is the price or cost of the new equipment installed in the plant.

15.1.4 Depreciation

This is an expenditure considered to be a loss of the value of the equipment with time and use. The life of the equipment is the time period between the time it was installed and the time it becomes useless and is scrapped. A simple definition of depreciation per annum is the difference between the first cost of the equipment less the scrap value (if any) divided by the life in years (the straight line method):

$$\text{Annual Depreciation} = \frac{(\text{Initial cost of the equipment} - \text{scrap value})}{\text{years of life}}.$$

15.2 COST OF A PLANT

A plant includes various processing and auxiliary equipment, the cost of a variety of materials, such as raw materials, utilities like steam, power, chemicals (other than raw materials), working personnels, and other expenses. The cost of a plant is the sum of the capital and the operating cost. Capital cost is the total cost of all the equipment installed. This is calculated as the total cost of the equipment including the cost of installation. Operating cost is discussed in the following section.

15.3 OPERATING COST

The operating cost is the yearly cost or expenses for running the plant. This includes the cost of the raw materials, chemicals, power consumed, salaries and wages of the manpower involved, overhead charges, such as depreciation for the capital cost, and other expenses.

15.4 PRODUCT COST

The product cost includes the cost of production until the raw materials are converted to finished products plus the cost of packing and filling, dispatching, after sales cost, etc. In fact, the *ex factory cost* of a product includes all the expenses incurred for making the finished product ready for despatch. Product cost is the sum of the ex factory cost plus the additional costs incurred to reach the consumer. Thus, typical elements of a product cost are listed as

1. Direct material cost
2. Direct wages
3. Direct expenses
4. Production overhead
5. Administration overhead
6. Packing and filling costs
7. Excise duty (which is paid before leaving the factory or plant)
8. Ex factory cost = (1 + 2 + 3 + 4 + 5 + 6 + 7)
9. Sales and distribution cost, including after sales service costs
10. Product cost = 8 + 9

15.5 PROFIT AND PRODUCT PRICE

The product price is the sum of the product cost plus the profit. Profit is thus the difference between the product price and the product cost. The product price is decided by the market price and profitability criteria, e.g., breakeven point, payout period, minimum rate of return (ROR), turnover ratio, trading rules, and various duties and taxes.

15.6 TAXES AND DUTIES

Various taxes and duties are charged by the state and central governments to the value of the product. Thus, product price includes all the duties and taxes. It is a fact that the entire burden of duties and taxes paid by the manufacturer and sellers are borne by the consumer. Details of the various taxes are available from the *Gazette of India* published in every financial year. These are also notified in newspapers during budget announcements.

Thus,

$$\text{Product price} = \text{product cost} + \text{profit} + \text{taxes and duties.}$$

15.7 BREAKEVEN POINT, PAYOUT PERIOD, AND RATE OF RETURN

The *breakeven point* is expressed in terms of the rate of production units when the value earned by selling the product equals the cost of producing it, i.e., while profit is zero. The total earnings for a certain volume of production that is sold is known as the sale volume, which is the product of the unit price of the product and the number of units of product sold. For example, if volume of product is n units in a year and the unit sale price of product is s, then sale volume = $n \times s$. Total expenses incurred are divided into variable expenses and fixed expenses. Variable expenses are the expenses that vary with the rate of production. Thus, if the variable cost per unit is c, then the total variable cost incurred for the production of n units will be $n \times c$. The fixed expenses will not vary with the units produced and remain unchanged for any number of units produced. Examples of fixed expenses are the salaries of the regular employees, depreciation of the plant and machineries, and rent of the land. The total expenditure will be the sum of the variable costs and the fixed costs per year. If the fixed cost of the plant is f, then profit = $ns - nc - f$.

At the breakeven point, profit = 0 and $n = f/(s - c)$. Hence, the number of units at the breakeven point is the fixed costs divided by the difference of the sales price and the variable cost per unit.

15.7.1 Payout Period or Payback Period

This is the time period within which the entire fixed investment on the plant and machineries are earned to recover or repay. If the initial investment is acquired by a loaned amount as I and the rate of earning is r per year, the payout period = I/r per year. In fact, the initial investment will include the additional expenses in getting the license fees, land development cost, etc., in addition to the cost of the plant and machineries and the installation costs. The rate of earning is the average profit and the depreciation per year.

15.7.2 Rate of Return

The ROR is part of the initial investment earned per year. This is calculated as the ratio of the profit per year divided by the investment, i.e.,

$$\text{ROR} = (\text{gross profit} - \text{tax over profit})/\text{initial investment} \times 100\%.$$

This is the ROR per year, which may vary from year to year depending on the variation in profit. If the investment is acquired by a loan at an interest of i per year, the ROR should be greater than i. In fact, the investor must earn greater than the interest payable to the lender.

Exercise 15.1

A plastic moulding plant with the following details has been planned for construction.

The rent of the land: Rs 30,000 per month
Land development cost: Rs 20 lakhs
Building: Rs 20 lakhs
Purchased price of moulding machines including installation charges: Rs 50 lakhs
Life of moulding machines: 10 years
Power requirement: 100 kW
Power cost: Rs 0.5 per kW-h
Maximum rate of production: 20,000 units
Raw material cost: Rs 10 lakhs
Water cost: Rs 20 lakhs

Chemicals costs: Rs 1 lakh
Salaries per year: Rs 10 lakhs
Insurance premium per year: Rs 5 lakhs
Exigency and safety arrangement: Rs 10 lakhs

Taxes and duties chargeable are 10% over the sales value. Assume tax on profit is 10%.

A bank loan is available charging 12% simple interest per annum.

Determine: (1) the initial investment, (2) the variable cost per unit, (3) the fixed cost chargeable per year, (4) the product cost, (5) the sales price of the product, (6) the breakeven point, (7) the ROR, and (8) the payout period.

Solution

1. Initial investment

Capital costs:	**Rs lakhs**
Land development cost	20
Building	20
Purchased price of moulding machines including installation charges	50
Total fixed capital cost	90

Working capital:	**Rs lakhs**
Rent of the land	3.6
Raw material cost	10
Water cost	20
Chemicals costs	1
Salaries per year	10
Insurance premium per year	5
Exigency and safety arrangement	10
Total of working expenses for a year	59.6

Hence, initial investment = fixed capital + working capital for a year = Rs 149.6 lakhs.

From the information, cost elements are segregated as fixed and variable as follows.

Fixed cost per year	**Rs lakhs**
Depreciation of moulding machines including installation charges (straight line method) (considering scrap value at the end of 10 years as nil)	5.0
Land development cost recovery per year in a 10-year span	1.0
Depreciation of safety appliances	1.0

(*continued*)

(CONTINUED)

Fixed cost per year	Rs lakhs
Depreciation of building	2.0
Salaries of the staff	10.0
Insurance premium	5.0
Rent of the land	3.6
Interest on investment (Rs 149.6 lakhs × 0.12)	17.952
Total	45.552

Variable cost per year	Rs lakhs
Raw material cost	10.0
Water cost	20.0
Chemicals costs	1.0
Power cost for 330 days/year continuous production in 24 hrs = 100 × 24 × 330 × 0.50	3.96
Total	34.96

2. Variable cost per unit is Rs 34.96 × 100,000/20,000: Rs 174.8 per unit
3. Fixed cost per year = Rs 45.552 lakhs
4. Product cost = (annual fixed cost + variable cost)/20,000 = (45.552 + 34.96) × 100,000/20,000 = Rs 402.56 per unit
5. Sales price of product = product cost + profit per unit
   Product price should be greater than the product cost. Let us assume that profit is 15% of the product cost, i.e., product sale price = 1.15*402.56 = Rs 462.944 per unit.
   It must be checked that the payback period should be less than 10 years. This will be done next.
6. Breakeven point:
   The number of units at BEP = annual fixed cost/$(s - c)$, where s = product price: Rs 462.95 and c = variable product cost = Rs 174.8.
   So, the rate of production to break even = 45.552 × 100,000/(462.95 − 174.8) = 15,808.4, i.e., 15,809 units in a year, which is less than the maximum rate of 20,000 units.
   Hence, the breakeven point will occur after 15,809/20,000 × 330 days = 261 days.
   Thus, the entrepreneur will start earning a profit after only 261 days.
7. ROR:
   Gross profit in the year = annual sales − annual cost = 20,000 × (s − c) − fixed cost = 20,000(462.95 − 174.8) − 45.552 × 100,000 = 12.11 lakhs
   Net profit after tax = 10.899 lakhs
   ROR = (profit after tax + depreciation)/Investment × 100 = (10.899 + 8) lakhs/149.6 lakhs × 100 = 12.6%, which is much higher than the bank interest rate.
8. Payout period:
   Payout period = initial investment/net profit per year = 149.6/10.899 = 13.72 years, i.e., about 14 years. If depreciation is also included with net profit, the payback period will be 149.6/(10.899 + 8) = 7.9 years.

Thus, the payback period will be between 8 and 14 years. However, if the existing market price of the product is known, the same should be used for re-determining the above entities.

15.8 LINEAR PROGRAMMING

Linear programming (LP) is a mathematical tool for determining the optimum rate of production, product mix, raw material mix, etc., based on certain objective function, usually profit is to be maximized or cost is to be minimized. These objective functions are the profitability criteria, such as maximum profit, maximum sales, maximum marginal profit, minimum cost, etc. The relations between the yield of products from each production unit and their costs, constraints of limitations of units, demand of products, supply of raw materials, and other miscellaneous constraints are applicable to determine the objective function. The relations and equalities are collected and arranged as a large number of simultaneous linear algebraic equations and solved with a computer. This is an essential tool for the management to make valuable decisions, such as increasing or decreasing the rate of production during a shortage of raw materials, fluctuation of demand of products, industrial strikes, any emergency situation, etc. This is explained in the following simple examples.

Exercise 15.2

A refinery processing two types of crude oils has the following data.

Type of crude	Cost Rs/MT
1	24,000
2	15,000

Product prices	Rs/MT
Gasoline	40,000
Kerosene	24,000
Fuel oil	21,000
Residue	10,000

Yield data:

Products	From crude 1 (vol %)	From crude 2 (vol %)	Maximum allowable rate of production (MT/day)
Gasoline	80	44	2400
Kerosene	5	10	200
Fuel oil	10	36	600
Residue	5	10	–
Refining cost (Rs/MT)	178	358	

Find the optimum profit at which crude oil blend should be processed.

Solution

Let x_1, x_2, x_3, x_4, x_5, and x_6 be the rates of crude oil 1, crude oil 2, gasoline, kerosene, fuel oil, and residue, respectively.

Hence,

Profit = p, the objective function to be maximised is given as, P = sale value of products – costs of crude oils – refining expenses.

$$= 40{,}000x_3 + 24{,}000x_4 + 21{,}000x_5 + 10{,}000x_6 - 24{,}178x_1 - 15{,}358x_2,$$

where, $x_3 \le 2400$, $x_4 \le 200$, $x_5 \le 600$, and x_6 has no restriction.

Using the yield data:

Gasoline: $x_3 = 0.80x_1 + 0.44x_2$,

Kerosene: $x_4 = 0.05x_1 + 0.10x_2$,

Fuel oil: $x_5 = 0.10x_1 + 0.36x_2$.

Residue: $x_6 = 0.05x_1 + 0.10x_2$.

Substituting these in the objective function, p, is

$$P = 11{,}622x_1 + 13{,}202x_2. \qquad \text{(a)}$$

Introducing slack variables, s_3, s_4, and s_5, in the constraints for the limits of production,

$$0.80x_1 + 0.44x_2 + s_3 = 2{,}400, \qquad \text{(b)}$$

$$0.05x_1 + 0.10x_2 + s_4 = 200, \qquad \text{(c)}$$

$$0.10x_1 + 0.36x_2 + s_5 = 600. \qquad \text{(d)}$$

There are three equations ($m = 3$), five unknowns ($n = 5$), possible solutions are therefore to be tried,

$$\frac{n!}{m!(n-m)!} = 10.$$

The following trials are carried out by using two of the unknowns as zeros and the objective function is evaluated.

Trial	x_1	x_2	s_3	s_4	s_5	p
1	0	5,454.54	0	–	–	7.2×10^7
2	3,000	0	0	–	–	3.39×10^7
3	2,096	689.6	0	0	–	3.96×10^7

4	1,500	1,250	–	0	0	3.39×10^7
5	2,495	983.6	0	–	0	4.16×10^7
6	0	0	–	–	–	0.00
7	0	2,000	–	0	–	2.64×10^7
8	0	1,667	–	–	0	2.2×10^7
9	4,000	0	–	0	–	4.64×10^7
10	6,000	0	–	–	0	6.97×10^7

From the solution it is found that p is maximum at $x_1 = 0$ and $x_2 = 5{,}454.54$ MT/day.

Exercise 15.3

A lube oil blending plant has received an order of 100 t of lube oil, which should contain 4% weight of a specific LOBS. The blender has no stock of the 4% wt LOBS product, but it has three stocks, A, B, and C, containing LOBS 4.5%, 3.7%, and 0% respectively. The cost price of these are Rs 2,000 and 1,500, respectively, per ton for A and B, whereas stock C has no value. It is also required that at least 10 t of stock A must be consumed. Determine the blending ratio of the stocks A, B, and C, at the minimum cost of the final product.

Solution

Let the blending in tons of A, B, and C be W_A, W_B, and W_C, respectively, then the following relations are set as

$$\text{Overall material balance: } W_A + W_B + W_C = 100, \quad \text{(a)}$$

$$\text{LOBS content balance: } 4.5W_A + 3.7W_B + 0W_C = 400, \quad \text{(b)}$$

$$\text{Constraints: } W_A \geq 10 \text{ or } W_A - s = 10, \quad \text{(c)}$$

$$W_B \geq 0, \quad \text{(d)}$$

$$W_C \geq 0, \quad \text{(e)}$$

$$\text{Total cost} = 2000W_A + 1500W_B. \quad \text{(f)}$$

Since there are three equations (a, b, and c) and four unknowns (W_A, W_B, W_C, and s), one unknown must be zeroed to get the solution. The number of possible solutions will be

$$\frac{4!}{3!(4-3)!} = 4.$$

Sol	W_A	W_B	W_C	s	Cost, Rs
1	10	95.95	−ve	0	Not feasible
*2	37.5	62.5	0	27.5	63,750
3	0	108.11	−ve	−10	Not feasible
4	88.89	0	11.11	78.89	177,980

*Lowest cost, and the blending weights are A = 37.5 tons and B = 62.5 tons. Stock C is not required.

15.9 MATERIAL AUDIT

This is a commercially accepted method of material accounting of any organisation (trading or manufacturing). This audit is a kind of material balances in terms of monetary value, reconciling the stocks (past and current), material purchased, and sold. Hence, it is inevitable that the materials must be identified by their category and price.

15.9.1 Category of Materials

Materials may be classified as capital and consumable materials. Capital materials are machines and equipment that are partly or wholly responsible for turning raw materials into finished goods. Consumable materials are identified into three broad classes: *raw material*, *intermediate* (or semi-finished goods), and *finished* goods.

Raw materials are the starting materials that are processed for separation or conversion or both separation and conversion to the desired components. For example, crude oil is the main raw material in refineries and naphtha in petrochemical plants. The quantity and price of the raw materials purchased must be recorded and stored in an identified storage location. *Intermediate* products, part of the finished product obtained through the processing of raw material(s), are also identified by code names and stored in an identified storage location. However, the price of intermediate products may not be assessed because of the complexity of the processing operation and the costs components. But, the quantity of these products produced must be recorded accurately. However, the cost of these products may be estimated and used in the *cost audit*. The *finished* products are the desirable materials finally produced and sold. These materials are also stored in identified storage locations before despatch for sales. The quantity and price must be ascertained before despatch. Finished products are also classified as the main products, joint products, co-products, and by-products. The *main products* are the high volume products, i.e., those that are generated in the largest *revenue* and that influence the organisational profit on a large scale. Of course, the main product is decided using other factors like national, social or political considerations. For instance, liquified petroleum gasoline (LPG), petrol, kerosene, diesel, and aviation turbine fuel (ATF) are the main products from refineries in India. The *joint products* are the materials produced simultaneously during the processing of the same raw material(s), e.g., LPG, naphtha, gasoline, kerosene, and diesel are the joint products from crude oil. The *co-products* are the materials of similar characteristics produced from different raw materials or different processing methods. For example, gasoline is a co-product from a reforming plant or a fluid catalytic cracking (FCC) plant. The by-products are the materials that are incidentally produced while the desirable finished products are obtained. For example, hydrogen is a by-product from a plat-reforming plant, whereas it is the main product from a steam reforming plant.

15.9.2 Papers to Be Maintained

Myriads of transactions take place, e.g., transfer of materials from one storage location to another, either within the same premises or to other locations in the organisation or to storage location or vessels in other organisations. All these transactions are always associated with paper records signed by the appropriate authorities of both the recipients and the suppliers.

The formats of these papers are designed and maintained uniquely and unanimously agreed by the organisations involved in the transactions. Some paper records are presented next as common examples. Two types of records are maintained—primary and secondary. In the primary records, dips and other details before and after each operation are maintained. Examples of primary records are the tank dip register and pass out vouchers. The secondary records are prepared based on the primary records. Daily stock, daily operation, daily pumping, monthly reports, annual reports, etc., are examples of secondary records.

15.9.2.1 Tank Dip Register

All the materials and liquid or liquified gases must be stored at identified locations with specific numbers. Dips of all these tanks are maintained every day, and every operation of receiving and transfer or despatch are recorded in a separate register known as the "tank dip register." Solid or semi-solid materials are also recorded in specific storages and are maintained in the same register.

15.9.2.2 Pass Out Vouchers

These are the valid gate passes of the vessels, wagons, trucks, etc., leaving after filling or loading, which contain the information, such as the name of the recipient organisation with the address, vessel number, product name, volume or mass loaded, temperature, density, etc. Such a voucher is displayed in Figure 15.1.

15.9.2.3 Tank Dip Memos

A tank dip memo is a record of the stock of any product after any receipt, transfer, or despatch. In order to have clarity in stock taking, it is essential to maintain the following modes of operation:

1. Receipts of any product should be kept preferably in a single tank from a distinct source, i.e., a particular supplier or company. The opening and closing dips with the temperature of the recipient tank must be taken before and after the receiving operation. The quantity received in the

PASS OUT VOUCHER

M/S XYZ Company Ltd Kolkata
(For removal of petroleum and petrochemicals from plant)

(For Tanker/Ship/Barge Despatches)

Date: 1/4/09

Time of issue: 0900 hrs

Address of owner: 99, Sovabazar lane, Kolkata-08

Address of consignee: M/S ABC Company, Chennai

Vessel no: WB0092V 05BB

Tank nos:	Product	Dips	Qty litres	Temp°C	Density at 15°C
A 2001	SRN	115.0	20000	20	0.6
B 2001	MS	120.5	25000	26	0.66
C 2007	SKO	122.0	18000	30	0.79
D 0081	HSD	130.5	20000	32	0.85

Signature of asst. engineer: M/S XYZ company ltd

Signature of the captain/officer vessel GAMA

FIGURE 15.1 Pass out voucher.

tank and the quantity transferred from the suppliers' vessel or tanks must be reconciled to be the same and agreed on. The dip memo is then jointly signed by the recipient and the supplier. It is also to be remembered that if multiple tanks are used for the receiving operation, a dip of all the tanks earmarked for receiving must be taken simultaneously before and after the receiving operations. Usually, this is applicable when a single product is received in a large quantity. No simultaneous receiving from any other source or despatch or transfer for processing is allowed during the receiving operation.

2. When any product is dispatched to a consignees' vessel or tank, a single tank is preferably used for the operation and the opening and closing dips with temperature must be taken before and after the despatch operation. No simultaneous receiving or transfer or despatch from other consignees is allowed. Multiple tanks may be used simultaneously for receiving large quantities of similar products provided that the opening and closing dips of all the tanks earmarked for this receiving operation is taken before and after the operation. This is also true for simultaneous dispatch operations.

3. During any transfer operation of a raw material and finished product, steps (1) and (2) must be followed and dip memos are prepared. However, dip memos are not required for the transfer operation of intermediate products not declared as finished products from one tank to another in the same premises of the organisation; of course, this is understood from the tank dip register as discussed earlier.

15.9.2.4 Daily Stock Report

The daily stock of raw materials and finished product are calculated based on tank dip register data and maintained on a separate sheet.

15.9.2.5 Daily Pumping Record

The daily processing rate of raw material and the rate of production of products (intermediate and finished) during 24-h (i.e., from 07:00 a.m. of the previous day to 07:00 a.m. the next day) are also recorded on a separate sheet.

15.9.2.6 Daily Operation Record

This is mainly a record of all despatch operations of products.

Exercise 15.4

A refinery has the processing scheme as shown in Figure 15.2.

From the daily tank dip register, the following stock data are available for a 24-h operation.

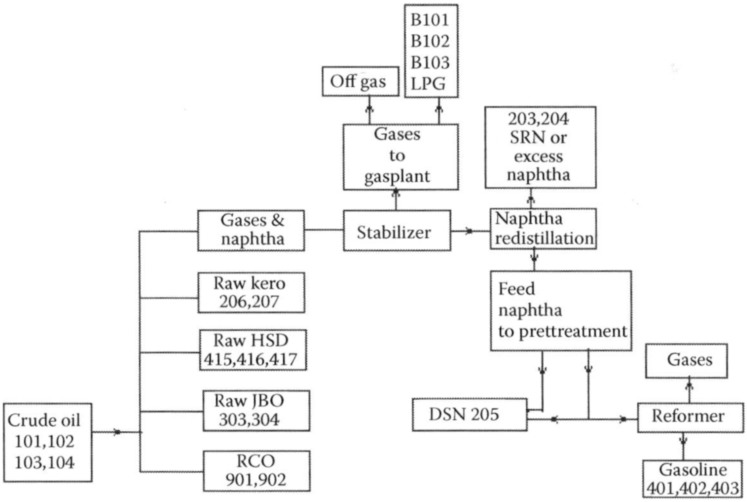

FIGURE 15.2 Processing scheme and storage locations of a typical refinery.

Dip Data as on February 9, 2007, at 7:00 a.m.

Tank No.	Product	Gross dip, cm	Water cut, cm	Temp, °C
101	Crude oil	900	50	30
102	Crude oil	850	10	32
103	Crude oil	700	10	30
104	Crude oil	750	20	30
B101	LPG	50	0	20
B102	LPG	50	0	20
B103	LPG	120	0	20
203	SRN	700	10	25
204	SRN	150	5	25
205	DSN	300	5	25
206	SRK	200	10	32
207	SRK	500	15	32
401	Gasoline	200	10	26
402	Gasoline	300	1	27
403	Gasoline	400	12	27
415	Diesel	200	20	30
416	Diesel	300	15	35
417	Diesel	400	10	32
303	JBO	300	12	32
304	JBO	300	10	35
901	RCO	200	0	120
902	RCO	200	0	105

Dip Data as on February 10, 2007, at 7:00 a.m.

Tank No.	Product	Gross dip, cm	Water cut, cm	Temp, °C
101	Crude oil	670.5	50	30
102	Crude oil	850	10	32
103	Crude oil	700	10	30
104	Crude oil	750	20	30
B101	LPG	210	0	20
B102	LPG	212	0	20
B103	LPG	163.6	0	20
203	SRN	700	10	25
204	SRN	355	5	25
205	DSN	505	5	25
206	SRK	460	10	32
207	SRK	500	15	32
401	Gasoline	280	10	26
402	Gasoline	300	1	27
403	Gasoline	400	12	27
415	Diesel	348	20	30

Dip Data as on February 10, 2007, at 7:00 a.m.

Tank No.	Product	Gross dip, cm	Water cut, cm	Temp, °C
416	Diesel	300	15	35
417	Diesel	400	10	32
303	JBO	442.5	12	32
304	JBO	300	10	35
901	RCO	604.5	0	120
902	RCO	200	0	105

Tank capacity factors (kL/cm) and volume reduction factor (VRF) to 15°C and density (ρ) at 15°C are listed below.

Tank No.	Product	kL/cm	Temp, °C	VRF	ρ at 15°C
101	Crude oil	38	30	0.9882	0.86
102	Crude oil	39	32	0.9812	0.86
103	Crude oil	37	30	0.9828	0.86
104	Crude oil	38	30	0.9820	0.86
B101	LPG	0.5	20	0.988	0.55
B102	LPG	0.5	20	0.988	0.55
B103	LPG	0.5	20	0.988	0.55
203	SRN	3.0	25	0.9854	0.65
204	SRN	3.0	25	0.9854	0.65
205	DSN	1.3	25	0.9854	0.65
206	SRK	9.0	32	0.9811	0.74
207	SRK	9.0	32	0.9811	0.74
401	Gasoline	8.0	26	0.9840	0.65
402	Gasoline	8.0	27	0.9825	0.65
403	Gasoline	8.0	27	0.9825	0.65
415	Diesel	8.0	30	0.9863	0.80
416	Diesel	8.0	35	0.9818	0.80
417	Diesel	8.0	32	0.9845	0.80
303	JBO	2.7	32	0.9864	0.85
304	JBO	2.7	35	0.9840	0.85
901	RCO	9.0	120	0.9248	0.90
902	RCO	9.0	105	0.9353	0.90

Determine the crude processed and the products generated, and make a reconciliation between the materials consumed and produced and also find the losses.

Solution

Crude throughput = (900 − 670.5) × 38 × 0.86 × 0.9882 = 7411.593 t,

LPG = (210 − 50) × 0.5 × 0.55 × 0.988 + (212 − 50) × 0.5 × 0.55 × 0.988
+ (163.6 − 120) × 0.5 × 0.55 × 0.988 = 99.33 MT,

$$SRN = (355 - 150) \times 3 \times 0.65 \times 0.9854 = 393.91365 \text{ MT},$$

$$\text{Gasoline} = (280 - 200) \times 8 \times 0.65 \times 0.9825 = 408.72 \text{ MT},$$

$$DSN = (505 - 300) \times 1.3 \times 0.65 \times 0.9854 = 170.69 \text{ MT},$$

$$SRK = (460 - 200) \times 9 \times 0.74 \times 0.9811 = 1{,}698.8728 \text{ MT},$$

$$HSD = (348 - 200) \times 8 \times 0.80 \times 0.9863 = 934.22336 \text{ MT},$$

$$JBO = (442.5 - 300) \times 2.7 \times 0.85 \times 0.9864 = 322.58979 \text{ MT},$$

$$RCO = (604.5 - 200) \times 9 \times 0.9 \times 0.9248 = 3{,}030.061 \text{ MT},$$

$$\text{Total production} = 99.33 + 393.91365 + 408.72 + 170.69 + 1{,}698.8728 + 934.22336 + 322.58979 + 3{,}030.061 = 7{,}058.400 \text{ t}.$$

Hence,

Crude processed = 7,411.593 MT,

Production = 7,058.400 t,

Loss and off gases (by balance) = 353.1924 t = 4.76% wt of crude.

(It is to be noted that the temperatures and water cuts were unchanged at the opening conditions, but variations of these will definitely occur in actual practice. As a result, the VRF and the densities will also change and the oil dip must be calculated by deducting the corresponding water cuts. Tank factors are not used in practice, but the calibration chart must be consulted for the respective tanks, to determine the actual volume corresponding to each dip.)

Exercise 15.5

The following data shows the receipts and issue of crude oil for processing in a refinery during a month.

Date	Transactions	Quantity in MT	Rate, $ per bbl
1/4/08	Opening stock	60,000	147
2/4/08	Received	20,000	148
2/4/08	Processed	10,000	
3/4/08	Processed	12,000	
4/4/08	Processed	12,000	
5/4/08	Processed	12,000	
6/4/08	Processed	10,000	
7/4/08	Processed	10,000	
8/4/08	Processed	12,000	
9/4/08	Received	35,000	45
9/4/08	Processed	12,000	

Date	Transactions	Quantity in MT	Rate, $ per bbl
10/4/08	Processed	12,000	
11/4/08	Received	20,000	40
11/4/06	Processed	10,000	
12/4/08	Processed	12,000	
13/4/08	Shut down	–	
14/4/08	Shut down	–	
15/4/08	Received	20,000	40
15/4/08	Shut down	–	
16/4/08	Shut down	–	
17/4/08	Processed	10,000	
18/4/08	Processed	12,000	
19/4/08	Shutdown	–	
20/4/08	Shutdown	–	
21/4/08	Shutdown	–	
22/4/08	Shutdown	–	
23/4/08	Shutdown	–	
24/4/08	Shutdown	–	
25/4/08	Shutdown	–	
26/4/08	Shutdown	–	
27/4/08	Received	30,000	50
28/4/08	Received	20,000	52
29/4/08	Processed	12,000	
30/4/08	Processed	12,000	
31/4/08	Processed	12,000	

Determine the stock of crude oil in quantity and price using

1. First in first out (FIFO) method
2. Last in first out (LIFO) method
3. Cost of crude oil processed during the month based on FIFO and LIFO methods

Solution

FIFO method of issue

Date	Balance b/d	Processed	Balance c/d
1/4/08	60,000@147	Nil	60,000@147
2/4/08	60,000@147	10,000@147	50,000@147
	20,000@148		20,000@148
3/4/08	50,000@147	12,000@147	38,000@147
	20,000@148		20,000@148
4/4/08	38,000@147	12,000@147	26,000@147
	20,000@148		20,000@148

(continued)

(Continued)

Date	Balance b/d	Processed	Balance c/d
5/4/08	26,000@147	12,000@147	14,000@147
	20,000@148		20,000@148
6/4/08	14,000@147	10,000@147	4,000@147
	20,000@148		20,000@148
7/4/08	4,000@147	4,000@147	14,000@148
	20,000@148	6,000@148	
8/4/08	14,000@148	12,000@148	2,000@148
9/4/08	2,000@148	2,000@148	25,000@45
	35,000@45	10,000@45	
10/4/08	25,000@45	12,000@45	13,000@45
11/4/08	13,000@45	10,000@45	3,000@45
	20,000@40		20,000@40
12/4/08	3,000@45	3,000@45	11,000@40
	20,000@40	9,000@40	
13/4/08	11,000@40		11,000@40
14/4/08	11,000@40		11,000@40
15/4/08	11,000@40		11,000@40
	20,000@40		20,000@40
16/4/08	11,000@40		11,000@40
	20,000@40		20,000@40
17/4/08	11,000@40	10,000@40	1,000@40
	20,000@40		20,000@40
18/4/08	1,000@40	1,000@40	9,000@40
	20,000@40	11,000@40	
19/4/08	9,000@40		9,000@40
20/4/08	9,000@40		9,000@40
21/4/08	9,000@40		9,000@40
22/4/08	9,000@40		9,000@40
23/4/08	9,000@40		9,000@40
24/4/08	9,000@40		9,000@40
25/4/08	9,000@40		9,000@40
26/4/08	9,000@40		9,000@40
27/4/08	9,000@40		9,000@40
	30,000@50		30,000@50
28/4/08	9,000@40		9,000@40
	30,000@50		30,000@50
	20,000@52		20,000@52
29/4/08	9,000@40	9,000@40	27,000@50
	30,000@50	3,000@50	20,000@52
	20,000@52		
30/4/08	27,000@50	12,000@50	15,000@50
	20,000@52		20,000@52
31/4/08	15,000@50	12,000@50	3,000@50
	20,000@52		20,000@52
Total		$110,377,740	

Total received with opening stock: 205,000 t ($118,433,520)
Total processed: 182,000 t ($110,377,740)
Balance stock: 23,000 t ($8,055,781)
Total of crude processed and the balance stock: 205,000 t ($118,433,521) (verified)
(*taking the API of crude as 36 for which 1 bbl of crude = 0.14772 t)

LIFO method of issue

Date	Balance b/d	Processed	Balance c/d
1/4/08	60,000@147	Nil	60,000@147
2/4/08	60,000@147	10,000@148	60,000@147
	20,000@148		10,000@148
3/4/08	60,000@147	10,000@148	58,000@147
	10,000@148	2,000@147	
4/4/08	58,000@147	12,000@147	46,000@147
5/4/08	46,000@147	12,000@147	34,000@147
6/4/08	34,000@147	10,000@147	24,000@147
7/4/08	24,000@147	10,000@147	14,000@147
8/4/08	14,000@147	12,000@147	2,000@147
9/4/08	2,000@147	12,000@45	2,000@147
	35,000@45		23,000@45
10/4/08	2,000@147	12,000@45	2,000@147
	23,000@45		11,000@45
11/4/08	2,000@147	10,000@40	2,000@147
	11,000@45		11,000@45
	20,000@40		10,000@40
12/4/08	2,000@147	10,000@40	2,000@147
	11,000@45	2,000@45	9,000@45
	10,000@40		
13/4/08	2,000@147		2,000@147
	9,000@45		9,000@45
14/4/08	2,000@147		2,000@147
	9,000@45		9,000@45
15/4/08	2,000@147		2,000@147
	9,000@45		9,000@45
	20,000@40		20,000@40
16/4/08	2,000@147		2,000@147
	9,000@45		9,000@45
	20,000@40		20,000@40
17/4/08	2,000@147	10,000@40	2,000@147
	9,000@45		9,000@45
	20,000@40		10,000@40
18/4/08	2,000@147	10,000@40	2,000@147
	9,000@45	2,000@40	7,000@45

(continued)

(Continued)

Date	Balance b/d	Processed	Balance c/d
	10,000@40		
19/4/08	2,000@147		2,000@147
	7,000@45		7,000@45
20/4/08	2,000@147		2,000@147
	7,000@45		7,000@45
21/4/08	2,000@147		2,000@147
	7,000@45		7,000@45
22/4/08	2,000@147		2,000@147
	7,000@45		7,000@45
23/4/08	2,000@147		2,000@147
	7,000@45		7,000@45
24/4/08	2,000@147		2,000@147
	7,000@45		7,000@45
25/4/08	2,000@147		2,000@147
	7,000@45		7,000@45
26/4/08	2,000@147		2,000@147
	7,000@45		7,000@45
27/4/08	2,000@147		2,000@147
	7,000@45		7,000@45
	30,000@50		30,000@50
28/4/08	2,000@147		2,000@147
	7,000@45		7,000@45
	30,000@50		30,000@50
	20,000@52		20,000@52
29/4/08	2,000@147	12,000@52	2,000@147
	7,000@45		7,000@45
	30,000@50		30,000@50
	20,000@52		8,000@52
30/4/08	2,000@147	8,000@52	2,000@147
	7,000@45	4,000@50	7,000@45
	30,000@50		26,000@50
	8,000@52		
31/4/08	2,000@147	12,000@50	2,000@147
	7,000@45		7,000@45
	26,000@50		14,000@50
Total		$109,572,160.00	$8,861,359.30

Cost of crude processed: $109,572,160.00, which is lower than the cost issued according to the FIFO method.

Total received with opening stock: 205,000 t ($118,433,521)

Total processed: 182,000 t ($109,572,160.00)

Balance stock: 23,000 t ($8,861,359.30)

Total of crude processed and the balance stock: 205,000 t ($118,433,519.30) (verified)

Exercise 15.6

A yearly report of a refinery presents the following information based on monthly reports.

Opening stock of crude: 50,000 MT as on April 1, 2007

Crude received during the year: 2,300,000 MT

Crude processed: 2,150,000 MT

Closing stock of crude is obtained from the crude tank dips taken on March 31, 2008:

Tank No.	Gross dip, cm	Water cut, cm	Temp, °C
101	900	10	30
102	950	5	32
103	900	5	32
104	850	0	30
105	800	0	30
106	750	0	31
107	800	10	32

Tank information and other data collected:

Tank no	kL/cm	Density at 15°C	Volume correction factor for temperature
101	38	0.86	0.9882
102	39	0.87	0.9812
103	37	0.87	0.9828
104	38	0.87	0.9880
105	39	0.85	0.9880
106	39	0.86	0.9882
107	40	0.87	0.9812

Solution

Tank No.	Kiloliter at tank temperature	Kiloliter corrected at 15°C	Metric ton, kiloliter at 15°C × density at 15°C
101	33,820	33,420.924	28,741.200
102	36,855	36,162.126	31,461.05
103	33,115	32,545.422	28,314.517
104	32,300	31,912.40	27,763.788
105	31,200	30,825.600	26,201.76
106	29,250	28,904.850	24,858.171
107	32,000	31,398.400	27,316.698
		Total	194,657.180

Closing stock: 194,657.180 MT

Closing stock should be = opening stock + receipts – processed = 50,000 + 2,300,000 – 2,150,000 = 200,000 MT.

Hence, loss of crude = 200,000 –194,657.180 = 5,342.82 MT, which is 0.24% of crude processed.

Exercise 15.7: The Price of Crude Oil and Products

From the following data, determine the different product costs (excluding taxes) of a refinery that processed 2,350,000 t per year.

	Rs/MT
Crude FOB (free on board) price	1700
Freight	95
Marine insurance	3
Ocean loss	11
Wharfage	20
Loading charges	35
Customs duty	600
Refining cost	75
Return on investment from market	70

Assume that the crude throughput for the refinery is 3,000,000 t per year.
Product yields:

Products	% wt yield
LPG	1.06
MS	4.33
Naphta	6.32
ATF	4.10
SKO	1.748
HSD	30.21
MTO	0.85
JBO	2.19
FO	15.65
CBFS	14.00
Bitumen	3.89
Slack wax	0.042
LOBS	5.61
Fuel and loss	10.00
Total	100.00

Considering the base product as kerosene, all the products costs are evaluated in terms of the kerosene equivalent. For this, the kerosene equivalence of all the other products is expressed in the following list where kerosene is taken as unity.

Products	Kerosene equivalent units
LPG	1.15
MS	0.95
Naphtha	0.90
ATF	1.10
SKO	1.00
HSD	0.95
MTO	1.10
JBO	1.00
FO	0.60
CBFS	0.80
Bitumen	0.70
Slack wax	0.70
LOBS	1.20

Solution

Expressing all the products rates and costs in terms of kerosene equivalent as listed in the table.

Products	Production rates, tons	Production rates in kerosene equivalent
LPG	25,000	28,750
MS	130,000	123,500
SRN	172,000	154,800
ATF	120,000	132,000
SKO	48,000	48,000
HSD	757,000	719,150
MTO	20,000	22,000
JBO	75,000	75,000
FO	509,000	305,400
CBFS	15,000	12,000
Bitumen	115,000	80,500
Wax	1,000	7,000
Lube stock	179,000	214,800
Total	2,166,000	1,922,900

Hence, the rate of production is 1,922,900 tons kerosene equivalent.
Total cost of crude oil is evaluated below,

	Rs/MT
Crude FOB price	1700
Freight	95
Marine insurance	3
Ocean loss	11
Wharfage	20
Loading charges	35
Customs duty	600
Refining cost	75
Return on investment from market:	70
Total	Rs 2490.89 per ton

Hence, the total cost is Rs 2490.89 × 2,350,000 = 5,853,591,500.

So, the cost per ton of equivalent kerosene = 5,853,591,500/1,922,900 = Rs 3,044.1.

The cost of all the products is then evaluated using this price of equivalent kerosene as the product.

Products	Price in Rs/ton
LPG	$1.15 \times 3044.1 = 3500.71$
MS	$0.95 \times 3044.1 = 2891.89$
Naphtha	$0.90 \times 3044.1 = 2739.69$
ATF	$1.10 \times 3044.1 = 3348.51$
SKO	$1.00 \times 3044.1 = 3044.10$
HSD	$0.95 \times 3044.1 = 2891.89$
MTO	$1.10 \times 3044.1 = 3348.51$
JBO	$1.00 \times 3044.1 = 3044.10$
FO	$0.60 \times 3044.1 = 1826.46$
CBFS	$0.80 \times 3044.1 = 2435.28$
Bitumen	$0.70 \times 3044.1 = 2130.87$
Slack wax	$0.70 \times 3044.1 = 2130.87$
LOBS	$1.20 \times 3044.1 = 3652.92$

Ans.

15.9.3 Material Audit of Capital Goods

So far, the previous discussions have been about consumable goods. Pumps, compressors, stirrers, motors, various drives, columns, and vessels are examples of capital goods. These are recorded in books giving details of specifications, values, date of installation, location, etc. Book values of these are maintained by deducting the amount of the annual depreciation value applicable for each of these capital goods. Small pumps, compressors, bearings, nut, bolts, screws, pipes, rods, motors, wheel keys, helmets, socks, gloves, safety shoes, recording papers, etc., are various materials that are also treated as consumable goods. These are kept in stores in specified drawers or a storage area, tagged with a bin card number containing details of the items, values, and quantity date wise. Valuation of these items are also carried out monthly based on daily receipts, issues, returns and balance of these items. Standard methods of issue, e.g., FIFO, LIFO, weighted average, etc., are applicable. A physical audit is also carried out monthly or annually to check the balance quantities with the book figures.

The maximum material (excluding the capital goods and spares) loss in terms of crude oil and its products is typically 1% of the crude throughput in a year.

15.10 ENERGY AUDIT

Energy is consumed in any plant in the form of electricity and heat, which are required to drive the machineries and to maintain the necessary operating conditions in the processing equipment. Electricity is either purchased from external power plants or generated within the plant (captive power plant). For refineries and petrochemical plants, a huge amount of steam is essential for the process operations. A captive power plant also serves the need for electricity and steam. Hence, the energy audit incorporates entities such as fuel, steam, and electricity. The lighting and motor drives of a plant consume a substantial amount of electricity. According to the Energy Conservation Act 2001, maximum limits are specified (depending on the size of the plant) above which a penalty has to be paid by the owners of the plant. Many industries are aware of the benefits of conservation measures and have been improving their productivity in recent years. A regular energy audit is carried out by their own team in order to assess the performance using various options of conservation, e.g., no cost, low cost, and high value options. At the no cost option, operating conditions are changed without any investment, which leads to savings in electricity or heat loss. In fact, simple tips of good housekeeping at no cost can save crores of rupees through savings on fuel and power. At the low cost option, less expensive methods can reduce power and heat consumption. Low cost modifications in the process and the replacement of old equipment may save an enormous amount of heat and power. The high value option will include major modifications or the replacement of an existing process plant. For example, switching a thermal cracking unit to a catalytic cracking unit, the change of a catalyst of an existing reactor, etc., can reduce power consumption. Another classic example of a high value option is replacing a traditional solvent dewaxing plant with a catalytic dewaxing plant, where power consumption of the former is about three times that of the latter.

15.10.1 Electricity Audit

A list of all the motor drives are recorded with their details, e.g., requisite supply voltage, frequency, rated consumption mentioned by the suppliers, day-to-day monitoring of the running hours, power consumption in terms of kilowatt-hour, and also the rate of current consumption. The no-load voltage or current must also be checked. Deviations of all these parameters must be checked daily. For audit purposes, total power consumption in a month or year must be noticed and checked against a stipulated limit. The total power consumption in the lighting of the plant and its buildings must be recorded and presented monthly or yearly. The electricity generation for a captive power plant per day, month, or year must be recorded. An inspection of the total generation must be done along with the consumption pattern.

15.10.2 Thermal Audit

Thermal energy is obtained primarily from the combustion of fuels in the furnaces, which include the process heaters (furnaces) and also the furnaces of the captive power plant. The heat generated from a chemical reaction is also obtained as waste heat. Hence, it is essential to record the amount of fuel consumed monthly or yearly. Total heat energy is then calculated by multiplying the net heating value (NHV) of the fuel and the waste heat generated by exothermic chemical reaction or combustion in catalyst regenerators (as in FCC or hydrocracker). Thermal energy consumption is then compared with the amount of steam generated and the heat recovered in the preheaters using steam and hot products. The difference is the loss. The maximum limit of loss of thermal energy is counted by the fuel consumption in a month or a year. For example, in a typical refinery, the maximum fuel consumption is 9% of the crude throughput in a year. Thus, the total fuel consumption and loss of consumable products is 10% of the throughput of a refinery.

15.10.3 Steam Balance

Steam generated in a power plant is high pressure (HP) superheated steam above a pressure of 50–60 kg/cm^2 and at a temperature of around 450°C. This steam is used to drive the turbine for power generation and is finally condensed to liquid water. Additional steam is generated in the boiler, which is converted to medium (MP) and low pressure (LP) steam. MP steam is used for stripping and distillation and also for driving small steam-driven pumps and compressors. LP steam is mainly used for heating purposes in the reboilers and steam-tracing lines. A balance of these HP, MP, and LP steam is carried out by using their respective enthalpies and quantities. It is to be mentioned that even though the mass of steam (possessed by the weight of water) of a certain quality (HP, MP, or LP) may be unchanged, the enthalpy may be changed and will cause loss of steam. Loss of steam may also occur materially to the surroundings. Hence, the steam balance is also associated with the condensate balance. Thus, the water condensed from the turbine, process heaters, reboilers and steam coils in the tanks and vessels, steam-tracing coils, etc., where steam flows in a closed system (conduits) and does not come in contact with the process fluids, is recovered and recycled as the boiler feed water. The difference between the amount of boiler feed water and the recycled condensates is the loss of boiler feed water. It is to be noted that the steam balance involves both the material and heat balance of water used in steam generation. A steam generation, distribution and consumption in a refinery is shown in Figure 15.3.

Exercise 15.8

From the following information, determine the annual electric consumption, fuel consumption, steam balance, condensate recovered, and makeup boiler feed water requirement.

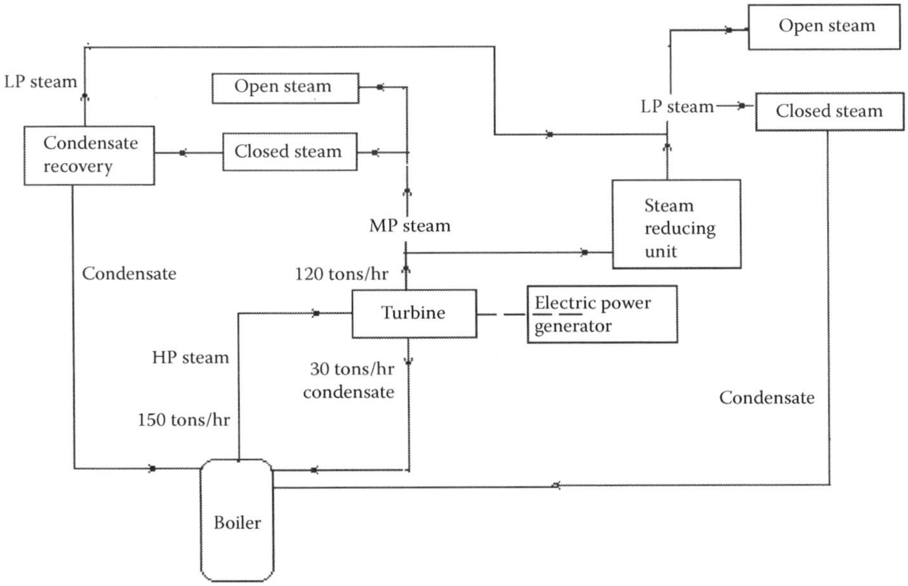

FIGURE 15.3 Steam generation, distribution, and consumption.

Annual Electric Consumption Data

No. of motors	Power per motor, kW	Running hours
10	20	7,000
20	35	6,000
40	25	5,000
5	75	6,000
10	100	5,000
No. of bulbs	**Power per bulb, W**	
100	100	3,000
200	60	4,000
50	250	1,000
No. of arc welding receptacles	**Power per point, W**	
100	1000	500

Annual Fuel Consumption Data

No. of furnaces	Fuel, kg/h per furnace	NHV, kJ/kg of fuel	Running hours
10	5	10,000	7,200
10	12	9,000	7,000
30	4	12,000	7,000

(continued)

Annual Fuel Consumption Data (Continued)

No. of furnaces	Fuel, kg/h per furnace	NHV, kJ/kg of fuel	Running hours
5	10	11,000	7,200
20	3	10,000	7,000
30	1	11,000	7,000

Annual Open Steam Consumption Data

Distillation, MP steam, ton/h	Stripping, MP steam, ton/h	Extraction, MP steam, ton/h	Quenching, MP steam, ton/h	Dewaxing unit, LP steam, ton/h
5.0	4.0	3.0	5.0	15.0

Annual Closed Steam Data for 7200 H per Year

Power plant boilers HP steam generation, ton/h	For power plant turbine, HP to MP, ton/h	Reboilers, ton/h LP steam to condensate	MP steam drives, ton/h to condensate	LP steam coils and tracing lines, ton/h to condensate
150	120	15	50	20

Solution

Annual Electric Consumption

No. of motors	Power per motor, kW	Running hours	kW-h
10	20	7,000	1,400,000
20	35	6,000	4,200,000
40	25	5,000	5,000,000
5	75	6,000	2,250,000
10	100	5,000	5,000,000
No. of bulbs	**Power per bulb, W**		
100	100	3,000	30,000,000
200	60	4,000	48,000,000
50	250	1,000	12,500,000
No. of arc welding receptacles	**Power per point, W**		
100	1000	500	50,000,000
		Total	1.42285×10^8

Hence, the annual electric power requirement = 1.42285×10^8 kW-h.

If the average running hours for a plant in a year is 7200 h, then the required power is $1.42285 \times 10^8/7200 = 19.76$ MW.

Annual Fuel Consumption

No. of furnaces	Fuel, kg/h per furnace	NHV, kJ/kg of fuel	Running hours	kJ
10	5	10,000	7,200	3.6×10^9
10	12	9,000	7,000	7.56×10^9
30	4	12,000	7,000	10.08×10^9
5	10	11,000	7,200	3.96×10^9
20	3	10,000	7,000	4.2×10^9
30	1	11,000	7,000	2.31×10^9
			Total	31.71×10^9

Hence, total thermal power consumption = 31.71×10^9 kJ per year.

Thus, the annual fuel consumption taking a standard fuel of NHV of 10,000 kJ/kg is $31.71 \times 10^9/10{,}000 = 3171$ tons.

Annual Open Steam Consumption

Distillation, MP steam, ton/h	Stripping, MP steam, ton/h	Extraction, MP steam, ton/h	Quenching, MP steam, ton/h	Dewaxing unit, LP steam, ton/h
5.0	4.0	3.0	5.0	15.0
Enthalpy, kcal/kg: 15 kg/cm², 260°C	Enthalpy, kcal/kg: 15 kg/cm², 260°C	Enthalpy, kcal/kg: 15 kg/cm², 260°C	Enthalpy, kcal/kg: 15 kg/cm², 260°C	Enthalpy, kcal/kg: 5 kg/cm², 160°C
Vap: 703; Con: 200	Vap:703; Con: 200	Vap:703; Con: 200	Vap:703; Con: 200	Vap: 661; Cond: 152

Annual Closed Steam Data for 7200 h per Year

Power plant boilers HP steam generation, ton/h	For power plant turbine, HP to MP, ton/h	Reboilers, ton/h, LP steam to condensate	MP steam drives, ton/h, to condensate	LP steam coils and tracing lines, ton/h, to condensate
150	120	15	50	20
Enthalpy, kcal/kg: 60 kg/cm², 460°C	Enthalpy, kcal/kg: 60 kg/cm², 460°C	Enthalpy, kcal/kg: 5 kg/cm², 160°C	Enthalpy, kcal/kg: 15 kg/cm², 260°C	Enthalpy, kcal/kg: 5 kg/cm², 160°C
Vap: 794; Cond: 288	Vap: 794; Cond: 288	Vap: 661; Cond: 152	Vap: 703; Con: 200	Vap: 661; Cond: 152

Mass balance of steam	
Steam used in reboilers	15 ton/h
Steam drives	50 ton/h
Steam for heating	20 ton/h
Total closed steam	85 ton/h
Process steam	
For distillation	5 ton/h
For stripping	4 ton/h
For extraction	3 ton/h
For quenching	5 ton/h
For dewaxing	15 ton/h
Total open steam	32 ton/h
Total open + closed steam	117 ton/h
Total of steam from power plant turbine	120 ton/h
Mass of steam loss	3 ton/h
Heat balance	
Steam used in reboilers (LP)	15 × (661 – 152) = 7,635 × 1,000 kcal/h
Steam drives (MP)	50 × (703 – 200) = 25,150,000
Steam for heating (LP)	20 × (661 – 152) = 10,180
Total closed steam	42,965,000 kcal/h
Process steam	
For distillation (MP)	5 × (703 – 200) = 2,515,000
For stripping (MP)	4 × (703 – 200) = 2,012,000
For extraction (MP)	3 × (703 – 200) = 1,509,000
For quenching (MP)	5 × (703 – 200) = 2,515,000
For dewaxing (LP)	15 × (661 – 152) = 7,635,000
Total open steam	16,186,000 kcal/h
Total open + closed steam	59,151,000 kcal/h
Total of steam from power plant turbine	120 × (703 – 200) = 60,360,000 kcal/h
Thermal loss	1,209,000 kcal/h
Equivalent loss of HP steam	1,209,000/(794 – 288) = 2,389 kg/h = 2.389 ton/h
Total condensate recovered	Condensate from turbine + condensate from the closed process steams = 20 + 75 = 95 ton/h
Total make up water (boiler feed) required	(150 – 95) = 55 ton/h
Annual steam loss	2.389 × 330 × 24 = 1,892,088 tons of HP steam
Annual water loss	3 × 330 × 24 = 23,760 ton
Annual make up water required for boiler	55 × 330 × 24 = 435,600 ton

It is assumed that condensates from the open steam from the process units are not suitable for recycling as boiler feed water, considering the possibility of contamination with hydrocarbons. If, however, high quality filtration is possible, a substantial amount of condensate could be recycled as boiler feed water.

Appendix

TABLE A1
Viscosity Data of Reference Oils for Viscosity Index Measurements

cSt at 98.8°C	H at 37.8°C	L at 37.8°C	cSt at 98.8°C	H at 37.8°C	L at 37.8°C
2	6.62	8.36	4.5	25.04	33.7
2.1	7.143	9.043	4.6	26	35.35
2.2	7.684	9.752	4.7	26.98	37.06
2.3	8.243	10.485	4.8	27.98	38.84
2.4	8.821	11.244	4.9	29	40.68
2.5	9.417	12.028	5	30.04	42.57
2.6	10.031	12.838	5.1	31.09	44.5
2.7	10.664	13.672	5.2	32.15	46.46
2.8	11.315	14.532	5.3	33.21	48.44
2.9	11.984	15.417	5.4	34.27	50.43
3	12.671	16.328	5.5	35.33	52.43
3.1	13.377	17.263	5.6	36.39	54.43
3.2	14.101	18.224	5.7	37.45	56.43
3.3	14.843	19.21	5.8	38.51	58.43
3.4	15.603	20.222	5.9	39.57	60.43
3.5	16.382	21.258	6	40.63	62.43
3.6	17.179	22.32	6.1	41.69	64.43
3.7	17.994	23.407	6.2	42.75	66.43
3.8	18.828	24.52	6.3	43.81	68.43
3.9	19.68	25.657	6.4	44.88	70.43
4	20.55	26.82	6.5	45.97	72.46
4.1	21.4	25.06	6.6	47.08	74.55
4.2	22.28	29.36	6.7	48.22	76.74
4.3	23.18	30.73	6.8	49.39	79.04
4.4	24.1	32.18	6.9	50.59	81.44
7	51.82	83.92	9.5	82.714	148.695
7.1	52.98	86.46	9.6	83.986	151.411
7.2	54.15	89.04	9.7	85.262	154.147
7.3	55.273	91.66	9.8	86.575	156.982
7.4	56.473	94.095	9.9	87.856	159.722
7.5	57.669	96.528	10	89.178	162.494
7.6	58.873	98.958	10.1	90.458	165.361
7.7	60.063	101.938	10.2	91.814	168.303
7.8	61.305	103.925	10.3	93.128	171.194
7.9	62.513	106.388	10.4	94.461	174.075

(continued)

TABLE A1 (CONTINUED)
Viscosity Data of Reference Oils for Viscosity Index Measurements

cSt at 98.8°C	H at 37.8°C	L at 37.8°C	cSt at 98.8°C	H at 37.8°C	L at 37.8°C
8	63.723	108.859	10.5	95.825	177.068
8.1	64.969	111.49	10.6	97.152	179.98
8.2	66.251	114.067	10.7	98.492	182.907
8.3	67.501	116.65	10.8	99.818	185.849
8.4	68.753	119.306	10.9	101.206	188.873
8.5	70.041	121.926	11	103.909	191.848
8.6	71.296	124.528	11.1	103.909	194.899
8.7	72.542	127.153	11.2	105.305	197.966
8.8	73.793	129.786	11.3	106.721	201.15
8.9	75.089	132.515	11.4	108.1	204.238
9	76.352	135.176	11.5	109.493	207.86
9.1	77.617	137.841	11.6	110.887	210.468
9.2	78.88	140.517	11.7	112.273	213.601
9.3	80.184	143.284	11.8	113.711	216.829
9.4	81.448	145.989	11.9	115.114	219.981
12	116.507	223.145	14.5	154.124	310.749
12.1	117.948	226.412	14.6	155.708	314.54
12.2	119.438	229.784	14.7	157.255	318.247
12.3	120.883	233.078	14.8	158.804	321.968
12.4	122.33	236.392	14.9	160.396	325.801
12.5	123.781	239.713	15	161.95	329.549
12.6	125.274	243.136	15.1	163.548	323.41
12.7	126.73	246.482	15.2	165.19	337.385
12.8	128.189	259.842	15.3	166.793	341.275
12.9	129.689	253.305	15.4	168.399	345.179
13	131.153	256.69	15.5	170.007	348.881
13.1	132.658	260.18	15.6	171.66	353.131
13.2	134.166	263.684	15.7	173.274	357.078
13.3	135.716	267.293	15.8	174.891	361.039
13.4	137.23	270.824	15.9	176.552	365.117
13.5	138.745	274.369	16	178.174	369.107
13.6	140.27	277.964	16.1	179.841	373.214
13.7	141.784	281.498	16.2	181.552	377.439
13.8	143.348	285.177	16.3	183.224	381.577
13.9	144.874	288.776	16.4	184.952	385.729
14	146.502	292.388	16.5	189.577	389.897
14.1	147.933	296.014	16.6	188.3	394.184
14.2	149.507	299.749	16.7	189.983	398.381
14.3	151.043	303.402	16.8	191.67	402.594
14.4	152.582	307.069	16.9	193.401	406.929
17	195.094	411.172	19.5	239.906	526.274
17.1	196.831	415.537	19.6	241.806	531.247

TABLE A1 (CONTINUED)
Viscosity Data of Reference Oils for Viscosity Index Measurements

cSt at 98.8°C	H at 37.8°C	L at 37.8°C	cSt at 98.8°C	H at 37.8°C	L at 37.8°C
17.2	198.571	419.917	19.7	243.666	536.164
17.3	200.357	424.421	19.8	245.529	541.075
17.4	202.104	428.832	19.9	247.44	546.12
17.5	203.853	433.26	20	249.31	551.07
17.6	205.605	437.704	20.1	253.1	561.12
17.7	207.361	442.161	20.4	256.86	571.13
17.8	209.162	446.745	20.6	260.59	581.08
17.9	210.923	451.237	20.8	264.64	591.94
18	213	455.743	21	269.26	601.66
18.1	214.35	460.266	21.2	272.35	612.67
18.2	216.268	464.915	21.4	275.99	622.52
18.3	218.042	469.469	21.6	280.1	633.67
18.4	219.818	474.039	21.8	284.22	644.8
18.5	221.597	478.625	22	287.9	654.94
18.6	223.423	483.339	22.2	292.05	666.3
18.7	225.208	487.956	22.4	296.22	677.75
18.8	226.996	492.59	22.6	299.96	687.98
18.9	228.831	497.332	22.8	304.13	699.57
19	230.625	502.017	23	308.34	711.24
19.1	232.466	506.812	23.2	312.09	721.67
19.2	234.354	511.739	23.4	316.32	733.47
19.3	236.201	516.568	23.6	320.57	745.35
19.4	238.052	521.413	23.8	324.36	755.98
24	328.63	768	29	438.1	1087
24.2	332.45	778.76	29.2	442.8	1101
24.4	336.75	790.92	29.4	446.9	1113.5
24.6	341.06	803.17	29.6	451.6	1127.6
24.8	345.4	815.49	29.8	456.2	1141.8
25	349.3	826.5	30	360.9	1156
25.2	353.6	839	30.5	472.8	1192
25.4	358	851.5	31	484.1	1226
25.6	362.4	864.1	31.5	496.1	1263.7
25.8	366.8	878.8	32	508.2	1301.1
26	371.2	889.6	32.5	520.4	1338.9
26.2	375.1	901	33	532.6	1377.2
26.4	379.6	913.9	33.5	544.9	1416
26.6	384	926.9	34	557.3	1455.3
26.8	388.5	939.9	34.5	569.9	1495
27	393	953.1	35	582.4	1535.2
27.2	397.5	966.3	35.5	595.8	1577.7
27.4	402	979.6	36	608.5	1618.9

(continued)

TABLE A1 (CONTINUED)
Viscosity Data of Reference Oils for Viscosity Index Measurements

cSt at 98.8°C	H at 37.8°C	L at 37.8°C	cSt at 98.8°C	H at 37.8°C	L at 37.8°C
27.6	406	991.4	36.5	621.4	1660.6
27.8	410.6	1004.9	37	643.3	1702.7
28	415.1	1018.4	37.5	647.4	1745.3
28.2	419.7	1032	38	660.5	1788.3
28.4	424.3	1045	38.5	674.4	1833.9
28.6	428.9	1059.4	39	687.7	1877.9
28.8	433.5	1073.2	39.5	701.1	1922.4
40	714.6	1967.4	52.5	1090.9	3280.4
40.5	728.3	2013.1	53	1107.1	3339
41	741.9	2058.7	53.5	1123.9	3399.9
41.5	756.1	2106.4	54	1140.6	3460.8
42	770.2	2154.1	54.5	1157.2	3521.4
42.5	784.2	2201.7	55	1173.7	3582
43	798.2	2249.3	55.5	1190.5	3643.6
43.5	812.7	2299	56	1207.2	3705.2
44	827.2	2348.6	56.5	1224.2	3767.8
44.5	842.2	2400.4	57	1241.1	3830.4
45	857.2	2452.1	57.5	1258.6	3895.4
45.5	872.1	2503.8	58	1258.6	3895.4
46	886.9	2555.4	58.5	1293.5	4025
46.5	902	2608.1	59	1310.8	4089.6
47	917.1	2660.7	59.5	1328.3	4155.2
47.5	932.7	2715.5	60	1345.8	4220.7
48	948.2	2770.3	60.5	1363.9	4288.8
48.5	963.7	2825	61	1382	4356.8
49	979.1	2879.6	61.5	1399.9	4424.4
49.5	994.8	2935.2	62	1414.7	4492
50	1010.4	2990.8	62.5	1435.8	4560.6
50.5	1026.2	3047.4	63	1453.9	4629.1
51	1042	3104	63.5	1472.6	4700.2
51.5	1058.4	3162.9	64	1491.2	4771.3
52	1074.7	3221.8	64.5	1509.7	4841.9
65	1528.1	4912.4			
65.5	1546.8	4984			
66	1565.4	5055.5			
66.5	1584.7	5129.7			
67	1603.9	5203.8			
67.5	1622.9	5277.4			
68	1641.9	5350.9			
68.5	1661.1	5425.5			
69	1680.3	5500			
69.5	1700.1	5577.2			

TABLE A1 (CONTINUED)
Viscosity Data of Reference Oils for Viscosity Index Measurements

cSt at 98.8°C	H at 37.8°C	L at 37.8°C
70	1719.9	5654.4
70.5	1739.5	5731
71	1759.1	5807.5
71.5	1779.3	5776.8
72	1799.5	5966
72.5	1819.5	6044.6
73	1839.4	6123.1
73.5	1860	6204.4
74	1880	6285.6
74.5	1901	6366.2
75	1921.3	6446.7

Note: H and L are the viscosities at 37.8°C of the oils whose VI are 100 and 0, respectively.

TABLE A2
Viscosity Blending Index

cSt	BI	cSt	BI	cSt	BI	cSt	BI	cSt	BI
2	20.54	46	30.56	180	34.93	900	38.85	18,000	44.15
4	22.78	48	30.71	190	35.08	1,000	39.07	20,000	44.3
6	24.08	50	30.86	200	35.22	1,200	39.45	25,000	44.63
8	25	55	31.2	220	35.48	1,400	39.76	30,000	44.89
10	25.73	60	31.51	240	35.71	1,600	40.03	35,000	45.1
12	26.315	65	31.79	260	35.92	1,800	40.26	40,000	45.59
14	26.811	70	32.04	280	36.11	2,000	40.46	60,000	45.83
16	27.24	75	32.27	300	36.29	2,500	40.88	70,000	46.03
18	27.62	80	32.48	320	36.45	3,000	41.21	80,000	46.21
20	27.96	85	32.68	340	36.61	3,500	41.49	90,000	46.63
22	28.26	90	32.87	360	36.75	4,000	41.73	100,000	46.49
24	28.54	95	33.04	380	36.88	4,500	41.93	150,000	47
28	29.04	110	33.49	400	37	5,000	42.11	200,000	47.34
30	29.26	120	33.76	420	37.12	6,000	42.42	250,000	47.61
32	29.5	130	34	480	37.44	7,000	42.68	300,000	47.82
34	29.67	140	34.21	500	37.53	8,000	42.89	400,000	48.15
36	29.85	150	34.41	600	37.95	12,000	43.53	500,000	48.39
38	30.02	160	34.6	700	38.3	14,000	43.77		
40	30.19	170	34.77	800	38.59	16,000	43.97		

TABLE A3
Dimensions of some Steel Pipes of 40 and 80 Schedule Numbers

Nominal Pipe size, in	Schedule No.	o.d., in	i.d., in	Wall Thickness, in
1/8	40	0.405	0.269	0.068
1/8	80	0.405	0.215	0.095
1	40	1.315	1.049	0.133
1	80	1.315	0.957	0.179
2	40	2.375	2.067	0.154
2	80	2.375	1.939	0.218
4	40	4.5	4.026	0.237
4	80	4.5	3.826	0.337
8	40	8.625	7.981	0.332
8	80	8.625	7.625	0.500
12	40	12.75	11.938	0.406
12	80	12.75	11.374	0.688

Index

A

Acrylonitrile butadiene styrene, 116
Addition polymerisation, 103
Adipic acid, 118
Air preheater, APH, 193
Alkenes. *See* olefins.
Amine washing of gases, 55
Anticline, 8
Antifoaming agent, 84
Aniline point, 38
Antioxidants, 83
API gravity, 6, 27
API separator, 140
Aromatics, 3
ASTM distillation, 34
ATF, xi, 39
Atmospheric distillation, 52
Atmospheric distillation control, 308
Automotive fuels, 32
Automotive lubricants, 84
Aviation fuels, 39
Aviation lubricants, 87

B

Batch reactor, 252
Basin, 9
Banbury mixer, 128
Batch distillation, 207
Batch extraction, 239
Battery limit, 131
Bearing lubricants, 85
Bitumen, 46
Bitumen blowing, 79
Biological treatment, 141
Bloom, 31
Blow moulding, 126
Boiling point diagram, 208
Bottom temperature, 312
Breakeven point, 345
BSW, 7, 28
Bullet, 136
Burners, 191

C

Cables, 297
Calcium grease, 91
Capacity ratio method of cost, 343
Caprock, 8
Catalytic processes for LOBS manufacture, 70
CDU startup, 324
Cetane number, 37
Centrifugal pumps, 281
Characterisation factor, 7, 27
CFR engine, 35
Char point, 31
Christmas tree, 15, 16
CI engine, 32
Circulating reflux, 230
Classes of products, 352
Cloud point, 43
Coal bed methane, 7, 22
Coking, 80
Components of lubricants, 83
Composition of crude oil, 1
Compressor lubricant, 86
Connate water, 13
Continuous distillation, 210
Continuous extraction, 240
Control hardware, 297
Control loops, 299
Control room, 305
Control software, 301
Copper strip corrosion, 32
Cost of equipment, 343
Cost of a plant, 344
Colburn number, 175
Compressors, 289
Compression cycle, 290
Condensation polymerisation, 103
Controllers, 299
Convection zone, 192
Cooling tower, 147
Corrosion protection, 276
Crankcase dilution, 34
Critical solution temperature, 238
Crown, 14
Crude conditioning and storage, 19
Crude oil analysis, 25
Crude oil receiving, 49
Crude oil transfer lines, 277
Crude throughput control, 305
CSTR reactor, 252
Cumene, 121

D

DEA, 55
Debutaniser, 62
De-ethaniser, 55, 105
Definition of terms related to extraction, 236
Definitions of petrochemicals, 101
Dehydrocyclisation, 61
Dehydrogenation, 61
Demetalation, 60
Demethaniser, 105
Denitrogenation, 60
Depreciation, 344
Depropaniser, 107
Derrick, 14
Desalter control, 306
Desalting of crude oil, 49
Desulfurisation, 59
Detergents, 83
Diesel hydrodesulfurisation (DHDS), 63
Diesel index, 37
Dispersants, 83
Distillation and stripping, 207
Distributed control system, 304
DMT, 113
Doctor's test, 30
Dodecyl lactum, 119
Domestic fuels, 28
DOW index, 338
Downstream units, 122
Draw plate, 310

E

Economic pipe diameter, 279
Energy audit, 367
Energy balance in a plant, 164
Enthalpy balance method, 219
Environmental impact of lubricants, 90
Equilibrium diagram, 209
Ethane–ethylene separator, 107
Ethylene glycol, 114
ETP, 140
Extended surface exchanger, 185
External refluxes, 229
Exploration techniques, 9
Extraction principle, 235
Extraction process, 236
Extrusion moulding, 124

F

Factories Act, 335
Fail-safe devices, 319
FCC reactor, calculations, 271
FCC start up, 328
FCC unit control, 314
Feedback control, 300
Feed plate, 314
Feed stock, 101
Final control elements, 299
Finished products, 352
Fire and explosion, 333
Fittings and supports, 275
Flame height, 38
Flare, 143
Flash point, 38
Flash zone pressure, 312
Floating roof tank, 50
Flow reactor, 255
Fluid catalytic cracking, 76
Fluidisation, 77, 273
Formaldehyde, 120
Fouling, 177
Friction modifiers, 83
Froud number, 95
Furfural extraction, 66
Furnace control, 312
Furnace fuels, 39

G

Gap and overlap, 230
Gas hydrates, 7, 21
Gas lift well, 16
Gas transfer lines, 280
Gaseous fuels, 39
Gear pumps, 288
Geochemical methods, 11
Geothermal gradient, 8, 13
Glass transition temperature, 123
Gravimetric method, 9
Greases, 88
Grease manufacture, 89

H

HAZAN, 337
HAZOP, 339
HDPE, 109
Heat exchangers, 171
Heat exchanger train, 186
Heater pipes, 190
Hexamethylene diamine, 119
Horton sphere, 138
HSD, xi, 36
Hydraulic lubricant, 86
Hydraulic pumping, 18
Hydrocracking, 71
Hydrofinishing, 69
Hydrogen generation, 74
Hydrogen separator, 76
Hydrogenation, 61

I

IFO circulation, 145
Igneous rock, 10
Industrial lubricants, 85
Injection moulding, 126
Intermediates, 101
Internal combustion engine, 33
Internal reflux, 230
Ion exchange resins, 148
Isomerisation, 61, 72

J

JBO, 44

K

Kerogen, 8
Kerosene, 30
Kerosene hydrodesulfurisation, 62
Kitchen, 8

L

Layout of petrochemical plant, 133
Layout of petroleum refinery, 132
LDPE, 108
Level control, 309
Linear programming, 349
Liquid fuels, 40
LLDPE, 110
Lobo and Evans method, 205
LOBS, 83
LPG, xi, 28
LPG splitter, 58
Lube blending, 89
Lubricating oils, 40

M

Magnetometric method, 10
Maintenance of plant and equipment, 332
Marine lubricants, 88
Material audit, 352
Material balance in a plant, 158
McCabe–Thiele method, 217
Measurement of quantity of crude oil and products, 151
Measurement of gases in closed vessels, 158
Melamine, 121
Melamine formaldehyde, 120
Meroxing and caustic washing, 56
Mercaptans, 4
Methanol, 120
Migration of oil, 8
Mild hydrocracking, 73
Minimum reflux, 219
Miscellaneous products, 44
Motor spirit, 32
Moulding of plastics, 124
MTO, 44

N

Naphtha cracking, 103
Naphtha pretreatment, 59
Naphtha pretreater startup, 327
Naphtha redistillation, 58
Naphtha reformer calculations, 269
Naphthenes, 3
Natural draft furnace, 194
Newtonian fluids, 13
NLGI specification, 90
NMP extraction, 66
NPSH, 284
Nusselt number, 174
Nylons, 117

O

Octane number, 34
Offsite facilities, 134
Oil field development, 14
Oil production processes, 16
Oil rig, 14
Oil shales, 7
Olefins, 1
Operation of a furnace, 193
Operating cost, 344
Origin of hydrocarbons, **7**
Overflash rate, 312
Oxidation stability, 36

P

Packies correlation, 231
Paraffins, 1
Partial solubility, 237
Partition coefficient, 236
Pay zones, 9
Payout period, 346
Penetration index, 46
Permissible limits of pollutants, 143
Petrochemical complex, 122
Petrochemicals' hub, 122
Petroleum coke, 47
Phase equilibrium in extraction process, 238
Phenol, 121
Phenol formaldehyde, 120
Phosgene, 122
Physical properties of crude oil, 6

Pig launching, 140
Pipe-still furnace, 188
Pipe-still furnace elements, 190
Pipes and tubes, 275
Plate type heat exchanger, 185
Polybutadiene, 116
Polycyclic hydrocarbons, 5
Polyethylene, 107
Polyethylene terepthalate, 111
Polynuclear aromatics, 4
Polypropylene, 111
Polystyrene, 115
Polyurethane, 121
Polyvinyl chloride, 115
Poly methyl metha acrylate, 117
Poly tetra flouro ethylene, 117
Porphyrins, 7
Porosity, 12
Possible reserve, 9
Pour point, 19, 38
Power and steam generating plant, 144
Power calculations for pumping and compression, 290
Power consumption in a blending tank, 94
Power number, 95
Power plant start up, 325
Prandtl number, 174
Pressure vessels, 136
Priming, 284
Probable reserve, 9
Process piping and instrumentation diagram, 301
Processes of distillation and stripping, 207
Processing of plastic, rubber, and fibre, 123
Processing units, 131
Product cost, 344
Product transfer lines, 279
Profit and product price, 345
Propane de-asphalting, 67
Propane–propylene separator, 107
Prospect, 9
Proven reserve, 9
PTFE, 117
Pumps and compressors, 281
Pump lubricant, 86
Pyridine, 6
Pyrophoric iron, 334
Pyrrole, 6

R

Radiant section, 192
Rate of reaction and kinetic equations, 250
Rate of return, 346
Reaction stoichiometry, mechanism, and pathways, 249
Reactor outlet temperature, 316
Reactors in refineries and petrochemical plants, 249
Reboiler, 213
Reclamation of used lubricants, 92
Rectification section, 212
Redwood viscosity, 41
Reflux drum pressure, 308
Reformer unit control, 314
Reforming of naphtha, 60
Reforming plant start up, 328
Refractories, 191
Regenerated catalyst, 77
Regeneration of catalyst of an NHDS unit, 330
Regeneration of reformer catalyst, 331
Remote sensing, 10
Reserves, 9
Reservoir, 8
Resource estimation, 11
Reynolds number, 278
Rotary pumps, 288
RVP, 29

S

SAE numbers, 84
Safety analysis, 337
Saybolt viscosity, 41
Schedule number, 275
Scrapped surface exchanger, 185
Screw pumps, 288
Sedimentary rocks, 7
Seismic survey, 10
Selectivity, 237
Separation of phases, 237
Shut down, 329
  of a CDU, 330
  of a naphtha pretreatment unit, 330
  of a reforming unit, 331
SI engine, 32
Silicone, 122
Silver strip corrosion, 39
Smoke point, 30
Softening point, 47
Solenoid cutoff switch, 320
Solvent dewaxing, 69
Solvent extraction, 66
Solvent to feed ratio, 237
Solvent power, 238
Solvent recovery, 237
Source rock, 8
Specific speed, 285
Spent catalyst, 78
Split range control, 309
Stabiliser, 54
Stabilisation, 54
Stack, 193

Standard signals in process control, 323
Start-up, 325
Stratigraphy, 11
Stripping section, 213
Stirred tank reactor, 252
Submersible pump lift, 17
Sucker rod lift, 17
Styrene butadiene rubber, 116

T

Tar sands, 7
Taxes and duties, 345
TBP distillation, 26
Terephthalic acid, 111
Theory of distillation, 208
Theory of heat exchange, 171
Thermoplastics, 123
Thickeners, 84
Toluene diisocyanate, 121
Top plate temperature, 310
Top reflux, 211
Transducers, 298
Transportation and metering of crude oil, 20
Trickle bed filter, 141
Tubular reactor, 256
Twin cell furnace, 190

U

Urea, 121
Urea formaldehyde, 120

V

Vacuum distillation, 64
Vacuum distillation control, 312
VI improvers, 83
Viscosity, 13, 38
Viscosity blend, 94
Viscosity index, 42
Vis-breaking, 79
Vulcanisation of rubber, 128

W

Water conditioning plant, 147
Wax, 48
Well logging, 16
Wilson, Lobo, and Hottel method, 194

Z

Ziegler catalyst, 107

Made in the USA
Monee, IL
12 November 2026

b9849be5-9389-48f2-974e-1c6fcb83de95R01